Contributions to Management Science

The series *Contributions to Management Science* contains research publications in all fields of business and management science. These publications are primarily monographs and multiple author works containing new research results, and also feature selected conference-based publications are also considered. The focus of the series lies in presenting the development of latest theoretical and empirical research across different viewpoints.

This book series is indexed in Scopus.

Saravanan Krishnan · Raghvendra Kumar ·
Swamynathan Ramakrishnan
Editors

Unveiling Blockchain Enabled Business Applications

Springer

Editors
Saravanan Krishnan
Department of Computer Science
and Engineering
Anna University, College of Engineering
Guindy
Chennai, Tamil Nadu, India

Raghvendra Kumar
Department of Computer Science
and Engineering
Gandhi Institute of Engineering
and Technology (GIET) University
Gobriguda, India

Swamynathan Ramakrishnan
Amity Business School
Amity University
Dubai, United Arab Emirates

ISSN 1431-1941 ISSN 2197-716X (electronic)
Contributions to Management Science
ISBN 978-981-95-2336-8 ISBN 978-981-95-2337-5 (eBook)
https://doi.org/10.1007/978-981-95-2337-5

This Springer imprint is published by the registered company Springer Nature Singapore Pte Ltd.
The registered company address is: 152 Beach Road, #21-01/04 Gateway East, Singapore 189721, Singapore

Preface

The rapid evolution of blockchain technology has transformed the digital landscape, enabling secure, transparent, and decentralized applications across multiple industries. From financial services to supply chain management and from healthcare to cybersecurity, blockchain is redefining the way businesses operate and interact in a trustless environment. The book *Unveiling Blockchain Enabled Business Applications* is a comprehensive compilation of research contributions that explore the diverse applications of blockchain technology in various sectors.

This edited volume consists of 19 meticulously curated chapters that shed light on the transformative potential of blockchain. The initial chapters provide foundational insights into blockchain applications in business and finance, with discussions on cryptocurrency as the new gold standard and smart contracts revolutionizing transactions. Subsequent chapters explore the integration of blockchain in critical sectors such as cybersecurity, healthcare, and supply chain management, demonstrating its role in enhancing transparency, efficiency, and security.

Several chapters delve into blockchain's impact on banking and governance, offering case studies and hybrid models that showcase its applicability in financial transactions and forensic accounting. Additionally, the book presents innovative discussions on how blockchain can contribute to green urbanization, industrial applications, and the secure exchange of digital assets, including non-fungible tokens (NFTs). The challenges and future directions of blockchain technology, including scalability, regulatory considerations, and cybersecurity frameworks, are also critically analyzed.

This book serves as an essential reference for researchers, industry professionals, policymakers, and students interested in exploring the evolving landscape of blockchain technology. By presenting cutting-edge research and real-world applications, it aims to bridge the gap between theoretical understanding and practical implementation.

We extend our sincere gratitude to all contributors, researchers, and experts whose valuable insights and expertise have made this book possible. We hope that this work will inspire further advancements and innovative applications of blockchain in the digital era.

Chennai, India — Saravanan Krishnan
Gobriguda, India — Raghvendra Kumar
Dubai, United Arab Emirates — Swamynathan Ramakrishnan

Contents

Blockchain for Business

Swamynathan Ramakrishnan, Saravanan Krishnan, Malini Mittal Bishnoi, Purnasha Chakrabarty, and Shanmugan Joghee

Abstract Blockchain is an open-source technology that makes it possible to record transactions over an ensemble of transparent and safe computing. In contrast to conventional databases that are controlled by a single entity, blockchain functions as a decentralized network in which every user, or node, has the information of the complete ledger. Because every transaction is confirmed by unanimity among the nodes and eternally recorded in blocks connected in a chronological chain, this guarantees that the data is transparent and unchangeable. Every block forms a safe and impenetrable record of all network activity, comprising a list of businesses, a timestamp, and a cryptographic link of the preceding block. Global reports indicate that the global blockchain market value is projected to grow from USD 20.1 billion in 2024 to USD 248.9 billion at a CAGR of 65.5% (Marketsandmarkets Research Private Limited 2024). Industry experts are estimating that the blockchain technology has potential to touch 137.8 billion USD by market size at the end of this decade. Especially, the Asia-Pacific region is leading the way with massive startups from China, Japan, and South Korea who are developing technologies for finance, healthcare, and supply management as the key focus (Hemant 2024).

S. Ramakrishnan (✉) · M. M. Bishnoi · P. Chakrabarty
Amity University Dubai, Dubai, UAE
e-mail: sramakrishnan@amityuniversity.ae

M. M. Bishnoi
e-mail: mbishnoi@amityuniversity.ae

P. Chakrabarty
e-mail: purnashaC@amitydubai.ae

S. Krishnan
COE, Anna University, Chennai, India
e-mail: saravanankrish@annauniv.edu

S. Joghee
School of Business, Horizon University College, Ajman, UAE
e-mail: jshanmugan@hu.ac.ae

S. Krishnan et al. (eds.), *Unveiling Blockchain Enabled Business Applications*,
Contributions to Management Science, https://doi.org/10.1007/978-981-95-2337-5_1

1 History and Evolution of Blockchain Technology

The first decentralized cryptocurrency, Bitcoin, was introduced in 2008 by an unidentified person known only as Satoshi Nakamoto (see "The Bitcoin Whitepaper"). The foundation of Bitcoin is blockchain technology. Nakamoto envisioned a peer-to-peer digital currency based on decentralized consensus and cryptographic evidence, eliminating the need for intermediaries like banks. The blockchain era began with the introduction of the Bitcoin whitepaper, which demonstrated how this technology has the potential to transform not only banking but also a number of other industries (see "The History of the Blockchain and Bitcoin | Freeman Law").

In 2015, Ethereum was developed as a result of developers and entrepreneurs exploring the broader possibilities of blockchain technology after the success of Bitcoin. Ethereum's introduction of smart contracts, which are self-executing agreements with the terms written directly into the code, enabled more complex and programmable transactions. This breakthrough facilitated the creation of numerous decentralized applications (DApps) and novel use cases including identity (digital) and decentralized finance (DeFi).

Blockchain has now evolved into a powerful tool for innovation in many other industries, having moved beyond a niche technology primarily associated with cryptocurrency. The potential of blockchain has been further enhanced by the introduction of new consensus mechanisms like Proof-of-Stake (PoS) and advancements in scalability and interoperability (see "The Evolution of Consensus Algorithms: From PoW to PoS and Beyond").

Figure 1.1 shows that research studies indicate that banking and financial sector, technology sector, government agencies, supply chain operations, healthcare, and education are the leading sectors implementing blockchain. Similarly, telecom and energy sectors are in initial stage of adoption to blockchain technology (Dehghani et al. 2023).

2 Key Characteristics: Decentralization, Transparency, and Security

The three main features of blockchain—security, transparency, and decentralization—set it apart from other systems and help explain why its use is expanding.

- **Decentralization:** Centralized systems is a single entity that owns the data but blockchain runs on decentralized systems. Every member, or node, has the same amount of power, and the consensus process ensures that everyone agrees that a transaction is genuine. By removing the intermediate channels and minimizing the possibility of failure, this decentralization gives users more control over their data and increases their sense of empowerment (Nakamoto 2008b).

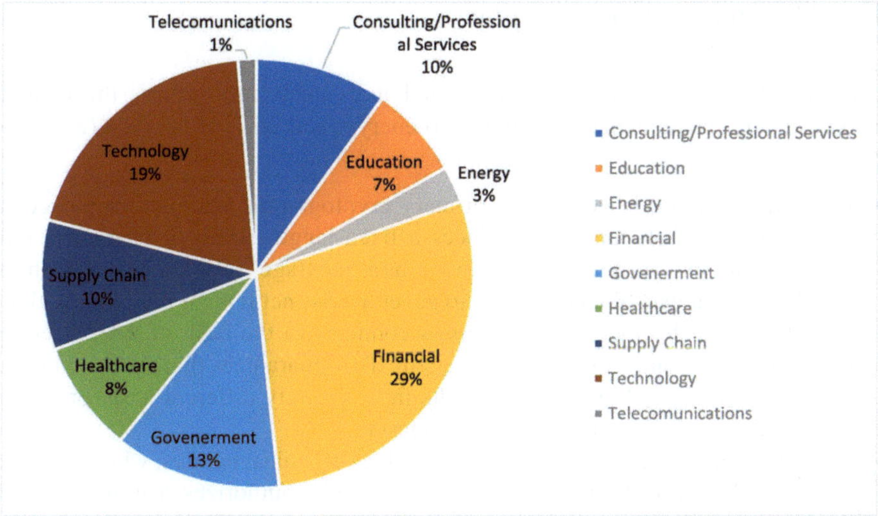

Fig. 1.1 Adaption of blockchain technology in industries. *Source* Dehghani et al. (2023)

- **Transparency:** Through inherent transparency, the network's participants can see every transaction that is registered on the blockchain. This open visibility allows anyone to confirm the validity and integrity of the data, which promotes user confidence (Tapscott and Tapscott 2016). This transparency is available to everyone on public blockchains, such as Bitcoin, but only authorized parties can access private or permissioned blockchains.
- **Security:** Cryptographic methods and consensus procedures are used in blockchain technology to ensure security. It is nearly impossible to alter any information without the network noticing since every single block is linked to the one before it via a cryptographic hash. Furthermore, because the blockchain is distributed, it is difficult to hack, as it would require an attacker to compromise a large number of network nodes simultaneously to change the data (Tapscott and Tapscott 2016). Because of these characteristics, blockchain provides a reliable and secure method of recording and verifying transactions, especially in environments where trust is a major issue (Nakamoto 2008b).

3 How Blockchain Works?

Blockchain technology is a groundbreaking tool for many businesses because of its distinct structure and underlying processes that guarantee integrity of information, transparency, and immutability. A thorough understanding of various essential elements, including cryptography, blockchain architecture, consensus processes, and smart contracts, is necessary to comprehend how blockchain functions.

3.1 The Role of Cryptography

Cryptography ensures that only authorized individuals may access the data, it involves using mathematical techniques to encrypt and decrypt the data (McNally 2017).

- **Hashing:** Hashing is one of the primary cryptographic methods employed in blockchain. A hash function produces a fixed-length string of characters that appears random from an input, also known as a "message" (Cachin 2016). Hashing is utilized in blockchain technology to generate distinct IDs for both transactions and blocks. Each block of the business connects to the hash of the preceding block and then connecting them in a chain. This guarantees that even a minimal modification to any block's data results in an entirely different hash, promptly alerting the network to any possible tampering.
- **Digital Signatures:** Cryptography also enables digital signatures, which are used to verify the authenticity of transactions. A user authorizes a transaction on the blockchain by recording their private key. The network then uses the corresponding public key to validate the signature, ensuring that the transaction has not been tampered with and that it originates from the rightful owner (Cachin 2016).
- **Public and Private Keys:** Blockchain utilizes a system of public–private key pairs to manage transactions and identity. The private key is used to sign transactions and prove ownership, while the public key serves as an address for receiving payments. This dual-key approach forms the foundation of blockchain's security architecture, enabling anonymous and secure transactions across the network (McNally 2017).

3.2 Blockchain Architecture: Blocks, Chains, and Nodes

Blockchain architecture uses its three main building blocks—blocks, chains, and nodes—to guarantee data confidentiality and integrity.

- **Blocks:** In a blockchain, a block is a group of related data that typically consists of the cryptographic hash of the previous block, a timestamp, a list of transactions, and a nonce (a random number applied to the proof-of-work process) (Swan 2015). Because each block uses this hash to connect to the one before it, a chain of blocks is created—hence the name "blockchain." The structure ensures that data uploaded to the chain and logged in a block are nearly unchangeable (Nakamoto 2008a).
- **Chains:** The blockchain is a series of chronological blocks each securely linked to the next by cryptographic hashes. This chaining approach guarantees data integrity, as it becomes computationally impractical to alter any block in the chain without also affecting all the blocks that follow it in a well-distributed network.
- **Nodes:** A node is an individual computer or other device that participates in the blockchain network. Each node independently verifies transactions and blocks, holding a complete clone of the blockchain. Nodes correspond with one another

to maintain consensus on the blockchain's state, ensuring that all copies of the ledger are accurate and synchronized (Nakamoto 2008a).

3.3 *Proofs for Work, Stake, and Consensus Mechanisms*

The dispersed chain of nodes in a blockchain must reach a consensus on the legitimacy of transactions and the ledger's current state for the blockchain to function. This is made possible via consensus mechanisms, which are essential to the software.

Proof-of-Work (PoW): For Bitcoin and other early blockchains, Proof-of-Work served as the original consensus method. PoW requires miners to compete by solving intelligent mathematical problems to validate transactions for addition of new blocks to the chain. The winning miner who solves the puzzle is awarded with cryptocurrency and gains the right to add the block (The Role of Consensus Mechanisms in Blockchain Networks 2023). Although PoW is secure and reliable, it is energy-intensive and demands significant computational power.

Proof-of-Stake (PoS): Proof-of-Stake (PoS) is a more energy-efficient alternative where validators build new blocks depending upon the number of coins they own and are cooperative to "stake" as collateral (What Are Consensus Mechanisms in Blockchain and Cryptocurrency 2023). Validators earn rewards for participating, while malicious actors risk losing the coins they have staked. PoS has gained popularity in Ethereum 2.0, due to its lower resource requirements compared to PoW.

In addition to PoW and PoS, other consensus mechanisms have been developed to address specific needs and challenges. For instance, Delegated Proof-of-Stake (DPoS) allows token holders to elect a limited number of delegates for reliable transactions and governance. Permissioned blockchains use a different approach called Practical Byzantine Fault Tolerance (PBFT), where adding newer blocks requires the agreement of a pre-approved group of participants (The Role of Consensus Mechanisms in Blockchain Networks 2023).

3.4 *Smart Contracts: Automation in Blockchain*

The self-executing agreements in alignment with the conditions of the contract which are encoded directly into the code are termed as smart contracts. By design, they intelligently carry out and implement the terms of a contract when predetermined clauses are met, thereby eliminating the need for human intervention and elimination of the risk of fraud or manipulation (McConaghy 2018).

Functionality: The smart contracts are designed with immutability and transparency and once deployed, their terms cannot be altered. Hence, when a product is delivered based on data inputs, a smart contract will automatically release their payment.

Applications: Smart contracts are utilized across various industries such as property management and real estate, where they streamline property transfers by automating title registration and payment processes. In monitoring the supplies, they can track the movement of goods and automatically initiate payments. In insurance, they can automate claims processing based on real-time data (Buterin 2013).

Pros and Cons: The benefits of smart contracts include efficiency, cost-effectiveness, and security. However, they also present challenges, particularly in legal enforcement, coding errors, and the complexity of translating legal language into executable code (McConaghy 2018).

4 Types of Blockchains

As blockchain technology has evolved over time, many kinds of blockchains that are suited to particular use cases and business requirements have been developed. Each type of blockchain—public, private, consortium, permissioned, and permissionless—offers a different set of benefits and is best suited for a certain application. To properly leverage blockchain in various circumstances, it is imperative to comprehend the differences between these types.

4.1 Public vs. Private Blockchains

Public Blockchains: Decentralized networks that allow participation from anyone are known as public blockchains. This type of blockchain is the most transparent, as all users are part of the network as nodes; they are involved in approving transactions and participate in the contractually agreed process. Every transaction is accessible to all users. Bitcoin is one of the most popular examples for public blockchains (What Is a Public Blockchain Beginner's Guide 2023).

Key Characteristics

- Decentralization
- Transparency
- Security
- Use Cases: Public blockchains are ideal for applications where trust and transparency are critical, such as cryptocurrency, decentralized finance (DeFi), and public record-keeping.

Private Blockchains: These are restricted networks where only authorized individuals can participate. These blockchains are usually managed by a distinct entity or a syndicate of organizations controlling the participation in the consensus process (What Is a Public Blockchain Beginner's Guide 2023; Public, Private, and Permissioned Blockchains Compared 2024).

Key Characteristics

- Centralization
- Confidentiality
- Efficiency
- Use Cases: Private blockchains are common in enterprise environments where control over the network and privacy are crucial. They are suitable for business operations requiring a private yet secure environment (Public, Private, and Permissioned Blockchains Compared 2024).

4.2 Consortium Blockchains

A hybrid approach that blends aspects of public and private blockchains, consortium blockchains—also referred to as federated blockchains—are managed by a group of organizations working together, with each member organization operating one or more nodes (Patnaik 2023).

Key Characteristics

- Semi-Decentralization
- Selective Transparency
- Collaborative Governance.
- Use Cases: Consortium blockchains are well-suited for industry-wide collaboration, where multiple organizations need to share and verify data without relying on a central authority. Examples include financial services (e.g., interbank settlements), healthcare (e.g., patient data sharing), and trade finance (What Is a Consortium Blockchain Features and Use Cases 2023).

4.3 Permissioned vs. Permissionless Blockchains

The distinction between permissioned and permissionless blockchains revolves around access control and participation rights within the network.

Permissionless Blockchains: These are open networks where anyone can participate without requiring permission; they are fully decentralized, permitting users to enter the network, confirm transactions, and partake in the agreed process. Examples of permissionless blockchains include Ethereum and Bitcoin (Permissioned Blockchain vs. Permissionless Blockchain: An In-Depth Comparison 2023).

Key Characteristics

- Open Participation
- Security
- Transparency.
- Applications: Permissionless blockchains are ideally suitable for open, transparent, and decentralized applications (DApps) and peer-to-peer networks (Permissioned and Permissionless Blockchains 2024).

Permissioned Blockchains: In contrast permissioned blockchains are suitable for private consortium settings, where pre-approval by the network's governing body is mandated for users since by nature, they restrict access and participation (Permissioned Blockchain vs. Permissionless Blockchain: Key Differences 2023).

Key Characteristics

- Controlled Access
- Efficiency
- Privacy
- Use Cases: Corporations and consortium environments where control, privacy, and efficiency are crucial especially the government applications, banking sector, and procurement management use permissioned blockchains (Permissioned Blockchain vs. Permissionless Blockchain: Key Differences 2023).

5 Blockchain Applications in Business

Blockchain technology is revolutionizing a number of global economic sectors at a rapid pace by providing solutions that improve efficiency, security, and transparency. Businesses are using blockchain more and more to solve difficult problems, improve workflow, and open up fresh avenues for creativity. The next segments go into significant blockchain implementations in several sectors, showcasing its adaptable and revolutionary capabilities.

5.1 Supply Chain Management

The complex nature of global supply chains, involving multiple stakeholders, geographies, and processes, and the challenges in transparency, traceability, and trust makes it one of the most promising areas for blockchain application.

Blockchain's Impact

- **Transparent and Traceable:** Each transaction on blockchain is a distributed ledger that is permanent and clear to all authorized parties, providing greater visibility across the value chain. By ensuring traceability at every stage, from procuring raw materials to delivering the finished product, blockchain eliminates the risk of fraud, counterfeiting, and inefficiencies (see "Using Blockchain to Drive Supply Chain Transparency").
- **Efficiency:** By automating verification processes and eliminating intermediaries, blockchain reduces the costs and delays associated with manual procedures. Smart contracts can also enhance productivity by automating payments, shipments, and other critical supply chain tasks according to predetermined criteria (see "The Role of Blockchain in Supply Chain Management (SCM)").
- **Use Case Example:** To improve food safety, companies like Walmart and IBM have implemented blockchain-based supply chain systems. This allows them to track the origin of food items and quickly identify and remove contaminated or counterfeit goods from the supply chain.

5.2 Finance and Banking

The finance and banking sector has been at the forefront of blockchain adoption, leveraging the technology to improve transaction speed, reduce costs, and enhance security.

Blockchain's Impact

- **Decentralized Finance (DeFi):** Thea decentralized financial services are the order of the day where there is no need for traditional intermediaries, such as trading, borrowing, and lending. This results in more accessible financial services and reduced transaction costs (Decentralized Finance (DeFi): What It Is and How It Works 2023).
- **Cross-Border Payments:** In order to eliminate an expensive, time-consuming, and error-prone process, blockchain allows instantaneous transactions at lower costs by eliminating the intervention of correspondent banks and other intermediaries.
- **Security:** Blockchain's cryptographic security measures make it highly resistant to fraud and hacking, thereby reducing the risk of financial crimes and protecting sensitive financial data.
- **Use Case Illustration:** Financial institutions use the blockchain-based payment protocol Ripple to enable real-time cross-border payments more quickly and at a lower cost than traditional banking systems.

5.3 Healthcare

Blockchain has transformed healthcare industry by modernizing patient data management, data sharing, and improved patient outcomes through efficient processes.

Blockchain's Impact

- **Data Security and Privacy:** By complying with data protection regulations like GDPR and reducing the risk of data breaches, blockchain ensures data security and authorized access (see "Blockchain Technology: A Potential Tool for the Management of Pharma Supply Chain").
- **Interoperability:** For advanced Medicare where patients need to receive care from multiple physicians or across international borders, blockchain ensures seamless exchange of medical records among different healthcare providers and receipt of accurate and consistent care.
- **Supply Chain Management:** Blockchain is implemented currently in the healthcare sector for follow-up of production schedules and distribution of medications and ensuring that counterfeit drugs are removed from the supply chain.
- **Use Case Example:** The MediLedger project uses blockchain technology to trace the movement of medications from manufacturers to consumers, ensuring that only genuine products are delivered to patients, thereby improving the pharmaceutical supply chain.

5.4 Real Estate

Blockchain technology is transforming the real estate sector by simplifying transactions, enhancing transparency, and reducing fraud.

Blockchain's Impact

- **Property Transactions:** Blockchain can streamline the buying and selling of real estate by using smart contracts which transfer the title of ownership automatically. This reduces role of intermediaries such as lawyers and agents, resulting in faster and more cost-effective transactions.
- **Land Registry:** Blockchain technology helps land registries to mitigate title disputes by providing an immutable record of property ownership. This is particularly beneficial in regions where land ownership records are disorganized or prone to corruption.
- **Tokenization:** The tokenization of real estate assets using blockchain technology allows for fractional ownership and increases market liquidity. This makes real estate investment more accessible by enabling the buying and selling of property fractions.
- **Use Case Example:** Propy, a real estate platform, uses blockchain technology to streamline international real estate transactions, providing a safe and improved

ways for buyers and sellers in transfer of property ownership without the need for traditional intermediaries (See "Propy: Revolutionizing Real Estate Transactions with Blockchain").

5.5 Digital Identity Verification

Digital identity verification is a critical challenge in the digital age, with increasing concerns over data privacy, security, and identity theft. Blockchain offers a solution by providing a secure, decentralized, and user-controlled identity management system.

Blockchain's Impact

- **Self-Sovereign Identity:** Blockchain allows individuals to create identities with full autonomy over their personal data, granting them complete sovereignty of information. Users can discreetly share information with third parties without intervention of centralized authorities.
- **Security and Privacy:** To eliminate identity theft and unauthorized access to private information, cryptographic security features of blockchain enables secured storage and authorized access.
- **Interoperability:** Blockchain-based identities facilitate a seamless and consistent identity verification process by enabling their use across different platforms and services.
- **Use Case Example:** Sovrin is a blockchain-based digital identity network that empowers individuals to monitor sharing of their identities and security of private information with organizations without relying on centralized identity providers (See "Sovrin: A Protocol and Token for Self-Sovereign Identity and Decentralized Trust").

5.6 Intellectual Property and Copyright Management

Blockchain is providing new ways to manage and protect intellectual property (IP) rights, ensuring that creators and rights holders are fairly compensated for their work.

Blockchain's Impact

- Blockchain makes it possible to create self-sovereign identities.
- **Immutable Records:** Blockchain technology provides an unchangeable, permanent record of intellectual property ownership, making it easier to track and verify ownership transfers. This is particularly useful for proving the originality of a work or in resolving copyright disputes.
- **Smart Contracts:** Smart contracts automate the processes of licensing and royalties' distribution, ensuring that creators are compensated whenever their work is used or shared.

- **Transparency:** Blockchain enables transparent tracking of the usage and distribution of protected works, reducing the risk of piracy and unauthorized use.
- **Use Case Example:** The music platform Mycelia utilizes blockchain technology to manage artists' rights and royalties, ensuring that musicians are fairly compensated for their work and that their rights are protected (see "How Blockchain is Transforming Music Royalties and Rights Management").

6 Real Time Applications of Blockchain in Business

Blockchain technology is already providing observable advantages in practical commercial applications, surpassing theoretical debates and experimental initiatives. Blockchain is transforming supply chain management, food safety, luxury products, and international trade by improving transparency, security, and efficiency. The aforementioned instances demonstrate how top businesses such as Walmart, IBM, De Beers, and Maersk are utilizing blockchain technology to tackle sector-specific issues and provide novel opportunities.

6.1 Walmart: Enhancing Supply Chain Transparency

One of the biggest retailers in the world, Walmart, has integrated blockchain technology to improve traceability and transparency throughout its extensive supply chain. The business realized that a more capable system was required to monitor the flow of commodities, especially in light of concerns about food safety.

Blockchain's Impact

- **Trackability:** Walmart can quickly determine the provenance of products—especially perishables—by utilizing blockchain technology. The blockchain ledger provides an unchangeable record of every transaction from farm to shelf, allowing Walmart to trace each item's journey through the supply chain (How Walmart Using Blockchain Technology—CoinCodeCap 2023).
- **Efficiency:** Blockchain technology has streamlined Walmart's supply chain processes, reducing the time it once took to trace a product's origin from days to just a few seconds. This enhanced efficiency ensures quicker response times during foodborne illness outbreaks and lowers supply chain management costs.
- **Trust:** Customers can validate the origin and journey of the products they consume in the blockchain, ensuring adherence to safety and quality standards. This transparency builds customer trust.
- **Use Case Example:** Walmart and IBM have collaborated to implement a blockchain-based system for tracking leafy greens. This system enables Walmart to quickly identify and remove contaminated products from stores, protecting customers and minimizing the impact of recalls (Walmart Blockchain 2022).

6.2 IBM Food Trust: Ensuring Food Safety

IBM Food Trust aims to increase security and transparency throughout the entire global supply chain. The platform connects growers, processors, distributors, and retailers in a transparent and secure network to track food products from farm to table.

Blockchain's Impact

- **Safety and Quality Assurance:** IBM Food Trust enhances food safety by providing a comprehensive and immutable history of a product. This traceability allows contamination sources to be quickly identified, enabling faster recalls and reducing the risk of widespread outbreaks.
- **Transparency:** The platform offers unprecedented transparency by allowing all participants in the food supply chain to access real-time data on the products they handle. This transparency ensures that food products meet safety standards, reduces waste, and prevents fraud (see "IBM Food Trust: A Blockchain-Based Solution for Food Safety").
- **Customer Trust:** IBM Food Trust increases customer confidence in the food they purchase by providing detailed information on the provenance and production processes of food products. Customers have the access product QR codes and tracking of the items' journey from farm to store.
- **Use Case Examples:** Major retailers and food manufacturers, including Walmart, Nestlé, and Unilever, have benchmarked IBM Food Trust to develop traceability, demonstrating the platform's effectiveness in improving transparency and trust in the global food supply chain (see "Supply Chain Insight: Inside IBM's Food Trust Blockchain System").

6.3 De Beers: Tracking the Provenance of Diamonds

The largest diamond firm in the world, De Beers, has used blockchain technology to monitor the origin of diamonds and ensure they are obtained ethically and without conflict. The Tracr program provides a secure and transparent trail of a jewel's journey from mines to retail stores.

Impact of Blockchain

- **Provenance Verification:** Tracr records each diamond's attributes, ownership, and history in a digital certificate created for each diamond using blockchain technology. This digital ledger is accessible to all parties, offering an immutable record that guarantees the legitimacy and ethical sourcing of diamonds. (De Beers Tracks 100 Diamonds through Supply Chain Using Blockchain 2023)

- **Transparency and Trust:** De Beers uses blockchain technology to enhance transparency throughout the diamond supply chain, fostering confidence among regulators, retailers, and consumers. Tracking a journey from origin to retail location ensures it is conflict-free and adheres to ethical standards.
- **Efficiency and Security:** Blockchain streamlines the process of verifying a diamond's provenance and thereby eliminates the delay and costs associated with traditional certification methods. The fool-proof and secure nature of blockchain ensures that there is transparency in recorded data.
- **Use Case Example:** To increase consumer trust and promote ethical sourcing methods in the diamond industry, major diamond producers and retailers, such as Signet Jewelers, have adopted Tracr. This technology provides consumers with verifiable documentation of a diamond's provenance (De Beers 2023)

6.4 Maersk: Improving Global Trade with Blockchain

In order to increase transparency, efficiency, and security in international commerce, Maersk, the biggest shipping business in the world, and IBM have teamed to create TradeLens, a blockchain-based platform. The platform addresses the challenges of managing enormous volumes of data and paperwork in global shipping.

Blockchain's Impact

- **Transparency:** With TradeLens, all parties involved in the supply chain—shippers, freight forwarders, port authorities, customs inspectors, and more—can access the same data in real time on a shared, immutable ledger of shipping transactions. This transparency reduces disputes, delays, and errors compared to traditional paper-based methods.
- **Efficiency:** TradeLens optimizes workflows, reduces processing times, and lowers costs by utilizing smart contracts to digitize and automate trade documents. The platform enhances coordination among all stakeholders involved in international trade, speeding up customs clearance.
- **Security:** Blockchain technology secures sensitive trade data, protecting it from alteration or any unauthorized entry. This security is the bedrock for the integrity of international shipments and preventing fraud.
- **Use Case Example:** Over 100 companies, including port operators, shipping lines, and customs authorities worldwide, have implemented TradeLens. The platform has demonstrated how blockchain technology can revolutionize international trade by significantly eliminating the risks associated with time and costs globally (see "TradeLens: How IBM and Maersk Are Sharing Blockchain to Build a Global Trade Platform").

7 Benefits of Blockchain for Business

Businesses in a variety of industries are beginning to realize and take advantage of the disruptive possibilities that blockchain technology provides. Some of the most important issues that businesses are currently facing are addressed by its special qualities especially decentralization and cryptographic security. The following advantages provide clarity on the role of blockchain technology for businesses.

7.1 Increased Transparency

The capacity of blockchain technology to offer unmatched transparency across corporate processes is one of its most important benefits. In traditional business operations, the visibility of transactions and data flow is often obscured by the many layers of intermediaries involved. Blockchain, with its distributed and decentralized ledger, ensures that every transaction is instantly recorded and available to all authorized parties.

Key Benefits

- **End-to-End Visibility:** Blockchain allows businesses to track the entire lifecycle of a product or transaction, from inception to completion. This visibility is particularly beneficial in supply chain management, where it is crucial to monitor the movement of goods, verify their origin, and ensure compliance with regulations.
- **Accountability:** Blockchain is permanent and time-stamped and easily auditable of transactions. This level of transparency fosters accountability among all parties involved and reduces the likelihood of disputes or mismanagement.
- **Consumer Trust:** Transparency builds trust with consumers, who can verify the provenance of products, ensuring they are ethically sourced, genuine, and meet quality standards. This is especially important in industries which deal with short shelf-life products where transparency is linked directly to brand integrity (see "Using Blockchain to Drive Supply Chain Transparency").

7.2 Enhanced Security

Security properties of blockchain are yet another important advantage for companies. Conventional systems are susceptible to fraud, data breaches, and cyber-attacks because they frequently store data in centralized databases. By using decentralized data storage and sophisticated cryptographic algorithms, blockchain significantly improves the security of corporate processes.

Key Benefits

- **Data Integrity:** Blockchain technology guarantees data integrity by maintaining data in an unchangeable ledger that is impervious to tampering or alteration. The cryptographic hash of the preceding block makes it nearly impossible to alter without permission from competent authorities.
- **Decentralization:** Blockchain mitigates the single point of failure common in centralized systems by distributing data across a network of nodes. This makes it nearly impossible for hackers to alter data, as they cannot compromise most nodes to do so.
- **Access Control:** Blockchain technology can enforce strict access controls using permissioned networks and private keys, assuring authenticity to modify data. This feature is particularly beneficial in sectors such as government, healthcare, and finance, where protecting private data from unauthorized access is crucial (see "Using Blockchain to Drive Supply Chain Transparency").

7.3 Improved Efficiency and Cost Reduction

A strong advantage for companies trying to maximize their operations is blockchain's capacity to simplify corporate procedures and lower operating expenses. Conventional corporate procedures frequently entail laborious human labor, third-party middlemen, and intricate workflows that can result in lost time and higher expenses. Many of these procedures are automated by blockchain, which decreases the need for middlemen and speeds up transaction processing.

Key Benefits

- **Automation Using Smart Contracts:** Smart contracts are automatically executing agreements encoded into their code. These contracts do not require middlemen such as attorneys or brokers, as they autonomously enforce and execute actions when predetermined conditions are met. This automation reduces the costs, errors, and delays associated with human processing (Effective Use of Blockchain for Business Process Management 2023).
- **Faster Transactions:** By removing the need for third-party verification and settlement processes, blockchain technology facilitates nearly instantaneous transactions. This is particularly advantageous in sectors where traditional transactions may take days to settle, including finance and international trade.
- **Cost Savings:** Blockchain reduces costs for firms by eliminating the need for intermediaries and optimizing workflows, which results in savings on transaction fees, administrative expenses, and the costs associated with fraud prevention and detection (Smart Contracts: Automating the Future 2023).

7.4 *Trust and Verification*

Any business transaction must be based on trust, and blockchain greatly strengthens that trust by offering a safe and independently verified transaction record. In conventional systems, intermediaries are frequently used to build confidence, which may be expensive and time-consuming. Blockchain provides a decentralized, transparent, and unchangeable record that all parties can rely on, removing the need for these middlemen.

Key Benefits

- **Immutable Record Keeping:** The immutable ledger ensures that transaction records are accurate and dependable for all parties involved (The Importance of Trust in Blockchain Technology 2023).
- **Decentralized Verification:** Blockchain technology enables decentralized transaction verification, distributing trust throughout the network rather than concentrating it in a single entity. This distribution reduces the likelihood of fraud and manipulation as no one party has total control over the system.
- **Enhanced Credibility:** By adopting blockchain technology, companies can demonstrate that their operations are secure, ethical, and transparent. This is particularly important in sectors such as banking, healthcare, and real estate, where trust is a crucial differentiator.

7.5 *Reduced Fraud and Counterfeiting*

Fraud and counterfeiting are persistent challenges across many industries, leading to significant financial losses and damage to brand reputation. Blockchain's inherent security features and transparency make it an effective tool for reducing fraud and counterfeiting.

Key Benefits

- **Authenticity Verification:** This is particularly useful in industries such as luxury goods, pharmaceuticals, and electronics, where counterfeit products are prevalent. Consumers can easily verify the authenticity of a product by scanning a QR code or accessing the blockchain record.
- **Supply Chain Integrity:** By providing end-to-end visibility in the process, blockchain ensures that every participant obeys the agreed-upon standards and practices. This eliminates the risk of counterfeit products in the supply chain (Will Blockchain Technology Mark a Turning Point in Fraud Prevention 2023).

- **Fraud Prevention:** Blockchain's decentralized and immutable nature prevents fraudsters to alter transactions through fake identities. This enables overall security and devoid of financial fraud.

8 Risks and Tasks

While blockchain technology offers numerous benefits, its adoption and implementation are accompanied by significant challenges and risks. As businesses and industries increasingly explore the potential of blockchain, they must also address these obstacles to fully realize the technology's advantages. Below are key challenges and risks associated with blockchain technology.

8.1 Scalability Issues

Scalability remains one of the most pressing challenges in the adoption of blockchain technology. Blockchain networks, particularly those that use Proof-of-Work consensus mechanisms. For example, in use case like Bitcoin, often a large volume of transactions is processed leading to doubts in efficiency. As a result, the network becomes Jammed in traffic, impacting processing times and costs too.

Key Challenges

- **Transaction Throughput:** Most blockchain networks currently have limited transaction throughput compared to traditional centralized systems. For example, Bitcoin can process around seven transactions per second, whereas Visa processers achieve thousands of transactions per second (" see Blockchain Scalability and its Challenges").
- **Network Congestion:** As more users and applications adopt blockchain technology, the likelihood of network congestion increases, which can lead to delays and inefficiencies in transaction processing.
- **Resource Intensity:** Scaling blockchain networks often requires significant computational resources, which can increase costs and limit the accessibility of blockchain technology to smaller organizations.
- **Scalability:** Addressing scalability is crucial for blockchain to support large-scale applications, particularly in banking sector, industrial supply chains, and international trade.

8.2 *Regulatory and Legal Challenges*

The legal landscape for blockchain technology is still in nascent stage, presenting numerous challenges for businesses seeking to implement blockchain solutions. Especially in areas such as cryptocurrencies and smart contracts regulatory bodies are struggling to classify and regulate blockchain-based activities.

Key Challenges

- **Uncertain Regulatory Environment:** The lack of clear and consistent regulatory frameworks across different jurisdictions creates uncertainty for businesses. This uncertainty can hinder the adoption of blockchain technology, as organizations may be reluctant to invest in blockchain solutions without a clear understanding of the regulatory implications.
- **Compliance Issues:** Compliance with anti-money laundering (AML) and customer (KYC) requirements are complex issues while implementing blockchain.
- **Legal Recognition of Smart Contracts:** For blockchain to be fully integrated into global legal and commercial systems, clear regulations and standards must be established to ensure that smart contracts are legally enforceable (see "Legal and Regulatory Compliance—WEF Blockchain Toolkit").

8.3 *Energy Consumption Concerns*

In networks that use Proof-of-Work (PoW), the computational energy required to solve complex cryptographic puzzles vast amounts of energy is consumed leading to environmental and sustainability concerns.

Key Challenges

- **High Energy Costs:** PoW-based blockchain networks, like Bitcoin, consume enormous amounts of electricity. This high energy consumption translates into significant operational costs, which can be a deterrent for businesses considering blockchain adoption.
- **Environmental Impact:** Businesses adopting blockchain must consider the sustainability of their operations and explore alternatives since the Proof-of-Stake (PoS) mechanisms are energy-intensive projects.
- **Public Perception:** The energy consumption associated with blockchain can negatively affect public perception, particularly among environmentally conscious consumers and stakeholders. Businesses must address these concerns to maintain a positive brand image and ensure sustainable practices (see "Blockchain Protocols and their Energy Footprint").

8.4 *Integration with Existing Software*

Most businesses rely on well-established IT infrastructures and processes that needs complete turnaround to become compatible with blockchain. This is a careful process of investment in technology, resources, and training.

Key Challenges

- **Compatibility Issues:** Blockchain networks may not easily integrate with legacy systems due to differences in architecture and data formats requiring technical expertise and re-engineering of existing systems.
- **Cost and Complexity:** The cost and complexity of integrating blockchain with legacy systems can be prohibitive for many organizations. Businesses must invest in new infrastructure, software, and training which are long-term expensive goals.
- **Disruption to Operations:** for integrating blockchain technology businesses must carefully manage the transition to avoid operational disruptions and ensure continuity of services (see "Blockchain Integration: Merging Legacy Systems with Blockchain").

8.5 *Privacy and Data Security Issues*

While blockchain is often touted for its security features, it also presents unique privacy and data security challenges. The immutable nature of blockchain means that once data is recorded, it cannot be altered in any form. This immutability, while beneficial for security, raises concerns about data privacy, particularly in light of regulations such as the General Data Protection Regulation (GDPR).

Key Challenges

- **Data Privacy:** Blockchain's transparency can conflict with privacy regulations, as personal data stored on the blockchain is accessible to all participants in the network. Ensuring compliance with data privacy laws while maintaining the integrity and transparency of the blockchain is a significant challenge.
- **Sensitive Information:** Storing sensitive information on a public blockchain can expose it to unauthorized access or misuse. Businesses must carefully consider what data to record on the blockchain and implement encryption and other security measures.
- **Data Ownership:** Blockchain complicates data ownership and control and determining who owns the data on a blockchain and how it can be used or shared requires clear policies and governance structures (Smita and Lata 2022).

9 Future Trends and Opportunities

As a part of industry 5.0, blockchain technology is reshaping industries and creating new opportunities. Emerging trends of blockchain technology will play a key role in transforming technologies and markets. Below are key future trends and opportunities that highlight the growing influence of blockchain across various sectors. Statista indicates that there is a huge futuristic potential for blockchain field (Fig. 1.2).

9.1 Internet of Things (IoT) and Blockchain

The synergy of the Internet of Things (IoT) along with blockchain will be the most efficient futuristic technology wherein IoT devices would facilitate bigdata management. Blockchain's role is to secure data exchange and enhancing the trustworthiness of IoT networks.

Key Opportunities

- **Secure Data Management:** Blockchain can secure data generated by IoT for industries such as healthcare, where sensitive data from medical devices needs to be protected.

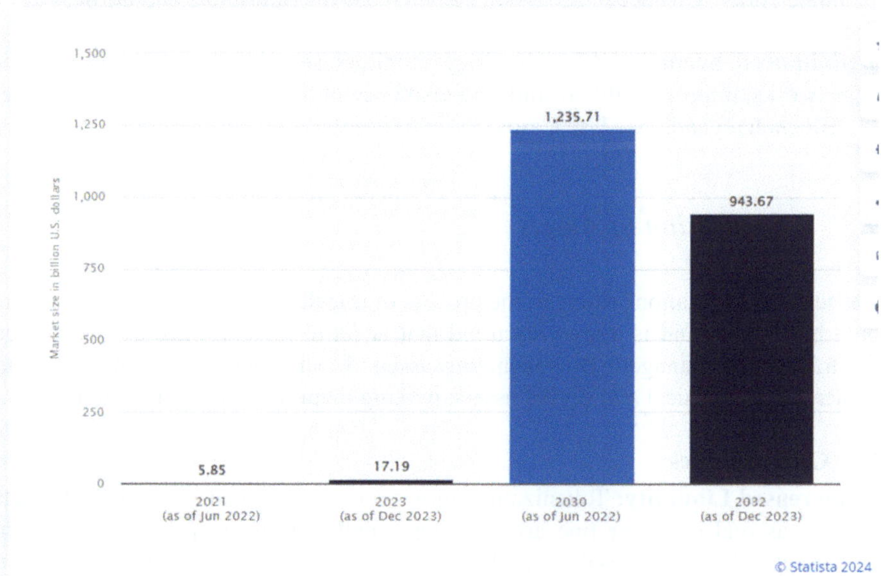

Fig. 1.2 Estimated market growth of blockchain. *Source* https://www.statista.com/statistics/1319369/global-blockchain-technology-market-size/

- **Decentralized IoT Networks:** Blockchain can facilitate the creation of decentralized IoT networks, reducing reliance on central authorities and increasing the resilience of these networks against cyber-attacks.
- **Efficient Automation:** Smart contracts within IoT ecosystems can execute transactions between devices and managing real-time logistics leading to greater operational efficiency (Convergence Between Blockchain and The Internet of Things 2021).

9.2 Integration with AI

The merge of blockchain with artificial intelligence (AI) is poised to unlock new possibilities for innovation and efficiency. Blockchain enhances transparent and secured AI models, while AI can optimize blockchain operations, making them more efficient and scalable.

Key Opportunities

- **Data Integrity:** Blockchain can be used to train AI models. This prevents data tampering resulting which AI-based decisions are made more accurate and reliable.
- **Decentralized AI:** Blockchain can enable the creation of decentralized AI models, where AI algorithms are distributed across a network rather than controlled by a single entity. This decentralization can increase the robustness and fairness of AI systems.
- **Enhanced Security:** AI technology monitors security threats in blockchain networks, improving the overall safe resilience of blockchain-based systems (see "Blockchain and AI—Use Cases").

9.3 Tokenization of Assets

The term "tokenization" refers to the process of translating assets into digital tokens on a blockchain and is a growing trend that is set to revolutionize the way assets are traded and managed. By tokenizing assets, blockchain enables the fractional ownership of physical and digital assets, making them more accessible and liquid.

Key Opportunities

- **Increased Liquidity:** Tokenization allows for the fractional ownership of assets, such as real estate or fine art, enabling smaller investors participation in the markets. This increases the liquidity of these assets and opens up new investment opportunities.

- **Efficient Trading:** Blockchain facilitates the instant and secure transfer of tokens, reducing the need for intermediaries and lowering transaction costs. This efficiency is particularly beneficial in markets such as securities trading, where speed and accuracy are critical.
- **Global Accessibility:** Tokenized assets are global scale, 24/7, without the constraints of traditional financial markets. This democratization of access to assets has the potential to attract a broader range of investors and increase market participation (see "Asset Tokenization: Complete Overview of Tokenized Assets").

9.4 The Rise of Decentralized Finance (DeFi)

Decentralized finance (DeFi) is offering a range of financial services that operate without traditional intermediaries such as banks or brokers. DeFi platforms leverage smart contracts on blockchain networks to provide services like lending, borrowing, trading, and insurance.

Key Opportunities

- **Financial Inclusion:** For the underprivileged populations, DeFi provides financial services. By removing intermediaries and enabling peer-to-peer transactions, DeFi can lower the barriers to financial inclusion.
- **Innovation in Financial Products:** DeFi enables the creation of new financial products and services that are more transparent, efficient, and accessible. For example, automated market makers (AMMs) in DeFi allow for decentralized trading without the need for order books.
- **Yield Opportunities:** DeFi platforms offer various opportunities for yield generation, such as staking, liquidity provision, and yield farming. These opportunities can provide higher returns compared to traditional financial products, attracting both retail and institutional investors (see "Decentralized Finance (DeFi): Revolutionizing the Financial Landscape Through Blockchain Technology").

9.5 Potential for Blockchain in Emerging Markets

Blockchain technology holds immense potential for driving economic growth and innovation in emerging markets. These regions often face challenges related to financial infrastructure, governance, and transparency, which blockchain can help address.

Key Opportunities

- **Financial Empowerment:** Blockchain can provide unbanked and underbanked populations in emerging markets with access to financial services, digital payments, transmittals, and microcredit finance. This can empower individuals and small businesses globally to grow.

- **Transparent Governance:** Blockchain can improve governance in emerging markets by providing transparent and immutable records of government activities, reducing corruption, and increasing accountability.
- **Supply Chain Transparency:** In emerging markets, blockchain can enhance supply chain transparency, ensuring that products are ethically sourced and comply with international standards. This can improve market access for local producers and increase consumer trust (Disrupt and Innovate: Harness the Power of Blockchain 2023).

10 Implementation Strategies for Businesses

The application of blockchain technology requires a well-thought-out strategy that aligns with the organization's goals and addresses its unique challenges. Businesses must navigate a complex landscape of technological, regulatory, and operational considerations to effectively harness the power of blockchain. Below are key strategies for implementing blockchain in business, with a focus on assessing needs, choosing the right platform, building a strategy, integrating effectively, and fostering collaborations.

10.1 Assessing Business Needs and Feasibility

Before diving into blockchain adoption, businesses must conduct a thorough assessment of their specific needs and the feasibility of implementing blockchain technology. This step is crucial for ensuring that blockchain is the right solution for the problems at hand.

Key Considerations

- **Identifying Pain Points:** Businesses should start by identifying the pain points and inefficiencies in their current processes. Blockchain is particularly valuable in scenarios where transparency, security, and decentralization can provide clear advantages.
- **Cost–Benefit Analysis:** A detailed cost–benefit analysis should be conducted to evaluate the potential return on investment (ROI) from blockchain implementation. This includes assessing both the initial costs of adoption and the long-term benefits.
- **Technical Feasibility:** Understanding the technical requirements and limitations of blockchain is essential. Businesses need to evaluate whether their existing IT

infrastructure can support blockchain and what modifications might be necessary (see "Identifying The Problem Or Pain Point That Your Blockchain Solution Aims to Address").

10.2 Choosing the Right Blockchain Platform

Selecting the appropriate blockchain platform is a critical decision that will impact the success of the implementation. The choice of platform should be aligned with the specific needs of the business and the use case in question.

Key Considerations

- **Public vs. Private Blockchains:** Businesses need to decide whether to use a public, private, or consortium blockchain. Public blockchains are largely transparent and decentralized; private blockchains are more controlled and secured. Consortium blockchains can be ideal for industry collaborations.
- **Platform Features:** The selected blockchain platform should offer the necessary features, such as smart contract functionality, scalability, and interoperability. Ethereum, Hyperledger, and Corda are popular platforms, each with its strengths and use cases.
- **Community and Support:** The strength of the platform's developer community and the availability of support resources are also important factors. A robust community can facilitate faster development and troubleshooting (see "Hyperledger vs Corda vs Ethereum: The Ultimate Comparison").

10.3 Building a Blockchain Strategy

A comprehensive blockchain strategy is essential for guiding the implementation process and ensuring alignment with the organization's overall objectives. This tailored strategy to the specific needs of the business is necessary to provide a clear roadmap.

Key Considerations

- **Defining Objectives:** The first step in building a blockchain strategy is to define clear and measurable objectives. These objectives should align with the business's long-term goals and address the identified pain points.
- **Stakeholder Engagement:** Engaging stakeholders from across the organization is crucial for building support and ensuring that the blockchain strategy meets the needs of all departments. This includes involving IT, legal, finance, and operations teams.

- **Risk Management:** A well-defined risk management plan should be included in the blockchain strategy. This plan should address potential risks, such as regulatory changes, security vulnerabilities, and integration challenges (see "A Strategist's Guide to Blockchain").

10.4 Best Practices for Integration and Adoption

Successful blockchain implementation requires careful integration with existing systems and processes. Following best practices can help businesses overcome common challenges and ensure a smooth adoption process.

Key Considerations

- **Incremental Implementation:** Businesses should consider implementing blockchain incrementally, starting with pilot projects or limited use cases. This approach allows for testing and refining the technology before scaling it across the organization (Blockchain Development Guide: Challenges and Best Practices 2024).
- **Training and Education:** Providing training and education to employees is critical for successful adoption. Employees need to understand how blockchain works and how it will impact their roles and responsibilities (How to Implement Blockchain in Business A Complete Guide 2024).
- **Change Management:** Effective change management is essential to address resistance to new technology. This includes clear communication of the benefits, addressing concerns, and providing ongoing support.

10.5 Collaborations and Partnerships

Blockchain's full potential often requires collaboration and partnerships, both within the industry and across sectors. Building strategic partnerships can accelerate blockchain adoption and create new opportunities for innovation.

Key Considerations

- **Industry Consortia:** Participating in industry consortia can provide businesses with access to shared resources, expertise, and best practices. Consortium blockchains, in particular, benefit from collaboration between multiple organizations (Blockchain Consortia Perspectives: The future of 'coopetition' 2023).
- **Technology Partnerships:** Partnering with technology providers, such as blockchain platform developers or consulting firms, can provide businesses with the expertise needed to implement and scale blockchain solutions.

- **Cross-Industry Collaboration:** Blockchain can enable cross-industry collaboration by creating shared platforms for data exchange and process automation. Businesses should explore opportunities for collaboration with partners in related industries (Industrial Blockchain Task Force Report: EU Consortia in Focus—Exploring Global Trends in Industrial Blockchain Projects 2023).

11 Conclusion

As industries across the globe increasingly recognize the benefits of blockchain, it is becoming clear that this technology will play a critical role in shaping the next era of digital transformation. Blockchain, with the potential to disrupt traditional business models and create new ones is enabling decentralized networks. This shift is particularly significant in industries such as finance, supply chain management, and healthcare, where trust and verification are paramount. As blockchain continues to evolve, it will enable more sophisticated applications, such as decentralized finance (DeFi), tokenization of assets, and smart contracts, further transforming how businesses operate and interact with stakeholders.

For business leaders, the adoption of blockchain technology presents both opportunities and challenges. The key takeaway is that blockchain is not just a technological innovation but a strategic enabler that can drive significant business value. To leverage blockchain effectively, business leaders must take a proactive approach, starting with a thorough assessment of their organization's needs and the feasibility of blockchain solutions. Choosing the right blockchain platform and building a comprehensive strategy are essential steps in ensuring successful implementation.

By participating in industry consortia and engaging with technology partners, businesses can access the expertise and resources needed to navigate the complexities of blockchain adoption. Furthermore, preparing the organization for blockchain integration through training, education, and change management is critical to achieving a smooth transition and maximizing the benefits of this technology.

As the world becomes increasingly blockchain-driven, businesses must be prepared to adapt and innovate. This preparation involves staying informed about the latest developments in blockchain technology, as well as understanding its implications for their industry. Business leaders should invest in building the necessary capabilities within their organization, including technical expertise, regulatory compliance, and data security measures. By fostering a culture of innovation and continuous learning, businesses can position themselves to thrive in a blockchain-driven world.

Blockchain had witnessed tremendous growth after the experience of the covid pandemics. World-wide, Corporations got their employees involved in businesses by creating work-from-home access. Cloud based platforms ensure security and efficiency. According to the World Economic Forum, BaaS (Blockchain as a Service) bridges the global trade financing gap, and thereby even the small businesses started using the technology enriching global supply chain operations. The rapidly increasing demand for Baas is witnessed in the recent years (Fortune Business Insights 2024).

In conclusion, blockchain technology represents a pivotal shift in the way businesses operate, offering significant opportunities for innovation, efficiency, and growth. Business leaders who recognize the strategic value of blockchain and take deliberate steps to integrate it into their operations will be well-positioned to lead in the digital economy of the future. As blockchain continues to evolve, its role in driving business transformation will only become more central, making it essential for businesses to embrace this technology and prepare for the changes it will bring.

Bibliography

A Strategist's Guide to Blockchain. Strategy+Business. [Online]. Available: https://www.strategy-business.com/article/A-Strategists-Guide-to-Blockchain

Asset Tokenization: Complete Overview of Tokenized Assets. ScienceSoft. [Online]. Available: https://www.scnsoft.com/blockchain/asset-tokenization

Blockchain and AI—Use Cases. Chainlink Blog. [Online]. Available: https://blog.chain.link/blockchain-ai-use-cases/

Blockchain Consortia Perspectives: The Future of 'Coopetition' (2023) Deloitte. [Online]. Available: https://www2.deloitte.com/us/en/pages/consulting/articles/blockchain-consortia-perspectives.html

Blockchain Development Guide: Challenges and Best Practices (2024) Nebulas Design. [Online]. Available: https://www.nebulasdesign.com/business/blockchain-development-guide-challenges-and-best-practices/. Accessed on 4 Sep 2024

Blockchain Integration: Merging Legacy Systems with Blockchain. Decubate. [Online]. Available: https://www.decubate.com/blog/blockchain-integration-merging-legacy-systems-with-blockchain

Blockchain Protocols and Their Energy Footprint. Adan. [Online]. Available: https://www.adan.eu/en/publication/blockchain-protocols-and-their-energy-footprint/.

Blockchain Scalability and its Challenges. Vezgo, [Online]. Available: https://vezgo.com/blog/blockchain-scalability/

Blockchain Technology: A Potential Tool for the Management of Pharma Supply Chain (2024-06-Jan) ScienceDirect. [Online]. Available: https://www.sciencedirect.com/science/article/abs/pii/S1551741124000792

Buterin V (2013) A next-generation smart contract and decentralized application platform. Ethereum White Paper. [Online]. Available: https://ethereum.org/en/whitepaper/

Cachin C (2016) Architecture of the hyperledger blockchain fabric. In 2016 18th international symposium on stabilization, safety, and security of distributed systems (SSS), Tokyo, Japan, pp. 1–16. [Online]. Available: https://ieeexplore.ieee.org/document/7916821

Convergence Between Blockchain and The Internet of Things (2021) International Journal of Technology, Innovation and Management. [Online]. Available: https://journals.gaftim.com/index.php/ijtim/article/download/11/5/34

De Beers Tracks 100 Diamonds Through Supply Chain Using Blockchain (2023) Supply Chain Digital. [Online]. Available: https://supplychaindigital.com/technology/de-beers-tracks-100-diamonds-through-supply-chain-using-blockchain

De Beers' Tracr Platform Now Open to the Industry (2023) Rapaport. [Online]. Available: https://rapaport.com/news/de-beers-tracr-platform-now-open-to-the-industry/

Decentralized Finance (DeFi): Revolutionizing the Financial Landscape Through Blockchain Technology. The Payments Association. [Online]. Available: https://thepaymentsassociation.org/article/decentralized-finance-defi-revolutionizing-the-financial-landscape-through-blockchain-technology/

Decentralized Finance (DeFi): What It Is and How It Works (2023) Investopedia. [Online]. Available: https://www.investopedia.com/decentralized-finance-defi-5113835

Dehghani M, Kennedy RW, Mashatan A, Rese A, Karavidas D (2023) High interest, low adoption. A mixed-method investigation into the factors influencing organisational adoption of blockchain technology. J Bus Res 393–411

Disrupt and Innovate: Harness the Power of Blockchain (2023) Agile Dynamics. [Online]. Available: https://gulfbusiness.com/new-blockchain-technology-research/

Effective Use of Blockchain for Business Process Management (2023) Protokol. [Online]. Available: https://www.protokol.com/insights/effective-use-blockchain-for-business-process-management/.

Fortune Business Insights (2024) Retrieved from https://www.fortunebusinessinsights.com; https://www.fortunebusinessinsights.com/industry-reports/blockchain-market-100072

Hemant (2024) Exploring the explosive growth potential of blockchain technology: forecasting a USD 137.8 billion market by 2030. Retrieved from https://www.linkedin.com/pulse/exploring-explosive-growth-potential-blockchain-technology-kolhe-i1ehf/

How Blockchain is Transforming Music Royalties and Rights Management. Music Industry Weekly. [Online]. Available: https://musicindustryweekly.com/how-blockchain-is-transforming-music-royalties-and-rights-management/.

How to Implement Blockchain in Business? A Complete Guide (2024) Apptunix. [Online]. Available: https://www.apptunix.com/blog/how-to-implement-blockchain-in-business-a-complete-guide/. Accessed on 4 Sep 2024

How Walmart Using Blockchain Technology?—CoinCodeCap (2023) CoinCodeCap. [Online]. Available: https://coincodecap.com/how-walmart-is-using-blockchain

Hyperledger vs Corda vs Ethereum: The Ultimate Comparison. 101 Blockchains. [Online]. Available: https://101blockchains.com/hyperledger-vs-corda-r3-vs-ethereum/

IBM Food Trust: A Blockchain-Based Solution for Food Safety. IBM. [Online]. Available: https://www.ibm.com/blockchain/resources/food-trust/food-logistics/

Identifying The Problem Or Pain Point That Your Blockchain Solution Aims to Address. Faster Capital. [Online]. Available: https://fastercapital.com/topics/identifying-the-problem-or-pain-point-that-your-blockchain-solution-aims-to-address.html

Industrial Blockchain Task Force Report: EU Consortia in Focus—Exploring Global Trends in Industrial Blockchain Projects (2023) INATBA. [Online]. Available: https://inatba.org/reports/industrial-blockchain-task-force-report-eu-consortia-in-focus-exploring-global-trends-in-industrial-blockchain-projects/

Legal and Regulatory Compliance—WEF Blockchain Toolkit. World Economic Forum. [Online]. Available: https://widgets.weforum.org/blockchain-toolkit/legal-and-regulatory-compliance/index.html

Marketsandmarkets Research Private Limited (2024) https://www.marketsandmarkets.com/. USA: marketsandmarkets research private limited. Retrieved from https://www.marketsandmarkets.com/Market-Reports/blockchain-technology-market-90100890.html?gad_source=1&gclid=Cj0KCQjwi5q3BhCiARIsAJCfuZnv9aQIkOgUgdXhniQ-BHNfqaQQp6sXk2jhQ-JFdt0TVZhjPok5_hYaApIMEALw_wcB

McConaghy SJ (2018) Smart contracts: the future of contract law. Stanf J Blockchain Law & Policy 1(1):1–28. [Online]. Available: https://stanford-jblp.pubpub.org/pub/smart-contracts-the-future-of-contract-law

McNally MJ (2017) Understanding blockchain technology: a guide to the basics. J Comput Commun 5(1):1–10, 2017. [Online]. Available: https://www.scirp.org/journal/paperinformation.aspx?paperid=80682

Nakamoto D (2008a) Bitcoin: a peer-to-peer electronic cash system. Bitcoin.org. [Online]. Available: https://bitcoin.org/bitcoin.pdf

Nakamoto S (2008b) Bitcoin: a peer-to-peer electronic cash system. Bitcoin.org. [Online]. Available: https://bitcoin.org/bitcoin.pdf

Patnaik M (2023) What is consortium blockchain?—a comprehensive guide. AlmaBetter. [Online]. Available: https://www.almabetter.com/bytes/articles/consortium-blockchain.
Permissioned and Permissionless Blockchains (2024) Freeman Law. [Online]. Available: https://freemanlaw.com/permission-and-permissionless-blockchains/
Permissioned Blockchain vs. Permissionless Blockchain: Key Differences (2023) Cointelegraph. [Online]. Available: https://cointelegraph.com/learn/permissioned-blockchain-vs-permissionless-blockchain-key-differences.
Permissioned Blockchain vs. Permissionless Blockchain: An In-Depth Comparison (2023) Spheron. [Online]. Available: https://blog.spheron.network/permissioned-blockchain-vs-permissionless-blockchain-an-in-depth-comparison.
Propy: Revolutionizing Real Estate Transactions with Blockchain. The big whale. [Online]. Available: https://www.thebigwhale.io/tokens/propy
Public, Private, and Permissioned Blockchains Compared (2024) Investopedia. [Online]. Available: https://www.investopedia.com/news/public-private-permissioned-blockchains-compared/
Smart Contracts: Automating the Future (2023) ArkSigner. [Online]. Available: https://blog.arksigner.com/en/smart-contracts-and-automation
Smita B, Lata R (2022) Challenges in making blockchain privacy compliant for the digital world: some measures. Sādhanā 47(168). [Online]. Available: https://link.springer.com/article/10.1007/s12046-022-01931-1
Sovrin: A Protocol and Token for Self-Sovereign Identity and Decentralized Trust. Sovrin Foundation. [Online]. Available: https://sovrin.org/wp-content/uploads/2018/03/Sovrin-Protocol-and-Token-White-Paper.pdf
Supply Chain Insight: Inside IBM's Food Trust Blockchain System. Supply Chain Digital. [Online]. Available: https://supplychaindigital.com/technology/supply-chain-insight-inside-ibms-food-trust-blockchain-system
Swan M (2015) Blockchain: blueprint for a new economy. O'Reilly Media. [Online]. Available: https://www.oreilly.com/library/view/blockchain-blueprint-for/9781491920499/
Tapscott M, Tapscott D (2016) Blockchain revolution: how the technology behind bitcoin is changing money, business, and the world. Penguin. [Online]. Available: https://www.penguinrandomhouse.com/books/317305/blockchain-revolution-by-don-tapscott-and-alex-tapscott/
The Bitcoin Whitepaper. Bitcoin.com.au. [Online]. Available: https://bitcoin.com.au/education/the-bitcoin-whitepaper/
The Evolution of Consensus Algorithms: From PoW to PoS and Beyond. Openware. [Online]. Available: https://www.openware.com/news/articles/the-evolution-of-consensus-algorithms-from-pow-to-pos-and-beyond
The History of the Blockchain and Bitcoin | Freeman Law. Freeman Law. [Online]. Available: https://freemanlaw.com/the-history-of-the-blockchain-and-bitcoin/
The Importance of Trust in Blockchain Technology (2023) Forbes. [Online]. Available: https://www.forbes.com/sites/bernardmarr/2023/01/30/the-importance-of-trust-in-blockchain-technology/
The Role of Blockchain in Supply Chain Management (SCM). Paltron. [Online]. Available: https://www.paltron.com/insights-en/the-role-of-blockchain-in-supply-chain-management-scm
The Role of Consensus Mechanisms in Blockchain Networks (2023) Smartsight. [Online]. Available: https://www.smartsight.in/technology/the-role-of-consensus-mechanisms-in-blockchain-networks/
TradeLens: How IBM and Maersk Are Sharing Blockchain to Build a Global Trade Platform (2018-27-Nov) IBM. [Online]. Available: https://www.ibm.com/blogs/think/2018/11/tradelens-how-ibm-and-maersk-are-sharing-blockchain-to-build-a-global-trade-platform/
Using Blockchain to Drive Supply Chain Transparency. Deloitte. [Online]. Available: https://www2.deloitte.com/us/en/pages/operations/articles/blockchain-supply-chain-innovation.html.
Using Blockchain to Drive Supply Chain Transparency. Deloitte. [Online]. Available: https://www2.deloitte.com/us/en/pages/operations/articles/blockchain-supply-chain-innovation.html
Walmart & Blockchain: New Era of Supply Chain Management (2022) Changelly Blog. [Online]. Available: https://changelly.com/blog/walmart-blockchain/

What Are Consensus Mechanisms in Blockchain and Cryptocurrency? (2023) Investopedia. [Online]. Available: https://www.investopedia.com/terms/c/consensus-mechanism-cryptocurrency.asp
What Is A Consortium Blockchain? Features and Use Cases (2023) Mudrex. [Online]. Available: https://mudrex.com/learn/what-is-consortium-blockchain/
What Is A Public Blockchain? Beginner's Guide (2023) 101 Blockchains, 2023. [Online]. Available: https://101blockchains.com/public-blockchain/
Will Blockchain Technology Mark a Turning Point in Fraud Prevention? (2023) Finance Magnates. [Online]. Available: https://www.financemagnates.com/cryptocurrency/education-centre/will-blockchain-technology-mark-a-turning-point-in-fraud-prevention/

Addressing Bottom of Pyramid Through Blockchain as a Service Application

Malini Mittal Bishnoi, Swamynathan Ramakrishnan, Purnasha Chakrabarty, and Saravanan Krishnan

Abstract The Bottom of the Pyramid (BoP) represents the largest but poorest socio-economic group, often lacking access to basic services and economic opportunities. Blockchain technology, with its decentralized and transparent nature, presents a promising solution to address the unique challenges faced by this demographic. This paper explores the potential applications of blockchain to enhance financial inclusion, supply chain transparency, and access to essential services for BoP communities. By leveraging blockchain's capabilities, we can facilitate secure and cost-effective financial transactions, thus integrating the unbanked population into the formal economy. Additionally, blockchain can improve supply chain efficiency and traceability, ensuring that essential goods reach their intended recipients without corruption or mismanagement. The paper also examines case studies where blockchain has been successfully implemented in BoP contexts, highlighting the transformative impact on local economies and social structures. Furthermore, the discussion extends to the challenges and limitations of deploying blockchain in low-resource settings, including technological literacy and infrastructure constraints. Through comprehensive analysis and empirical evidence, this study underscores the potential of blockchain technology to drive sustainable development and economic empowerment for the Bottom of the Pyramid, fostering a more inclusive and equitable global economy.

M. M. Bishnoi (✉) · S. Ramakrishnan · P. Chakrabarty
Amity University Dubai, Dubai, UAE
e-mail: mbishnoi@amityuniversity.ae

S. Ramakrishnan
e-mail: sramakrishnan@amityuniversity.ae

P. Chakrabarty
e-mail: purnashaC@amitydubai.ae

S. Krishnan
COE, Anna University, Chennai, India
e-mail: saravanankrish@annauniv.edu

S. Krishnan et al. (eds.), *Unveiling Blockchain Enabled Business Applications*, Contributions to Management Science, https://doi.org/10.1007/978-981-95-2337-5_2

1 Introduction

Blockchain technology has the potential to significantly impact poverty alleviation efforts, particularly for marginalized communities often referred to as the "Bottom of the Pyramid" (BoP). By providing a decentralized, transparent, and secure framework for transactions, blockchain can address various socio-economic challenges faced by low-income populations. Traditional financial systems frequently exclude these communities due to high costs, lack of trust, and inadequate infrastructure. However, blockchain offers innovative solutions that enhance financial inclusion, improve access to essential services, and foster economic empowerment.

One of the most promising applications of blockchain in this context is its ability to facilitate financial services for the unbanked and underbanked. Initiatives like Mojaloop, an open-source payment platform, aim to create a more inclusive financial ecosystem by enabling seamless transactions across different financial providers, thus reaching millions of people who lack access to banking services. Furthermore, blockchain can enhance the efficiency and transparency of social programs, ensuring that aid reaches its intended beneficiaries without corruption or mismanagement. The World Food Programme's Building Blocks initiative exemplifies this by utilizing blockchain to streamline food assistance distribution in conflict-affected regions.

Additionally, blockchain's capacity for secure digital identity management allows individuals to establish verifiable identities, which is crucial for accessing financial services and government support. Projects such as Agri-Wallet in Kenya demonstrate how blockchain can empower smallholder farmers by providing them with secure payment solutions and access to credit. As these applications continue to evolve, they hold the promise of transforming the socio-economic landscape for BoP communities, driving sustainable development, and reducing poverty on a global scale.

2 Review of Literature

The application of blockchain technology has garnered significant attention as a potential solution for addressing the challenges faced by the Bottom of the Pyramid (BoP) populations. These communities, often characterized by extreme poverty and limited access to essential services, stand to benefit from the unique attributes of blockchain, including transparency, security, and decentralization. Various studies have explored how blockchain can facilitate financial inclusion, enhance governance, and empower marginalized groups, thereby contributing to poverty alleviation.

One of the primary areas where blockchain shows promise is in improving financial access for the unbanked. According to Kshetri and Voas, blockchain can help reduce fraud and corruption, thereby increasing legal property titles, which are crucial for economic empowerment in developing countries.

Initiatives like Mojaloop and Agri-Wallet exemplify this potential by providing decentralized financial services that enhance the ability of individuals to engage in economic activities without the need for traditional banking infrastructure. Mojaloop, developed by the Gates Foundation, aims to connect various financial services, enabling millions of individuals to access banking solutions that were previously unavailable.

Moreover, blockchain's capacity for enhancing transparency in governance is critical for BoP populations, who often face corruption and mismanagement in public services. Projects such as the Abuja Electronic Government Records System in Nigeria utilize blockchain to ensure that public records are tamper-proof, thus increasing accountability and reducing corruption. This transparency is essential for ensuring that resources intended for poverty alleviation reach the intended beneficiaries, as highlighted by the World Food Programme's "Building Blocks" initiative, which has successfully improved food distribution in conflict-affected areas.

The integration of blockchain with big data technologies has also been proposed as a means to enhance targeted poverty alleviation efforts. By standardizing personal information and utilizing blockchain for secure data management, organizations can better identify and assist those in need. However, challenges such as the need for digital literacy, infrastructure development, and regulatory frameworks remain significant barriers to effective implementation (Kshetri and Voas 2018).

Despite these challenges, the literature suggests that the potential benefits of blockchain for BoP communities are substantial. For instance, blockchain can facilitate access to microloans and crowdfunding opportunities, allowing entrepreneurs in underserved areas to secure funding for their ventures. Furthermore, initiatives like SOLshare in Bangladesh demonstrate how blockchain can enable peer-to-peer energy trading, providing affordable electricity to communities without reliable access.

However, the successful implementation of blockchain solutions requires a nuanced understanding of local contexts and the specific needs of BoP populations. As noted by Hye et al., the adaptation of blockchain applications to fit the socio-economic realities of these communities is crucial for their success (Kshetri and Voas 2018). Additionally, the need for interdisciplinary training and capacity building among local stakeholders is essential to ensure that blockchain technology is effectively integrated into existing systems.

In conclusion, the literature indicates that while blockchain technology presents transformative opportunities for addressing the challenges faced by the Bottom of the Pyramid, significant hurdles must be overcome. Collaborative efforts among governments, NGOs, and the private sector are necessary to harness the full potential of blockchain for poverty alleviation, ensuring that its benefits reach the most vulnerable populations.

3 Chapter Overview

Investigating the transformative potential of blockchain applications for addressing the Bottom of the Pyramid (BoP), this study employs a secondary qualitative methodology. The research is anchored in an extensive review of current literature, comprising peer-reviewed academic journals, authoritative reports, and reputable online sources. Key sources include publications from leading institutions and industry experts such as the World Bank, IBM, and blockchain-focused research centers, selected for their credibility and relevance. This approach ensures the accuracy and reliability of the findings by drawing on high-quality references in the fields of blockchain technology and socio-economic development. Through a meticulous analysis of theoretical and practical implications, this methodology provides a thorough understanding of how blockchain can enhance financial inclusion, transparency, and efficiency for BoP communities. Adhering to scholarly research standards, the study's conclusions are derived from a rigorous examination of existing knowledge.

3.1 Blockchain for Financial Inclusion

Blockchain technology has emerged as a transformative solution for enhancing financial inclusion, particularly for the Bottom of the Pyramid (BoP) populations. These communities, often marginalized from traditional banking systems, face significant barriers to accessing financial services. Blockchain's decentralized, secure, and transparent nature can bridge these gaps, facilitating access to essential financial resources.

One prominent example is "Mojaloop", an open-source payment platform developed by the Gates Foundation. Mojaloop aims to connect various financial services, enabling individuals without bank accounts to engage in digital transactions. By providing a framework for interoperability among different payment systems, Mojaloop has the potential to reach millions of unbanked individuals across Africa, allowing them to access mobile money services and participate in the digital economy.

Another noteworthy initiative is "Agri-Wallet", which addresses the financial challenges faced by smallholder farmers in Kenya. This blockchain-based mobile application allows farmers to receive payments securely and access microloans, thereby improving their cash flow and enabling them to invest in their agricultural activities. Within a year of its launch, Agri-Wallet successfully onboarded approximately 4,000 farmers, demonstrating the viability of blockchain in enhancing financial access for rural communities.

"Wala" a blockchain firm operating in Africa, focuses on creating a financial ecosystem tailored to low-income individuals. By eliminating intermediaries and reducing transaction costs, Wala enables users to send and receive money efficiently, access credit, and manage their finances through a mobile app. This approach not only fosters financial inclusion but also empowers users to build a credit history, which is crucial for accessing larger loans in the future (Kshetri, n.d.).

The "World Food Programme's (WFP) Building Blocks" initiative exemplifies how blockchain can enhance humanitarian aid and financial inclusion. In Pakistan, WFP implemented a blockchain system to streamline cash transfers to vulnerable populations, ensuring that aid reaches its intended recipients without loss or corruption. This system has significantly reduced administrative costs and improved the efficiency of aid delivery, demonstrating blockchain's potential in addressing financial exclusion in crisis situations.

Furthermore, "Grassroots Economics", funded by UNICEF's Innovation Fund, employs blockchain to create community currencies that facilitate local trade and enhance access to credit. By allowing communities to trade goods and services using a decentralized currency, Grassroots Economics empowers individuals who lack traditional credit histories, enabling them to participate in the economy and access essential services (Grassroots Economics, n.d.).

Despite these promising developments, challenges remain in the widespread adoption of blockchain for financial inclusion. Issues such as digital literacy, regulatory frameworks, and infrastructure development must be addressed to maximize the impact of blockchain solutions (Blockchain's Impact in Fostering Global Financial Inclusion, n.d.). Nonetheless, the potential for blockchain to transform financial services for BoP populations is substantial, offering a pathway to greater economic empowerment and poverty alleviation.

3.2 *Blockchain for Supply Chain Transparency*

Blockchain technology has emerged as a transformative solution for enhancing supply chain transparency, particularly in the context of Bottom of the Pyramid (BoP) communities. By providing a decentralized, secure, and immutable ledger for recording transactions, blockchain enables traceability and trust throughout the supply chain, benefiting producers, consumers, and intermediaries alike.

One prominent example is the "Provenance" initiative in Indonesia, which uses blockchain to track the origin and sustainability of fisheries. By providing a transparent record of the supply chain, Provenance empowers small-scale fishermen, promotes environmental stewardship, and ensures that consumers have access to ethically sourced seafood (Provenance | Sustainability Marketing Technology, n.d.). Another notable case study is "Agri-Wallet" in Kenya, which leverages blockchain to provide smallholder farmers with secure payment solutions and access to credit, thereby enhancing their ability to participate in the supply chain and improve their livelihoods.

The "Mitsubishi Chemical Group" has also embraced blockchain technology to drive circularity in its supply chain. By using blockchain to track material flows and ensure responsible sourcing, the company can guarantee the authenticity of its products and demonstrate its commitment to sustainability to consumers and regulators.

Similarly, "BASF", in collaboration with SAP, has developed a blockchain-based mass balance approach called Green Token to enhance traceability in the chemical industry, enabling companies to verify the origin and sustainability of raw materials.

In the pharmaceutical industry, blockchain has been used to combat the growing threat of counterfeit drugs. A pilot project involving KPMG, Merck, Walmart, and IBM demonstrated the potential of blockchain to trace prescription drugs from manufacturer to patient, reducing the time for traceability from 16 weeks to just two seconds. This level of transparency is crucial for ensuring the safety and efficacy of medicines, particularly in regions where counterfeit drugs are prevalent.

The IBM Food Trust initiative exemplifies how blockchain can enhance transparency and trust in the food supply chain. By bringing together a diverse group of growers, distributors, and retailers, IBM Food Trust has delivered real business results and boosted confidence in the world's food products and their supply chains (Grassroots Economics, n.d.). This collaborative approach is particularly relevant for BoP communities, where fragmented supply chains and lack of trust can hinder access to markets and fair prices.

While blockchain technology holds significant promise for enhancing supply chain transparency, challenges remain in terms of scalability, interoperability, and regulatory frameworks. Successful implementation requires collaboration among stakeholders, investment in infrastructure, and the development of common standards and protocols. However, as blockchain continues to evolve and demonstrate its potential, it holds the key to unlocking more inclusive, sustainable, and resilient supply chains that benefit BoP communities and the global economy (IBM Supply Chain Intelligence Suite—Food Trust, n.d.).

3.3 Blockchain for Identity Management

Blockchain technology offers a revolutionary approach to identity management, particularly for marginalized populations at the Bottom of the Pyramid (BoP). Traditional identity verification systems often exclude these individuals due to a lack of documentation, centralized control, and vulnerability to fraud. Blockchain's decentralized and secure nature can empower these communities by providing them with self-sovereign identities that enhance access to essential services.

One significant example is the "ID2020 Alliance", which aims to provide digital identities to individuals in need, particularly refugees and those without formal identification. By leveraging blockchain technology, ID2020 enables individuals to control their own identity data, reducing dependency on centralized authorities and minimizing the risk of identity theft. This initiative has garnered support from organizations like Accenture and the World Food Programme, which have implemented blockchain-based identity solutions to streamline aid distribution and improve service access for vulnerable populations.

The "World Food Programme (WFP)" has also successfully utilized blockchain for identity management in its humanitarian efforts. In a pilot project in Jordan,

WFP implemented an Ethereum-based identity system that allowed Syrian refugees to access food assistance without the need for traditional identification documents. This system not only improved the efficiency of aid distribution but also empowered refugees by giving them control over their identity data, thus enhancing their dignity and agency in the aid process.

Another notable case is "Civic", a blockchain-based identity verification platform that allows users to manage their personal information securely. Civic enables individuals to create a digital identity that can be used across various services without the need for multiple logins or passwords. This is particularly beneficial for BoP populations, who often lack access to reliable identity verification methods. By providing a secure and user-friendly platform, Civic facilitates access to essential services such as banking, healthcare, and education.

"NEC Corporation" has developed a blockchain-based identity management system that enhances security and privacy while allowing users to maintain control over their personal data. This system is particularly relevant for BoP communities, where individuals often lack formal identification and are at risk of exclusion from vital services. NEC's solution enables users to verify their identity through a decentralized platform, which can be particularly beneficial in regions with limited infrastructure (Blockchain-for-digital-identity, n.d.).

Moreover, the "IBM Blockchain" platform offers a robust framework for managing digital identities. By allowing individuals to curate their profiles and control data sharing, IBM's solution empowers users while ensuring that their information remains secure and private. This is crucial for BoP populations, who may have limited access to traditional identity verification methods and are often vulnerable to data breaches (Blockchain for Digital Identity and Credentials I IBM, n.d.).

Despite the potential benefits, challenges remain in the widespread adoption of blockchain for identity management. Issues such as digital literacy, infrastructure development, and regulatory frameworks must be addressed to maximize the impact of these solutions. However, as blockchain technology continues to evolve, it holds the promise of transforming identity management for BoP communities, enhancing their access to essential services and empowering them to participate fully in society.

3.4 Impact on Economic Empowerment

Blockchain technology is increasingly recognized for its potential to drive economic empowerment, particularly among the Bottom of the Pyramid (BoP) populations. By enhancing access to financial services, creating job opportunities, and fostering entrepreneurial ventures, blockchain can significantly improve the economic landscape for marginalized communities.

One of the most significant impacts of blockchain is its ability to facilitate financial inclusion. Traditional banking systems often exclude individuals in low-income communities due to high transaction costs and lack of formal identification.

Blockchain technology mitigates these barriers by enabling secure, low-cost transactions, and providing a platform for microfinance. For instance, "M-Pesa", a mobile money service in Kenya, has successfully lifted approximately 194,000 households out of extreme poverty within its first six years by allowing users to conduct financial transactions via mobile phones, demonstrating the potential of digital financial services in enhancing economic stability and growth.

Moreover, blockchain enables "Microtransactions", which are vital for low-income individuals who may not have access to traditional banking services. By allowing small-scale transactions, blockchain empowers users to engage in global commerce. For example, "BitPesa", a blockchain-based payment platform, enables businesses in Africa to conduct cross-border transactions efficiently, reducing costs and improving access to international markets (Varghese 2018). This not only facilitates trade but also creates job opportunities as businesses can expand their operations.

In addition to financial inclusion, blockchain fosters entrepreneurial opportunities by providing a transparent and efficient means for startups to raise capital. Platforms like "Kickstarter" have evolved to include blockchain-based fundraising, allowing entrepreneurs to reach a global audience without relying on traditional venture capital. This democratization of funding sources is particularly beneficial for BoP entrepreneurs, who often struggle to secure financing through conventional channels.

Case studies further illustrate the transformative potential of blockchain in economic empowerment. For instance, the "One Acre Fund" in Kenya utilizes blockchain to streamline loan repayments for farmers. By allowing farmers to repay loans via mobile money, the program has reduced payment leakages by 85% and significantly improved financial management for low-income farmers. This not only enhances their economic stability but also encourages agricultural productivity, which is crucial for food security and income generation in rural areas.

Additionally, blockchain can create job opportunities in emerging sectors. The rise of blockchain technology has led to the creation of new roles in software development, data analysis, and cybersecurity. As more businesses adopt blockchain solutions, the demand for skilled professionals in these areas is expected to grow, providing employment opportunities for individuals in BoP communities who acquire the necessary skills.

Despite the promising potential of blockchain, challenges remain in its implementation, including the need for infrastructure development, digital literacy, and regulatory frameworks. However, as blockchain technology continues to evolve, it offers a pathway to economic empowerment for BoP populations, fostering job creation and entrepreneurial opportunities that can significantly improve their livelihoods.

3.5 Social and Cultural Impacts

Blockchain technology has the potential to significantly transform social and cultural dynamics, especially at the Bottom of the Pyramid (BoP), where access to essential services is often limited. The following sources provide insights into the social impacts of blockchain, particularly in promoting financial inclusion, building trust, and addressing cultural challenges.

Social Inclusion and Trust-Building: Blockchain enables marginalized populations to participate in the formal economy by providing access to financial services. For instance, platforms like BitPesa in Kenya allow individuals without traditional bank accounts to engage in cross-border transactions, fostering trust within communities by facilitating secure and low-cost transactions without intermediaries. This capability enhances financial accessibility and reduces reliance on corrupt institutions, thereby strengthening community ties and economic resilience (Realizing Blockchain's Potential for Social Impact, n.d.).

Cultural Adoption and Resistance: Despite its benefits, the adoption of blockchain technology faces cultural resistance in some communities. In rural India, skepticism toward digital platforms can hinder acceptance. Initiatives like the eSewa platform in Nepal demonstrate the importance of community engagement and education in overcoming these barriers. By introducing blockchain in a culturally sensitive manner, these efforts have led to greater acceptance and integration of technology into daily life, highlighting the need for tailored approaches to technology adoption (Realizing Blockchain's Potential for Social Impact, n.d.).

Impact on Social Structures: Blockchain can democratize access to resources and information, particularly in areas with disputed land ownership. Projects like Bitland in Ghana utilize blockchain to create transparent records of land ownership, helping to resolve conflicts and strengthen community relationships. This transparency not only reduces disputes but also promotes a more equitable distribution of resources, enabling individuals to assert their rights and ownership claims effectively (Mhlanga 2023).

3.6 Environmental Impact

The deployment of blockchain technology at the Bottom of the Pyramid (BoP) has significant environmental implications, including both potential benefits and drawbacks. The following sources provide insights into the environmental impact of blockchain, particularly regarding energy consumption, sustainable practices, and waste management.

Energy Consumption and Environmental Strain: Blockchain networks, especially those using Proof-of-Work (PoW) mechanisms, are known for their high energy consumption. For instance, Bitcoin mining consumes more electricity than some small countries, raising concerns about sustainability. In resource-constrained regions like sub-Saharan Africa, this energy-intensive technology could exacerbate existing challenges, leading to increased carbon emissions and environmental degradation. The ecological impact of blockchain technologies must be carefully considered to ensure they do not worsen energy scarcity and reliance on non-renewable sources (Salehi 2023a, b).

Promotion of Sustainable Practices: Conversely, blockchain can enhance environmental sustainability by improving supply chain transparency. In agriculture, for example, platforms like Bext360 utilize blockchain to trace the origins of products, ensuring that sustainable practices are followed. This transparency not only promotes eco-friendly practices but also empowers farmers by allowing them to verify and market their sustainable methods, potentially attracting premium prices for their products (Dupuis 2024).

Reducing Waste and Improving Resource Management: Blockchain technology can significantly contribute to environmental conservation by improving waste management. Initiatives like Plastic Bank incentivize waste collection by offering digital tokens for plastic waste. In Haiti, this program has successfully reduced plastic pollution while providing income to local communities (Piddubna 2024). Such applications illustrate how blockchain can foster environmental stewardship and economic empowerment simultaneously.

3.7 Technological Feasibility and Infrastructure

Deploying blockchain technology at the Bottom of the Pyramid (BoP) requires addressing unique technological challenges and infrastructure needs to ensure scalability and sustainability. The following sources provide insights into the technical feasibility and infrastructure requirements for blockchain adoption in low-resource settings.

Technical Requirements and Infrastructure Needs: Effective blockchain deployment at the BoP necessitates robust technical infrastructure, including reliable internet connectivity, sufficient computational power, and secure digital devices. However, many BoP communities, particularly in rural areas, lack these basic necessities. For instance, in sub-Saharan Africa, internet penetration remains below 30% in some regions, and access to electricity is still a significant barrier (Narayanan et al. 2019). To overcome these challenges, lightweight blockchain solutions and hybrid models that combine on-chain and off-chain processing are being explored. For example, the BanQu platform employs a mobile-first approach, requiring minimal data usage and allowing users to access services even with basic smartphones, reducing the dependency on high-end infrastructure (Crosby et al. 2016).

Scalability Issues: Scalability is a significant challenge when implementing blockchain at the BoP. Public blockchains, such as Ethereum, can face scalability issues due to high transaction volumes, leading to slower processing times and increased costs (Nakamoto, n.d.). To address these concerns, solutions like sidechains or Layer 2 protocols, which process transactions off the main blockchain and then anchor them to the main chain, can be employed. These solutions help reduce congestion on the main blockchain and ensure faster, more affordable transactions (Bada et al. 2021).

Sustainability of Blockchain Technologies: The sustainability of blockchain technologies in BoP contexts is closely linked to their energy consumption and environmental impact. High energy costs associated with certain blockchain consensus mechanisms, such as Proof-of-Work (PoW), are unsustainable in low-resource settings where electricity is both scarce and expensive (Crosby et al. 2016). To mitigate this, blockchain projects targeting the BoP are increasingly adopting energy-efficient consensus mechanisms, such as Proof-of-Stake (PoS) or Proof-of-Authority (PoA), which significantly reduce energy consumption. For example, the Celo blockchain, designed for mobile-first financial services in developing countries, uses a PoS consensus mechanism, making it more sustainable and feasible for use in BoP communities (Narayanan et al. 2019).

3.8 Financial Mechanisms and Microfinance

Blockchain technology has the potential to revolutionize microfinance by providing innovative financial mechanisms that enhance economic inclusion for individuals at the Bottom of the Pyramid (BoP). The following sources provide insights into the role of blockchain in microfinance, decentralized finance (DeFi) applications, and the impact on remittances.

Role of Blockchain in Microfinance: Traditional microfinance institutions often face challenges such as high operational costs and inefficiencies in tracking funds. Blockchain technology addresses these issues by enabling peer-to-peer transactions and ensuring transparency in fund flows. For example, the Kiva Protocol in Sierra Leone creates secure digital identities for borrowers, allowing for efficient loan disbursement even in areas without formal banking infrastructure (Contributors 2023). This streamlining of the loan process enhances financial inclusion by extending services to more individuals at the BoP.

Decentralized Finance (DeFi) Applications for the BoP: DeFi platforms operate without centralized intermediaries, providing direct access to financial services through smart contracts. This is particularly beneficial for BoP populations who often lack access to traditional banking. Platforms like Aave and Compound allow users to borrow and lend cryptocurrencies, with options for microloans in stablecoins to mitigate volatility (Sahay 2023). These platforms can be designed with low transaction fees and mobile accessibility, making them practical for low-resource settings.

Remittances and Financial Inclusion: Blockchain technology significantly improves remittance processes, which are crucial for many families in the BoP. For instance, platforms like Coins.ph in the Philippines facilitate low-cost, near-instant cross-border money transfers using stablecoins like USDC. This reduces the costs and time associated with traditional remittance services, enhancing financial stability for recipients and enabling greater economic participation (Moreira and Silva 2020).

3.9 User Experience and Accessibility

Ensuring that blockchain applications are accessible and user-friendly is crucial for their successful adoption at the Bottom of the Pyramid (BoP). The following sources provide insights into the design of user-friendly interfaces, educational strategies for blockchain literacy, and the importance of accessibility in blockchain applications.

Designing User-Friendly Interfaces: To make blockchain technology accessible to BoP users, interfaces must be simple, intuitive, and culturally relevant. Many individuals in BoP communities rely on basic mobile phones, necessitating optimization for low-end devices with limited data capabilities. The M-Pesa platform in Kenya exemplifies a successful user-centric design with its simple text-based interface, which can inform the development of blockchain applications like Celo that cater to similar user needs (Nest 2023).

Educational Strategies for Blockchain Literacy: Comprehensive educational strategies are essential for building blockchain literacy among BoP users, who may be unfamiliar with concepts like digital wallets and cryptocurrencies. Tailored initiatives, such as video-based learning used by organizations like Digital Green, can effectively educate communities about blockchain. Programs like Binance Academy also offer free resources in multiple languages, helping to empower individuals with the knowledge needed to use blockchain applications confidently (Salehi 2023a, b).

Accessibility in Blockchain Application: Accessibility is a key consideration in the design of blockchain applications. Research indicates that many existing crypto wallets lack accessibility features for users with disabilities, highlighting the need for focused design efforts (Zhou et al. 2023). By implementing features that enhance usability for all users, including those with visual impairments, blockchain applications can improve user experience and security, ensuring that technology is inclusive and beneficial for a broader audience (Mohammad and Vargas 2022).

3.10 Partnerships and Ecosystem Development

Deploying blockchain technology to address the needs of the Bottom of the Pyramid (BoP) requires robust partnerships between governments, non-governmental organizations (NGOs), and the private sector. These collaborations are essential for creating a supportive ecosystem that fosters the adoption and sustainable development of blockchain solutions tailored to the unique challenges faced by BoP communities.

Role of Partnerships in Deploying Blockchain for BoP: Partnerships between diverse stakeholders are crucial in overcoming the infrastructural, regulatory, and educational barriers that often impede the deployment of blockchain technology in

BoP regions. Governments play a key role in establishing regulatory frameworks that encourage innovation while protecting vulnerable populations. For example, the government of India has partnered with the private sector and NGOs to pilot blockchain projects aimed at improving transparency in public distribution systems and ensuring that subsidies reach the intended beneficiaries.

NGOs bring valuable on-the-ground experience and an understanding of local contexts, which are essential for designing blockchain applications that are relevant and accessible to BoP communities. For instance, the World Food Programme (WFP) partnered with blockchain technology firm Parity Technologies to develop the Building Blocks project, which uses blockchain to manage and distribute aid more efficiently to refugees in Jordan. The private sector, particularly tech companies and financial institutions, contributes innovation, capital, and technical expertise. Companies like IBM and Microsoft are actively involved in blockchain projects that target financial inclusion and supply chain transparency in BoP markets. IBM's partnership with Kenya's Twiga Foods uses blockchain to optimize the agricultural supply chain, helping small-scale farmers access markets more efficiently (Dahir 2022).

Building Ecosystems for Blockchain Innovation: Developing a broader ecosystem that supports blockchain innovation is critical. In regions like Southeast Asia, blockchain hubs have emerged where stakeholders can share knowledge, pool resources, and scale successful pilot projects across different communities. Accelerators and incubators focused on social impact are also playing a crucial role in nurturing blockchain startups that address BoP issues. Programs like the UNICEF Innovation Fund provide early-stage financing and technical support to blockchain startups developing solutions for children in low-resource settings.

3.11 Policy and Regulatory Implications

The deployment of blockchain technology to address the needs of the Bottom of the Pyramid (BoP) requires supportive regulatory frameworks and an understanding of legal challenges. The following sources provide insights into the policy and regulatory implications of blockchain initiatives, focusing on the need for effective regulation, legal challenges, and opportunities for economic empowerment.

Regulatory Frameworks to Support Blockchain Initiatives: Effective regulatory frameworks are essential for encouraging innovation while ensuring consumer protection and financial stability. For example, India's regulatory sandbox for fintech innovations allows startups to test blockchain applications under regulatory oversight, providing a controlled environment for innovation while helping regulators understand potential risks and benefits (Asak 2024). Tailored regulations that focus on data privacy and digital identities are also crucial, as seen in Estonia's approach to legitimizing blockchain-based identities, which can help bridge access gaps for BoP individuals.

Legal Challenges and Opportunities: Deploying blockchain at the BoP presents legal challenges related to jurisdiction, compliance with financial regulations, and managing decentralized systems. Ensuring compliance with anti-money laundering (AML) and know-your-customer (KYC) requirements can be particularly challenging in regions with weak regulatory systems, potentially exposing users to fraud risks (Asak 2024). However, blockchain also offers opportunities for legal empowerment, such as creating immutable records of land ownership, as demonstrated by the Bitland project in Ghana, which enhances property rights and economic security for BoP individuals.

Smart Contracts and Compliance: Smart contracts provide a way to enforce agreements without intermediaries, making them useful in contexts where legal systems are slow or inaccessible. In microfinance, for instance, smart contracts can automatically enforce loan repayment terms, reducing risks for both lenders and borrowers (Moreira and Silva 2020).

3.12 Ethical Considerations and Data Privacy

Ethical considerations and data privacy are crucial aspects to address when deploying blockchain technology at the Bottom of the Pyramid (BoP). The following sources provide insights into the ethical issues surrounding blockchain use for vulnerable populations and the data privacy concerns and solutions.

Ethical Issues in Blockchain Use for Vulnerable Populations: One of the primary ethical concerns is the potential for exploitation of vulnerable individuals with limited digital literacy. They may unknowingly expose themselves to risks such as loss of funds due to private key mismanagement or fall victim to scams and fraudulent schemes. Additionally, the introduction of blockchain can inadvertently exacerbate existing inequalities if only a subset of the population has access to the technology or understands how to use it. Ethical blockchain deployment must include measures to ensure broad accessibility and education to benefit all community members equally (DuPont 2020).

Data Privacy Concerns and Solutions: Blockchain's inherent transparency can conflict with the need to protect personal information, exposing sensitive data such as financial transactions or personal identifiers. To address these concerns, blockchain developers are exploring privacy-enhancing technologies like zero-knowledge proofs (ZKPs), which allow the verification of transactions without revealing the underlying data. Projects like Zcash have applied this technology to enable private transactions while maintaining the integrity of the blockchain (Zerocash: Decentralized Anonymous Payments from Bitcoin 2014).

Another approach is the use of permissioned blockchains, where access to data is restricted to trusted entities. This model is particularly relevant for applications such as digital identity systems or healthcare records in BoP communities. The ID2020 initiative explores the use of permissioned blockchains to provide secure and private digital identities for refugees and stateless individuals, ensuring that their data is protected from unauthorized access while enabling them to access essential services (ID2020 2023).

3.13 Impact Measurement and Evaluation

Measuring the impact of blockchain applications on Bottom of the Pyramid (BoP) communities is essential for understanding their effectiveness and guiding future developments. The following sources provide insights into the metrics and frameworks for evaluating the impact of blockchain, as well as the importance of long-term studies and ongoing monitoring strategies.

Metrics and Frameworks for Evaluating Impact: Developing appropriate metrics and frameworks is crucial for evaluating the impact of blockchain on BoP communities. Quantitative metrics might include the number of people gaining access to financial services or the volume of transactions conducted through blockchain platforms. For example, a blockchain-based remittance platform could be evaluated based on the percentage decrease in fees compared to traditional services and the increase in remittances received by BoP families (Realizing Blockchain's Potential for Social Impact, n.d.). Qualitative metrics, focusing on user satisfaction and perceived economic empowerment, are also important for a comprehensive evaluation.

Frameworks for impact evaluation, such as the Theory of Change (ToC), can help outline expected pathways through which blockchain benefits BoP communities and identify key indicators along these pathways (Mohammad and Vargas 2022). This approach allows for a nuanced understanding of how blockchain contributes to long-term development goals, such as poverty reduction and financial inclusion.

Long-Term Impact Studies and Ongoing Monitoring Strategies: Long-term impact studies are crucial for understanding the sustained effects of blockchain applications. These studies should track the same indicators over time to determine whether benefits are maintained or diminish after initial implementation (Chang et al. 2020). For instance, long-term studies on blockchain-based microfinance initiatives could examine whether increased access to credit leads to sustained improvements in income levels among BoP entrepreneurs.

Ongoing monitoring strategies, enabled by the transparency of blockchain, can continuously assess the performance of blockchain applications. Real-time data analytics can identify potential issues and inform data-driven adjustments to improve service delivery (Gromyko et al., n.d.). Additionally, participatory monitoring approaches, involving BoP users in the evaluation process, can provide deeper insights into the real-world impact of blockchain, ensuring that the perspectives of those directly affected are considered (Mohammad and Vargas 2022).

3.14 Challenges and Limitations

Blockchain technology presents a transformative opportunity for addressing the challenges faced by the Bottom of the Pyramid (BoP) populations. However, its implementation is fraught with challenges and limitations that must be navigated to realize its full potential. These challenges can be categorized into technical barriers, regulatory and policy issues, and socio-economic challenges.

- **Technical Barriers:** One of the primary technical challenges of blockchain is "Scalability". Many blockchain networks, particularly those utilizing proof-of-work consensus mechanisms, struggle to handle increasing transaction volumes, leading to congestion and high transaction fees. For instance, Bitcoin and Ethereum have faced significant delays and costs during peak usage periods, which can deter users from adopting these technologies for everyday transactions. Solutions such as off-chain transactions and layer-2 scaling solutions are being explored, but widespread implementation remains a work in progress. Another technical limitation is "Interoperability" among different blockchain platforms. Currently, various blockchain networks operate in silos, which complicates data sharing and transaction processing across platforms. This lack of interoperability can hinder the development of comprehensive solutions that address the needs of BoP communities, as users may be locked into specific ecosystems without the ability to leverage broader blockchain capabilities.
- **Regulatory and Policy Issues:** Regulatory uncertainty poses significant challenges for blockchain applications aimed at BoP populations. Many countries lack clear frameworks for blockchain technology, leading to confusion among developers and users alike. For example, in India, the regulatory environment surrounding cryptocurrencies and blockchain remains ambiguous, creating hesitance among potential users and investors. Such uncertainty can stifle innovation and limit the growth of blockchain solutions designed to enhance financial inclusion and economic empowerment. Moreover, compliance with existing regulations, such as anti-money laundering (AML) and know-your-customer (KYC) requirements, can be cumbersome for blockchain applications. These regulations often require extensive documentation and verification processes, which can be particularly challenging for individuals in BoP communities who may lack formal identification or access to necessary resources.
- **Socio-Economic Challenges:** In addition to technical and regulatory barriers, socio-economic challenges also limit the effectiveness of blockchain applications in BoP contexts. Many individuals in these communities may lack digital literacy, making it difficult for them to navigate blockchain platforms and understand their benefits. For instance, a study found that while mobile money services like M-Pesa have improved financial inclusion in Kenya, a significant portion of the population remains unaware of how to use these technologies effectively. Furthermore, infrastructure limitations, such as unreliable internet access and inadequate technological support, can hinder the adoption of blockchain solutions. In rural areas, where many BoP populations reside, these infrastructural challenges can create significant barriers to accessing blockchain-based services.

While blockchain technology holds promise for addressing the needs of BoP populations, it is essential to recognize and address the numerous challenges and limitations that accompany its implementation. By focusing on enhancing scalability, improving regulatory clarity, and addressing socio-economic barriers, stakeholders can work toward creating a more inclusive and effective blockchain ecosystem that empowers marginalized communities.

3.15 Future Prospects and Recommendations

Blockchain technology holds significant promise for addressing the challenges faced by the Bottom of the Pyramid (BoP) populations, particularly in enhancing financial inclusion and economic empowerment. As the technology matures, its future prospects hinge on overcoming existing challenges and leveraging opportunities for scalability, regulatory frameworks, and areas for further research.

- **Potential for Scalability:** Scalability remains a critical factor in the successful implementation of blockchain solutions for BoP communities. Current blockchain networks often face limitations in transaction throughput, which can hinder their ability to serve large populations effectively. To address this, innovations such as layer-2 solutions and sharding are being explored to enhance transaction speeds and reduce costs. For instance, the "Lightning Network" for Bitcoin allows for faster and cheaper transactions by enabling off-chain transactions, which can significantly benefit users in low-income areas where affordability and speed are paramount.
- **Policy and Regulatory Recommendations:** The regulatory landscape for blockchain technology is still evolving, and clear policies are essential to foster innovation while ensuring consumer protection and compliance. Governments should develop comprehensive regulatory frameworks that encourage the adoption of blockchain solutions while addressing concerns related to fraud, money laundering, and data privacy. For example, the "Central Bank of Nigeria's Project GIANT" aims to leverage blockchain for identity verification, which can facilitate access to financial services for underserved populations. Such initiatives highlight the need for collaboration between regulators and technology providers to create an enabling environment for blockchain applications. Moreover, governments should consider implementing pilot programs that allow for the testing of blockchain applications in real-world scenarios. This approach can provide valuable insights into the practical challenges and benefits of blockchain technology, informing future policy decisions.
- **Areas for Further Research:** Research into blockchain applications for BoP populations should focus on several key areas. First, studies examining the long-term impacts of blockchain on economic empowerment and poverty alleviation are essential. For instance, the "World Food Programme's Building Blocks initiative" has shown promising results in improving aid distribution, but further research is needed to quantify its impact on beneficiaries' livelihoods over time. Additionally, exploring the integration of blockchain with other emerging technologies, such as artificial intelligence and the Internet of Things, can yield innovative solutions for BoP communities. For example, combining blockchain with IoT

can enhance supply chain transparency, ensuring that agricultural products reach consumers efficiently and sustainably. Finally, addressing the digital divide is crucial for the successful adoption of blockchain technologies. Research should investigate strategies for improving digital literacy and access to technology in BoP communities, ensuring that all individuals can benefit from blockchain applications.

In conclusion, while blockchain technology presents transformative opportunities for addressing the needs of BoP populations, its future prospects depend on overcoming scalability challenges, establishing robust regulatory frameworks, and pursuing targeted research initiatives. By fostering collaboration among stakeholders and focusing on practical implementations, blockchain can significantly contribute to economic empowerment and poverty alleviation for marginalized communities.

4 Conclusion

Blockchain technology offers transformative potential for addressing the needs of the Bottom of the Pyramid (BoP) by fostering financial inclusion, enhancing supply chain transparency, and empowering marginalized communities. However, the successful implementation of blockchain solutions in these contexts requires careful consideration of technological, social, and environmental factors.

One of the most significant advantages of blockchain is its ability to provide financial services to the unbanked and underbanked populations. By enabling decentralized finance (DeFi) and microfinance platforms, blockchain can reduce the costs and barriers associated with traditional banking systems. This opens up new opportunities for economic participation among BoP individuals, allowing them to access credit, savings, and insurance services that were previously out of reach. The integration of blockchain into remittance services also promises to enhance the economic stability of BoP families by reducing transaction costs and processing times, thus increasing the net income from cross-border transfers.

Blockchain's impact extends beyond financial inclusion. By improving supply chain transparency, blockchain can ensure that products are sourced sustainably and ethically, benefiting both producers and consumers. For instance, small-scale farmers in BoP communities can use blockchain to verify the origins of their produce, gaining access to global markets and commanding fair prices for their goods. This not only boosts their income but also promotes sustainable agricultural practices, contributing to environmental conservation.

Nevertheless, the deployment of blockchain technology at the BoP is not without challenges. The technical feasibility of implementing blockchain in low-resource settings is a critical concern, given the limitations in infrastructure, such as unreliable internet connectivity and inadequate access to electricity. Addressing these issues requires innovative solutions, such as lightweight blockchain applications and mobile-first platforms, which can operate efficiently on basic devices. Moreover, the scalability and sustainability of blockchain technologies must be considered to ensure that they can handle the transaction volumes and energy demands of widespread adoption without exacerbating existing resource constraints.

Furthermore, ethical considerations and data privacy are paramount in the deployment of blockchain at the BoP. Vulnerable populations must be protected from exploitation, and their data must be securely managed to prevent misuse. Establishing robust regulatory frameworks and fostering partnerships between governments, NGOs, and the private sector are essential steps toward creating an ecosystem where blockchain can thrive.

In conclusion, while blockchain technology holds great promise for addressing the challenges faced by BoP communities, its success depends on a holistic approach that combines technological innovation with social responsibility and environmental sustainability. By carefully navigating these complexities, blockchain can become a powerful tool for driving inclusive growth and empowering the world's most vulnerable populations.

Bibliography

Asak R (2024) Crypto regulation: governments shaping blockchain's Future. coinpaper. https://coinpaper.com/4619/crypto-regulation-governments-shaping-blockchain-s-future

Bada AO, Damianou A, Angelopoulos CM, Katos V (2021) Towards a green blockchain: a review of consensus mechanisms and their energy consumption. In Computing & informatics. Bournemouth University. https://eprints.bournemouth.ac.uk/36968/1/GREEN_BLOCKCHAIN.pdf

Blockchain's impact in fostering global financial inclusion. (n.d.-b). PwC. https://www.pwc.com/us/en/services/digital-assets/blockchain-financial-inclusion.html

Buterin V, Illum J, Nadler M, Schär F, Soleimani A (2024) Blockchain privacy and regulatory compliance: towards a practical equilibrium. Blockchain Research and Applications 5(1):100176. https://doi.org/10.1016/j.bcra.2023.100176

Chang V, Baudier P, Zhang H, Xu Q, Zhang J, Arami M (2020) How blockchain can impact financial services—the overview, challenges and recommendations from expert interviewees. Technol Forecast Soc Chang 158:120166. https://doi.org/10.1016/j.techfore.2020.120166

Chen Y, Bellavitis C (2020) Blockchain disruption and decentralized finance: the rise of decentralized business models. J Bus Ventur Insights 13. https://www.bitget.com/wiki/how-emerging-markets-and-blockchain-can-bring-an-end-to-poverty

Contributors F (2023) Blockchain to the rescue: a new era of microfinance. Financial and business news | finance magnates. https://www.financemagnates.com/cryptocurrency/education-centre/blockchain-to-the-rescue-a-new-era-of-microfinance/

Crosby M, Nachiappan PP, Verma S, Kalyanaraman V (2016) BlockChain technology: beyond bitcoin. In Applied innovation review (2). https://scet.berkeley.edu/wp-content/uploads/AIR-2016-Blockchain.pdf

Dahir AL (2022) Blockchain technology is being used in the early disruption of Kenya's agribusiness. Quartz. https://qz.com/africa/1281840/blockchain-machine-learning-and-ibm-are-disrupting-kenyas-agriculture-and-farming

DuPont Q (2020b) Guiding principles for ethical cryptocurrency, blockchain, and DLT research. https://doi.org/10.21428/58320208.a8364373

Dupuis A (2024, April 4) Blockchain agriculture: revolutionizing farming with Eden green | Eden green. Eden green. https://www.edengreen.com/blog-collection/blockchain-technology-in-agriculture

Forbes (2023) The 5 biggest problems with blockchain technology everyone must know about. Forbes. Retrieved from https://www.forbes.com/sites/bernardmarr/2023/04/14/the-5-biggest-problems-with-blockchain-technology-everyone-must-know-about/

Georgetown University (n.d.) Blockchain and financial inclusion. Georgetown University. Retrieved from https://d3h0qzni6h08fz.cloudfront.net/reports/blockchain-and-financial-inclusion.pdf

Grassroots Economics (n.d.) Grassroots economics: community currencies for empowerment. Grassroots economics. Retrieved from https://grassrootseconomics.org/

GreenToken by SAP (2022) Material traceability for increased circularity in the chemical industry: A blockchain-based mass balance approach using greentoken by SAP. GreenToken by SAP. Retrieved from https://www.sap.com/documents/2022/05/d4f5a6d5-0b7d-0010-bca6-c68f7e60039b.html (ND)

Gromyko A, Luca P, Chaijitwanitkul SMP, Rozowicz R, Shaheen T, Stelea MC, Tao S (n.d.) Blockchain technology and its implications | OxJournal. https://www.oxjournal.org/blockchain-technology-and-its-implications/

Harvard Business Review (2020) Building a transparent supply chain. Harvard business review. Retrieved from https://hbr.org/2020/05/building-a-transparent-supply-chain

IBM (n.d.) Blockchain for digital identity and credentials. IBM. Retrieved from https://www.ibm.com/blockchain-identity

ID2020 (2023) Digital identity alliance. https://id2020.org/

1Kosmos (n.d.) Blockchain identity management: a complete guide. 1Kosmos. Retrieved from https://www.1kosmos.com/blockchain/blockchain-identity-management-a-complete-guide/

Kshetri N, Voas J (2018) Blockchain in developing countries. IEEE IT Prof 20(2), 11–14. Retrieved from https://libres.uncg.edu/ir/uncg/f/N_Kshetri_Blockchain_Developing_2018.pdf

LinkedIn (n.d.) Blockchain technology—challenges and limitations. LinkedIn. Retrieved from https://www.linkedin.com/pulse/blockchain-technology-challenges-limitations-polyd-2c

LinkedIn (n.d.) How blockchain will help to eradicate global poverty. LinkedIn. Retrieved from https://www.linkedin.com/pulse/how-blockchain-help-eradicate-global-poverty-chant%C3%A9-joseph

MarketLinks (n.d.) How blockchain can eradicate poverty in developing countries. MarketLinks. Retrieved from https://www.marketlinks.org/blogs/how-blockchain-can-eradicate-poverty-developing-countries

Mhlanga D (2023) Block chain technology for digital financial inclusion in the industry 4.0, towards sustainable development? Front Blockchain 6. https://doi.org/10.3389/fbloc.2023.1035405

Mohammad A, Vargas S (2022) Challenges of using blockchain in the education sector: a literature review. Appl Sci 12(13):6380. https://doi.org/10.3390/app12136380

Moreira J, Silva D (2020) Can blockchain technology reduce the cost of remittances? OECD Development Co-operation Working Papers. https://doi.org/10.1787/d4d6ac8f-en

Nakamoto S (n.d.) Bitcoin: a peer-to-peer electronic cash system. https://bitcoin.org/bitcoin.pdf

Narayanan A, Bonneau J, Felten E, Miller A, Goldfeder S, Clark J (2019) Bitcoin and cryptocurrency technologies. https://d28rh4a8wq0iu5.cloudfront.net/bitcointech/readings/princeton_bitcoin_book.pdf

NEC Corporation (n.d.) Blockchain for digital identity. NEC Corporation. Retrieved from https://www.nec.com/en/global/solutions/blockchain/blockchain-for-digital-identity.html

Nest (2023) The application layer of blockchain: powering user-friendly solutions. https://www.linkedin.com/pulse/application-layer-blockchain-powering-user-friendly-solutions-jhs4c

Pisa M, Juden M (n.d.) Blockchain and economic development: hype vs. reality. Center for Global Development. Retrieved from https://www.cgdev.org/publication/blockchain-and-economic-development-hype-vs-reality

Piddubna A (2024) How to apply blockchain technology in the agriculture supply chain? Intellias. https://intellias.com/blockchain-in-agriculture-supply-chain/

Rasmussen M (2022) Top 7 blockchain projects for poverty. The Borgen Project. Retrieved from https://borgenproject.org/top-7-blockchain-projects-for-poverty/
Realizing Blockchain's Potential for Social Impact (n.d.-b) UNDP. https://www.undp.org/blog/realizing-blockchains-potential-social-impact
Realizing Blockchain's Potential for Social Impact (n.d.) UNDP. https://www.undp.org/blog/realizing-blockchains-potential-social-impact
ResearchGate (n.d.) Blockchain technology: applications and challenges. ResearchGate. Retrieved from https://www.researchgate.net/publication/351346456_Blockchain_Technology_Applications_and_Challenges
Sandeep Kumar M, Venkat M, Jayagopal P, Mani P, Jothikumar R (2021) Deep learning and edge computing solutions for high performance computing. In: Applying blockchain in agriculture: a study on blockchain technology, benefits, and challenges, pp 167–181. https://doi.org/10.1007/978-3-030-60265-9_11
Sahay V (2023) The benefits of decentralized finance (DeFi) for the unbanked—1Ansah technologies. 1Ansah Technologies. https://1ansah.in/blog/the-benefits-of-decentralized-finance-defi-for-the-unbanked/
Salehi A (2023a) Best practices for designing a user-friendly blockchain interface. Cronj. Retrieved from https://www.cronj.com/blog/best-practices-for-designing-a-user-friendly-blockchain-interface/
Salehi M (2023b) Blockchain is revolutionizing the waste management industry. Towards a circular economy: a case study by carbon copilot. https://www.linkedin.com/pulse/blockchain-revolutionizing-waste-management-industry-towards-salehi
Schuller M (2021) 4 ways blockchain technology aids poverty alleviation in Nigeria. The Borgen Project. Retrieved from https://borgenproject.org/poverty-alleviation-in-nigeria/
ScienceDirect (n.d.) The future of blockchain technology: opportunities and challenges. ScienceDirect. Retrieved from https://www.sciencedirect.com/science/article/abs/pii/B9780128144473000100
Technology Innovation Management Review (2021) Blockchain implementation challenges in developing countries. Technol Innov Manag Rev 11(2):30–39. Retrieved from https://timreview.ca/article/1479
The Borgen Project (2021) 4 ways blockchain technology aids poverty alleviation in Nigeria. The Borgen Project. Retrieved from https://borgenproject.org/poverty-alleviation-in-nigeria/
The Borgen Project (2022) Agri-Wallet: empowering farmers with blockchain. The Borgen Project. Retrieved from https://borgenproject.org/agri-wallet-empowering-farmers-with-blockchain/ (ND)
The Conversation (n.d.) Can blockchain technology help poor people around the world? The Conversation. Retrieved from https://theconversation.com/can-blockchain-technology-help-poor-people-around-the-world-76059
Varghese S (2018) BitPesa is putting ancient payment systems on the blockchain. WIRED. https://www.wired.com/story/bitpesa-blockchain-payment-systems-monzo-revolut/
Zerocash: Decentralized Anonymous Payments from Bitcoin (2014) IEEE conference publication | IEEE Xplore. https://ieeexplore.ieee.org/document/6956581
Zhou Z, Sharma T, Emano L, Wang Y (2023) Iterative design of an accessible crypto wallet for blind users. In Proceedings of the nineteenth symposium on usable privacy and security. https://www.usenix.org/system/files/soups2023-zhou.pdf

Cryptocurrency: The New Generation Gold Standard

Shaista Hasan, M. Safir, and Sameer Ahmad

Abstract Economists are more inclined to analyze how a monetary system is being built on Bitcoin compared to a gold-standard monetary system. Since its inception in 2009, Bitcoin, created by the enigmatic figure Satoshi Nakamoto, has spurred endless debates and discussions about the notional value of cryptocurrency based on Blockchain. Contrary to being an everyday currency like fiat, Blockchain-based currencies don't serve well for daily transactions due to their instability. The paper places Bitcoin as a supranational central bank reserve currency, eliminating existing design flaws by evaluating the similarities and differences between the gold and Bitcoin standards. The research further highlights the proposed "Bitcoin Standard" or "Gold 2.0," Bitcoin's challenges, such as latency and scalability issues, which have become inconsequential because of the growing acceptance of the currency. The research is important to give insight to policymakers, academicians, and economists in understanding the future of the monetary system.

S. Hasan (✉)
Amity Business School, Amity University, Dubai, UAE
e-mail: salvi@amityuniversity.ae

M. Safir
Davis, USA

S. Ahmad
Dubai, UAE

S. Krishnan et al. (eds.), *Unveiling Blockchain Enabled Business Applications*,
Contributions to Management Science, https://doi.org/10.1007/978-981-95-2337-5_3

1 Introduction

1.1 Evolution of Cryptocurrency

Cryptocurrencies represent a real-world application of Blockchain technology, serving as a digital currency that operates independently of intermediaries such as banks. In this context, transactions are authenticated using cryptographic protocols and systematically documented in a decentralized public ledger. This decentralized nature eliminates the need for traditional financial institutions and offers a peer-to-peer electronic payment system. (Barbara Matthews 2019).

The idea for cryptocurrency first emerged in 1983 when American cryptographer David Chaum published a conference paper that was part of his dissertation thesis entitled 'Blind Signatures for Untraceable Payments', which outlined an early form of anonymous cryptographic currency that could be sent untraceable and in a manner that did not require centralized entities such as banks. The first cryptocurrency, Bitcoin, was invented by an unknown person or group of people using the name Satoshi Nakamoto and was released as open-source software in 2009 (Krishnan 2020), Satoshi Nakamoto initiated the inaugural transaction on 10 January 2009 by transferring 10 bitcoins to Hal Finney (Hayes 2019). In February 2011, a significant milestone in the cryptocurrency landscape unfolded with the establishment of Mount Gox, a pioneering platform that gained widespread recognition as the first to facilitate the exchange of Bitcoin for traditional fiat currencies. This development marked a pivotal moment in the evolution of the cryptocurrency market, introducing a tangible bridge between the digital realm of Bitcoin and established financial systems based on fiat currencies. (Wątorek 2021).

The crypto market has gone through boom and bust cycles since Bitcoin's launch in 2009, with rapid price swings driven by speculation and hype. However, in 2022, the overall crypto market capitalization stabilized around $900 billion after declining from all-time highs (Chen et al. 2023). Bitcoin remains the largest cryptocurrency, accounting for about 40% of the total value of the crypto market. As of February 20, 2023, Bitcoin's price sits around $23,000, down significantly from its all-time high of around $69,000 in November 2021 (CoinDesk 2023). Ethereum is the second largest cryptocurrency, with a market share of about 20%. Its native token Ether has also seen price declines in 2022–2023 amid the "crypto winter", but remains an important base layer for decentralized apps and NFTs (Laurinaitis and Štreimikienė 2022). Bitcoin remains the largest with a market cap of $418 billion, followed by Ethereum at $182 billion. (CoinMarketCap 2023) Cryptocurrency adoption continues to increase globally, an estimated 446 million people worldwide owned cryptocurrencies as of January 2022, up from 295 million in 2021 (Foley et al. 2018).The usage is highest in emerging markets such as Vietnam, India, and Nigeria. However, regulatory uncertainty persists in many countries regarding taxation, securities treatment, and anti-money laundering (Auer et al. 2023). The widespread embrace of cryptocurrency is escalating globally, boasting a staggering user base exceeding 300 million worldwide, according to recent research (Hameed and Mian 2022). In Q3

2022, 312 million people owned cryptocurrencies worldwide, up from 295 million in Q2 2022 (Crypto.com 2022).As of February 2023, there are more than 21,000 cryptocurrencies in existence, with a total market capitalization of more than $1 trillion (CoinMarketCap 2023). Cryptocurrencies are transitioning from speculative investments toward more utility and real-world use cases. Payment usage has risen particularly in emerging markets with high inflation or unreliable banking (Ngongang et al. 2022). While volatile, cryptocurrency markets have shown increasing correlations with equity markets and macroeconomic trends (Umar et al. 2022). This suggests deeper integration with mainstream finance, although major regulatory uncertainty remains regarding cryptocurrencies' future.

The expansion of decentralized finance (DeFi) has been a key trend, growing from $15 billion in 2020 to more than $247 billion in 2023 (Chen et al. 2023). The term Decentralized Finance (DeFi) encompasses a dynamic ecosystem of financial applications constructed on Blockchain networks. Within this realm, most DeFi platforms manifest themselves as decentralized applications (dApps), leveraging smart contracts to streamline and automate financial transactions. The overarching aim of DeFi is to establish decentralized frameworks for cryptocurrency borrowing, lending, and the provision of banking services. Among the notable projects within this domain are Chainlink, Wrapped Bitcoin, Uniswap, Maker, and Compound, which collectively represent pioneering efforts in reshaping traditional financial paradigms through decentralized technologies. (Wątorek 2021). DeFi provides financial services such as lending, trading, insurance, and more using blockchain technology and smart contracts without intermediaries. Nonfungible tokens (NFTs) have also seen rapid adoption, with $11.5 billion in sales volume in Q3 2022 (Kumar et al. 2023). Although speculative investment remains dominant, payment usage is growing from nearly nothing in 2015 to $162 billion in transactions in 2022 (Foley et al. 2018). The value locked in DeFi grew from $41 billion in October 2021 to $82 billion in October 2022 indicating growing engagement (Sadhukhan et al. 2023). Decentralized finance (DeFi) based on public blockchains is also growing as a new paradigm for financial services like lending, trading, and borrowing. More than $100 billion is now locked in DeFi protocols, although concerns exist about transparency and regulation (Khwaja et al. 2022).

1.2 Monetary Characteristics of Early Cryptocurrencies

One of the most groundbreaking innovations of cryptocurrencies was the creation of a decentralized ledger system known as blockchain. Unlike traditional ledgers maintained by banks or financial institutions, blockchains operate through a distributed network of computers with no central authority (Nakamoto 2008). This allows cryptocurrency transactions to be recorded and validated through a transparent, immutable, and censorship-resistant public ledger without the need for third-party intermediaries (Catalini and Gans 2021). The decentralized and open-source nature of Blockchains enabled cryptocurrencies to function as peer-to-peer payment systems that eliminate financial intermediaries (Harwick 2016).

Furthermore, most of the early cryptocurrencies had a maximum capped supply. For example, the total supply of Bitcoins is limited to 21 million (Nakamoto 2008). This fixed supply is in contrast to fiat currencies that have unlimited supply controlled by central banks. Capped supply introduces digital scarcity and provides a hedge against inflationary monetary policies of central banks that can devalue fiat currencies through money printing (Van Alstyne 2014). The limited supply supports the view of early cryptocurrencies as 'digital gold', scarce digital assets that can serve as stores of value.

Cryptocurrencies also introduced a novel method for currency issuance and distribution through a process called mining. Instead of being issued by central authorities, new cryptocurrency units are created and rewarded to miners around the world who use specialized computing equipment to validate transactions and add new blocks to the Blockchain (Narayanan et al. 2016). This decentralized creation and distribution of money presents an alternative to centralized control of currency supplies by governments and banks.

In terms of transaction capabilities, cryptocurrencies allow direct digital transfers of value between parties without intermediaries. Users can send and receive crypto payments globally through pseudonymous Blockchain addresses without regard for national borders or banking requirements (Folkinshteyn et al. 2015). This facilitates nonreversible cross-border payments and international remittance transfers that may be cheaper, faster, and more accessible than traditional banking methods (Abramova and Böhme 2016).

Although groundbreaking in many aspects, early cryptocurrencies also had some limitations that prevented wider mainstream adoption as monetary systems. Volatile and speculative price swings made cryptocurrencies risky as units of account and stores of value (Cheah and Fry 2015). Scalability issues led to delays in transaction processing times and high fees that hindered everyday payments (Croman et al. 2016). The pseudonymous nature of crypto transactions also raised concerns about illegal uses (Foley et al. 2018). The decentralized nature of cryptocurrencies made them subject to security risks such as exchange hacks, scam projects, and loss of private keys (Moore and Christin 2013).

The monetary innovations of early cryptocurrencies presented both opportunities and challenges. Their decentralized, transparent, and censorship-resistant transaction capabilities offered new models of digital money and value transfer. However, the outstanding technical hurdles related to scalability, security, and price volatility prevented greater circulation as functional monetary instruments. Nevertheless, these pioneering cryptocurrencies introduced groundbreaking economic properties that continue to evolve through ongoing innovation and research.

1.3 Gold Monetary Standard

For centuries, gold has played a vital role in monetary systems as a standard of value and a medium of exchange backed by its intrinsic physical properties. The gold standard refers to financial systems where paper currency is backed by and can be freely converted into gold coins or bullion (Eichengreen 2019). Under the classic gold standard prevalent in the late nineteenth and early twentieth centuries, governments guaranteed to redeem notes issued by central banks for their value in physical gold upon demand.

The gold standard emerged because of gold's natural qualities that make it well-suited to serve monetary functions such as acting as a store of value and facilitating transactions. Specifically, gold has consistent rarity, divisibility, durability, uniformity, widespread acceptability, and portability throughout history and cultures (Rockoff 2015). The limited global supply and the difficult extraction process of gold prevent sudden devaluations. At the same time, gold can be reliably measured, tested for purity, and divided into small coins or quantities without losing its metallic qualities or value. These attributes enabled gold to evolve as a trusted exchange and accounting unit.

By backing paper currency to fixed amounts of gold, the gold standard provides stability in foreign exchange markets and international trade while restricting governments from running perpetual deficits and excessive money printing. All major global powers used a gold standard during the late 1800s up until 1914 before transitioning briefly back to it from 1925 to 1931 (Eichengreen 2019). Under the gold standard, exchange rates were kept relatively fixed as central banks guaranteed convertibility between domestic currencies and gold, facilitating lower-cost global trade and capital flows.

However, the gold standard also has some disadvantages and shortcomings. Tying money supply growth directly to gold reserves can generate volatility during gold discoveries and shortages (Rockoff 2015). The gold standard can also worsen economic downturns as central banks lose the flexibility of monetary policy to stimulate growth when necessary. These issues came to the fore during the Great Depression, causing countries to abandon their gold standards. Additionally, carrying costs are required to transport and protect large physical gold reserves. These factors led global economies to transition to floating exchange rate fiat currencies backed by more diverse asset reserves.

In 1944, the Bretton Woods system introduced a partial gold standard where only the U.S. dollar was directly convertible into gold while other currencies maintained fixed exchange rates with the dollar (Eichengreen 2019). However, this modified gold standard also proved unsustainable due to fiscal costs, balance-of-payment deficits, and other macroeconomic pressures. President Nixon unilaterally ended the partial U.S. dollar-gold convertibility in 1971, marking the final fall of the gold standard.

Today, no global currency operates on a gold standard. However, gold remains an important reserve asset held by central banks and institutions. It functions as a long-term store of value during financial crises when confidence in fiat money and credit weakens (Reinhart and Rogoff 2014). Private investors also continue to use gold as an inflation hedge and portfolio diversification asset. Although the gold standard is no longer used officially, gold thus retains monetary relevance through central bank reserves and private ownership. Discussions around reintroducing aspects of the gold standard even occur occasionally due to concerns about reckless government spending and debt.

The classical gold standard historically served a pivotal role in anchoring currency values and exchange rates to gold reserves. Although the inflexibility of gold constrained monetary policy responses, it provided long-term stability. However, economic evolutions ultimately made the rigid gold standard unsustainable for modern economies. Gold remains nevertheless embedded in the global financial architecture both officially and unofficially as a trusted asset of intrinsic value. The gold standard and its collapse-shaped current floating exchange rate regimes are backed by more flexible central bank management rather than gold reserves.

Economists show a growing interest in dissevering the construction of a monetary system anchored on Bitcoin, contrasting it with the traditional gold-standard monetary system. The comparison between cryptocurrency and the gold standard is complex, and while there are some similarities, it is essential to consider the distinct characteristics and implications of each in the context of the monetary system. By exploring existing literature, our objective is to shed light on the potential of Bitcoin as a new gold standard, addressing the gaps in current research.

The structure to flow here onwards is the next section is a literature review that provides a literature survey. The third section is about Bitcoin, a world monetary standard, which is followed by the fourth section's conclusion.

2 Literature Review

Several recent papers have continued to analyze the potential for cryptocurrencies to take on gold-like roles in the monetary system. Raiborn and Sivitanides (2015) argue that Bitcoin shares many properties with gold, such as the scarcity and durability that made gold a historical standard. However, the authors note issues such as the anonymity of Bitcoin, questions around the security of the system, and lack of authoritative control as challenges that must be overcome for Bitcoin to become a viable monetary standard. Glaser et al. (2014) conducted an empirical analysis of Bitcoin's ability to serve as a medium of exchange and store of value, two key functions of a monetary standard. They found that Bitcoin exhibited high levels of volatility in its early years, reducing its current effectiveness as a unit of account or a store of value. The authors conclude that Bitcoin in its current form falls short of gold and existing fiat currencies in stability and adoption metrics. In contrast, Baur et al. (2018) argue that cryptocurrencies can fulfill the roles of money, having advantages in transmission speed, international divisibility, and storage compared to precious metals. Their models show that Bitcoin prices are mainly driven by speculative investment rather than use in trade currently. However, they suggest that cryptocurrencies may displace gold in the coming generations as investment assets and hedges against stocks. Hikmat (2022) and Dudukalov (2021) both highlight the growing use of cryptocurrencies as a means of payment and their potential to transform the digital economy. Mattke (2020) provides empirical evidence that people perceive cryptocurrencies, particularly Bitcoin, as fulfilling the core functions of money, including the medium of exchange, the store of value, and the unit of account. Athey et al. (2016) argue that Bitcoin in particular shares many features with gold, most notably the scarcity of control of supply, durability, portability, and the difficulty of counterfeiting. They state that Bitcoin's source code establishes digital scarcity reliably and that its decentralized nature could provide benefits over centrally controlled fiat currencies. Foley et al. (2018) debated that Bitcoin has emerged as a digital analog to gold given its absolute scarcity and how it is mined and valued. They show that Bitcoin's market value largely coincides with the available stock of Bitcoin and the cost of mining it. However, they note that regulatory risk is still a major impediment to broader adoption. Bariviera (2022) analyzed the informational efficiency of Bitcoin markets compared to gold and found that while Bitcoin is more efficient in weak-form tests, gold remains more efficient than Bitcoin in semi-strong-form tests. This implies that Bitcoin markets still need further maturity and adoption to rival gold's information efficiency. Meanwhile, Auer et al. (2023) discuss how CBDCs could integrate some cryptocurrency benefits with traditional monetary policy. We will examine academic perspectives on comparing cryptocurrencies to previous monetary assets, especially gold, on key attributes of money including scarcity, stability, legitimacy, and efficiency as exchange mechanisms.

2.1 Scarcity

Several researchers have compared the scarcity of commodities like gold to the supply controls embedded in cryptocurrencies like Bitcoin. Van Wijk (2013) notes that the algorithm governing Bitcoin's supply has rules similar to the physical constraints around gold mining, giving both assets scarcity. It is similarly stated that gold and bitcoin are both intentionally scarce assets, with grasping either requiring substantial effort or energy input. Glaser et al. (2014) argue that Bitcoin's ultimate quantity limit provides superior scarcity predictability over gold, removing uncertainty about future discovery and extraction technologies. However, Klein et al. (2018) counter that the physical scarcity and tangibility of gold make it less hypothetically replicable by competitors than blockchain-based cryptocurrencies designed by human programmers.

2.2 Stability

Several studies have analyzed and contrasted the stability and volatility dynamics of gold against cryptocurrencies such as Bitcoin. Dyhrberg (2016) empirically finds that Bitcoin has volatility clustering and hedging abilities comparable to gold using GARCH models, though with substantially wider swings in value. More recently, Klein et al. (2018) determined that Bitcoin has a volatility that is more than four times higher than that of gold, negatively impacting its ability to store value. Hayes (2017), however, projects that Bitcoin volatility will naturally decline over time as its user base and market capitalization grow, mirroring the historical trajectory of gold. On the other hand, most current literature indicates that cryptocurrencies exhibit far less stability than precious metals in their short histories.

2.3 Legitimacy

Research comparing social and institutional recognition of novel cryptocurrencies against long-standing monetary metals like gold and silver is growing but still limited in scope. Van Wijk (2013) contends that Bitcoin and other cryptocurrencies still lack firm grounding both legally and practically compared to established precious metal trade, though blockchain technology shows promise in enabling decentralized yet secure transactions. It similarly classifies the prevailing academic skepticism around cryptocurrencies achieving lasting high-trust roles in economies where gold is deeply culturally embedded through history. But Glaser et al. (2014) cite the strengthening momentum in Bitcoin adoption to argue that both government regulation and financial innovation could cement cryptocurrencies alongside gold in formal monetary systems.

2.4 Efficiency

Analyses that weigh the logistic efficiency of transferring and storing gold against electronic cryptocurrency transactions have generated fairly consistent assertions about the higher speed and lower friction inherent in blockchain-based digital payments (Hayes 2017). Van Wijk (2013) notes that cryptocurrencies can facilitate global peer-to-peer transfers without relying on intermediating authorities, which often impede cross-border gold trading. Hayes (2017) quantifies one aspect of this advantage, estimating that Bitcoin transactions incur less than 1% of the costs of retail commodity transfers using precious metals. However, Folkinshteyn et al. (2015) highlight cybersecurity threats surrounding cryptocurrencies that do not endanger physically held gold, elucidating a countervailing efficiency issue unique to digital currencies. Cirillo (2022) similarly identifies the "digital gold" properties of Bitcoin but also acknowledges the threats posed by quantum computing advancements. The author recommends diversification into quantum-secure cryptocurrencies as a future hedge.

2.5 Complementary Roles

Ametrano (2016) proposes a solution to the price instability of cryptocurrencies, suggesting a dynamic rebasing of the monetary base. The emerging academic literature generally indicates potential complementary roles rather than outright competition between cryptocurrencies and gold in serving monetary functions in the future. Klein et al. (2018) formally support diversification rationales for investors holding both Bitcoin and gold to combine scarcity with stability. It is too early to consider cryptocurrencies as successors to the enduring status of gold, but foresee parallel enduring functions for both assets as stores of value and media of exchange. Constructing formal models, Hayes (2017) ultimately projects innovative digital currencies that augment and circulate alongside precious metal coins rather than completely replacing them (Table 3.1).

3 Bitcoin Monetary Standard—Gold 2.0

This research examines Bitcoin's viability as a new monetary standard using a mixed quantitative and qualitative methods approach supported by empirical market data as well as conceptual analyses of Bitcoin's underlying technology and surrounding community. Athey et al. (2016) effectively used similar mixed-method analyses in their seminal research comparing Bitcoin and gold. By integrating various lenses, a well-versed perspective on Bitcoin's sound money properties emerges. Analyzing both statistical trends reflecting Bitcoin's real-world utility alongside fundamental attributes of its decentralized system design allows a fuller understanding of Bitcoin's sound money properties compared to scarce assets like gold.

Cryptocurrency, particularly Bitcoin, has been compared to gold as a potential new gold standard due to their shared characteristics of scarcity, decentralization, and use as a store of value. Bitcoin's security, limited supply, and global acceptance contribute to its potential as a store of value, similar to gold. Bitcoin wasn't even the first cryptocurrency—some obscure ones preceded it. There are some widely used tokens like Ethereum and upstarts like Dogecoin, etc. But none have reached Bitcoin's prominence or demonstrated the same staying power and hardcore principles that underpin true sound money. Bitcoin stands apart in dominance and recognition as uniquely positioned to fulfill a digital gold or global standard money role (Foley et al. 2018; Klein et al. 2018) Bitcoin stands apart as the most prominent

Table 3.1 Literature summary

Scarcity	– Bitcoin and gold share scarcity traits – Bitcoin's algorithm ensures predictable supply limits – Gold's physical scarcity is seen as more tangible	Raiborn and Sivitanides (2015), Van Wijk (2013), Glaser et al. (2014), Klein et al. (2018)
Stability	– Bitcoin exhibits high volatility, reducing its effectiveness as a store of value – Gold has more stable value – Bitcoin's volatility is expected to decrease over time with growing user base and market capitalization	Glaser et al. (2014), Dyhrberg (2016), Klein et al. (2018), Hayes (2017)
Legitimacy	– Cryptocurrencies lack the legal and practical grounding compared to gold – Growing adoption and potential regulatory support could enhance cryptocurrency legitimacy – Bitcoin shows promise in decentralized secure transactions	Van Wijk (2013), Glaser et al. (2014)
Efficiency	– Cryptocurrencies offer higher speed and lower transaction costs compared to gold – Peer-to-peer transfers are more efficient with cryptocurrencies – Cryptocurrencies face cybersecurity threats not applicable to gold – Potential for quantum computing to impact digital currencies	Hayes (2017), Van Wijk (2013), Folkinshteyn et al. (2015), Cirillo (2022)
Comparative Roles	– Cryptocurrencies and gold may serve complementary roles rather than competing directly – Both can act as stores of value and media of exchange – Innovative digital currencies may augment precious metals rather than replace them	Ametrano (2016), Hayes (2017), Klein et al. (2018)

Source Authors own tabulation

cryptocurrency discussed as a potential monetary standard due to several key technical and economic attributes. As the first widely adopted decentralized cryptocurrency, Bitcoin pioneered innovations including blockchain-based consensus, elimination of centralized intermediaries, and incentive schemes that mathematically enforce scarcity essential for sound money (Nakamoto 2008). Bitcoin was the breakthrough that introduced key innovations like decentralized governance, a permissionless blockchain ledger infrastructure, and built-in crypto-economic incentives that allow global value transfer to occur trustlessly between untrusted parties without any company or state or third-party intermediary needed. It was the first to solve the "double spend" problem that plagued previous attempts at digital cash. And unlike early progenitors like eCash or DigiCash, Bitcoin didn't just create tokens—it bootstrapped an entire network for issuing and securing those tokens on its own, independent from legacy financial rails. These hardened attributes distance Bitcoin from alternative cryptocurrencies more susceptible to governance changes that erode scarcity guarantees critical to sound money.

To evaluate the potential of Bitcoin as a new monetary standard, it is instructive to examine the gold standard that existed before government-issued fiat money was dominated. Under the gold standard, paper notes issued by governments or central banks represented a claim on underlying gold reserves (Eichengreen 2019). Most developed nations adhered to the gold standard in some form during the late nineteenth and early twentieth centuries until it was abandoned in the 1970s in favor of fiat money.

To further examine Bitcoin's potential as a monetary standard, it helps to evaluate the history of monetary regimes. For centuries, commodity monies such as gold and silver served as mediums of exchange, stores of value, and units of account in many societies. The natural scarcity of precious metals provided an inherent limitation on supply that formed the basis for the value of early coins. When paper money was later introduced, it consisted of claim checks representing underlying precious metal deposits. The gold standard prevailed in the late 1800s into the early 1900s when major economies fixed the value of their currencies to a specified amount of gold. The global gold standard limited the discretionary inflation powers of central banks and promoted global currency stability and trade.

However, the gold standard was subject to strains such as speculative attacks on fixed exchange rates and growth limitations due to scarce gold reserves. England was forced off the gold standard in the early 1930s during the Great Depression, followed by the United States in 1933 when convertibility was suspended. The 1944 Bretton Woods agreement instituted a modified gold standard, with the U.S. dollar becoming the global reserve asset valued at 1/35th an ounce of gold. Other nations fixed their currencies to the dollar rather than directly to gold. But this partial gold standard also eventually broke down due to fiscal costs of increased government spending and military expenditures, such as the Vietnam War. By 1973, the major developed countries had transitioned to floating exchange rates and fiat money systems no longer tied to the gold standard.

Under fiat monetary regimes, central banks have discretionary control over the supply of money unconstrained by gold reserves. Some economists argue that this enables more flexibility for monetary policy to promote full employment and price stability. However, critics contend that unsound monetary policies have led to recurrent boom-bust cycles, inflationary spikes, financialization of the economy, and exacerbated income inequality. Advocates of hard-money standards like gold assert that rules-based systems limiting money supply growth and currency debasement impose fiscal discipline and macroeconomic stability.

Enter Bitcoin in 2009, billed as "digital gold" and a decentralized peer-to-peer electronic cash system. The Bitcoin protocol established the issuance of new Bitcoin to occur on a fixed schedule, capped at 21 million Bitcoins. Some proponents believe that this digital scarcity makes Bitcoin well-suited to serve as the basis for a new monetary standard. They contend that the transparency and algorithmic control of

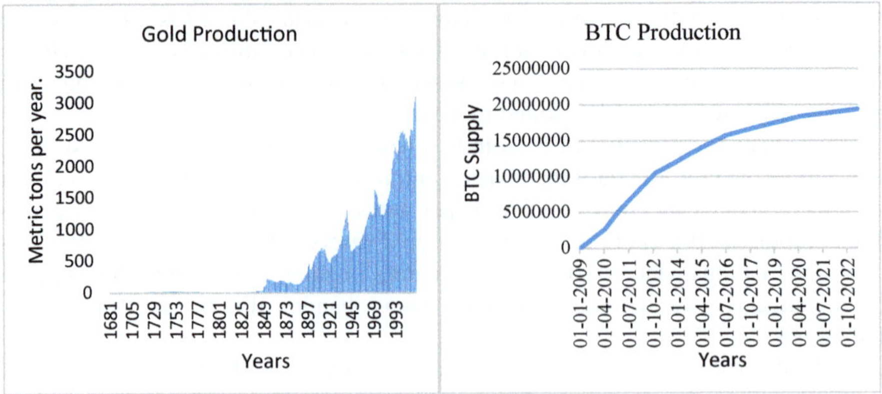

Fig. 3.1 Gold and Bitcoin supply. *Source* Authors compilation

Bitcoin monetary policy could restrict unrestrained inflationary powers of central bankers that erode currency values over time. In addition, Bitcoin's independence from nation-state control provides the opportunity for a truly global and apolitical monetary standard facilitating international trade and commerce.

According to the *Controlled Supply—Bitcoin Wiki*, (n.d.), *Gold production.* (n.d.). In Fig. 3.1 we can see that the existing global supply of gold experiences a gradual increase each year due to gold mining activities worldwide. This augmentation is responsive to and stabilizes the purchasing power of gold. Additionally, non-monetary demands for gold are influenced by its price, helping to mitigate fluctuations in its overall value. On the contrary, the creation of Bitcoin follows a completely programmed trajectory. Bitcoin's mining supply grew at an infinite pace in 2009 when the currency burst into existence. It remains unresponsive to its purchasing power and no external commodity demands influence its production.

Unlike traditional monetary regimes, Bitcoin utilizes a proof-of-work consensus paradigm to enable secure, censorship-resistant transactions without centralized intermediaries (Nakamoto 2008). Analyses of Bitcoin's game theoretical incentives and cryptography (Bonneau et al. 2015; Böhme et al. 2015) demonstrate robustness against common failure modes facing centralized institutions like banks. This antifragility and elimination of single points of control strengthen Bitcoin's long-term viability as a neutral global monetary network resistant to cooption or politicization attempts frequently observed throughout history under gold standards and at central banks.

Bitcoin's strictly limited token supply stands in contrast to unlimited fiat money printing susceptible to hyperinflation effects under discretionary political management as historically observed in various countries. This results from the inability of central banks to inflate obligations by manipulating Bitcoin's algorithmically fixed supply. Hyperinflation crises suffered under some fiat regimes with central banks printing unlimited money could potentially be prevented under a Bitcoin Standard that limits such discretionary expansion of the money supply. Deepening global commerce could occur through low-cost cross-border Bitcoin transactions, further integrating world markets and enhancing trade compared to limitations of government-issued currencies geographically limited in their circulation (Bordo and Levin 2017). Strengthened property rights protections could arise from Bitcoin's cryptographic security making it difficult for governments to arbitrarily confiscate wealth, as has occurred historically under various fiat regimes.

By capping total issuance at 21 million bitcoins released on a fixed, auditable schedule (Nakamoto 2008), Bitcoin mimics gold's scarcity while acting as a disinflationary complement to existing monetary regimes (Fernández-Villaverde and Sanches 2019). Analyses demonstrate this provable scarcity preserves purchasing power better than some of the fiat currencies, with Bitcoin increasingly behaving as digital gold for the twenty-first century (Klein et al. 2018). Proponents argue that Bitcoin could eliminate flaws in fiat money and improve the gold standard as a decentralized system with a fixed supply intended to counter inflation, enable borderless transactions, and enhance security (Ammous 2018).

Critics originally faulted Bitcoin's lack of real economy usage, but recent analyses highlight rapid adoption for payments, transactions, and remittances – especially across borders. Over 30,000 merchants accept Bitcoin payments (Mae 2023) as trust in decentralized money grows. Emergent microeconomies around Bitcoin mining and support services also exhibit endogenous growth uncorrelated with legacy systems (Foley et al. 2018). Together with fascination from institutions like central banks, Bitcoin demonstrates expanding utility and integration with mainstream commerce and finance. Some countries have begun adopting Bitcoin as a legal tender, providing real-world examples of its use as a monetary standard. El Salvador adopted Bitcoin as a legal tender in 2021 (BBC News 2022), and some economic benefits have been touted, like increasing financial inclusion for those lacking access to traditional banking services. Remittances sent home from abroad can be converted to Bitcoin instead of having to pay high fees for traditional money transfer services. The adoption of Bitcoin in El Salvador as a legal tender provides real-world insights into the use of it as standard money. Some touted benefits are increasing financial inclusion, with about one-third of Salvadorans downloading the government's Bitcoin wallet. Remittances in Bitcoin avoid fees amounting to millions in savings. This expands financial access. Bitcoin's growing acceptance and the innovations it has inspired in decentralized finance demonstrate its expanding role in the financial system. As a decentralized, borderless, and digital monetary system, Bitcoin could shape the future by fundamentally transforming financial services, banking, and commerce globally. The Central African Republic became the second country to adopt bitcoin as legal tender alongside the existing CFA franc currency. (BBC News 2022). Some countries that have taken steps to regulate and integrate cryptocurrencies include Australia, Canada, Malta, The Netherlands, Singapore, Slovenia, Estonia, Germany, Denmark, Japan, Switzerland, Spain, and the United Kingdom. However, it is important to note that the regulatory landscape with respect to cryptocurrencies is constantly evolving and governments are still exploring the potential of cryptocurrencies as a monetary system (B2BinPay 2023; Weston 2023; Bajpai 2023).

Additionally, Bitcoin's rise has inspired innovations in decentralized finance (DeFi) and nonfungible tokens (NFTs) built atop its blockchain platform. These provide further opportunities to improve efficiency, transparency, access, and overall functioning of the global financial system if Bitcoin becomes the predominant monetary standard as Bitcoin Monetary Standard—Gold 2.0. DeFi aims to recreate traditional financial services such as lending and derivatives trading in a decentralized architecture without intermediaries. NFTs enable the digital ownership and exchange of unique assets and collectibles across borders. Such financial innovations that leverage blockchain technology could accelerate if Bitcoin becomes the global monetary base asset.

However, economists continue to debate whether Bitcoin has the right properties to serve as an effective monetary standard. Price volatility has been one major criticism, although some argue that volatility would decline with increased adoption and as bitcoin markets mature. The risk was highlighted in El Salvador like exchange rate volatility, with bitcoin's dramatic price swings versus the US dollar, which is also legal tender there. Goods can be priced one day and then see the value of the Bitcoin exchange versus the dollar plummet the next day, reducing real wages and purchasing power. The bitcoin price crash in 2022 led to mass protests in El Salvador against the bitcoin law. However, President Bukele remains committed to the experiment. Other developing countries adopting Bitcoin as legal tender could demonstrate more pros and cons as monetary standards evolve. If adopted more widely, Bitcoin could impose greater fiscal and monetary discipline on governments compared to fiat regimes (Bordo and Levin 2017).

Security and operational risks associated with its supporting blockchain and cryptographic infrastructure also raise concerns about the reliability of the safety of stored value. Furthermore, the vast energy needed for mining Bitcoin has implications for environmental sustainability. Although built-in supply controls prevent arbitrarily high inflation, a bitcoin standard could also impose harmful deflationary effects on the economy if demand wanes.

Additional adoption hurdles relate to Bitcoin, which still represents a niche asset class with limited real-world transactional usage compared to established fiat currencies. Some economists argue its algorithmically fixed supply could engender excessive price deflation that stunts economic growth, reducing living standards over time as wages and prices fall across the board (Schilling and Uhlig 2019). This derives from monetarist notions of the deleterious impacts of deflation on output and employment, which occurred during historical periods like the Great Depression when gold standard constraints hampered the central bank's abilities to expand the money supply. However, others contend that this fear is overstated. Mainstream acceptance has been relatively slow, despite the increasing interest of institutional investors. Transitioning global trade and banking systems to rely on Bitcoin as a unitary monetary standard appears to be extremely challenging both practically and politically. However, in an increasingly digitalized economy, the decentralized and borderless design principles underlying Bitcoin offer a modernized foundation for an evolutionary step toward digitized programmable money as an alternative to traditional currency as Bitcoin Monetary Standard—Gold 2.0.

However, challenges remain around technological readiness, volatility, cybersecurity, privacy, and resistance to adoption from incumbent institutions. Much maturation would be required for Bitcoin to become a formal standard. The backdrop of prolonged economic turbulence and political turmoil experienced by the war-torn nation, the Central African Republic, held considerable expectations for Bitcoin and its associated venture, Sango. This initiative, the country's Blockchain project, aimed at tokenizing land and natural resources, was viewed with optimism as a potential remedy for the prevailing challenges. The inability of Bitcoin to gain acceptance as legal tender in the Central African Republic (CAR) was ascribed to the nation's challenging economic circumstances, skepticism regarding its intentions, limited technological accessibility, and unmet ancillary commitments. The political figures advocating for a new economic paradigm through Bitcoin often had a track record of unfulfilled political assurances, fostering a pervasive skepticism surrounding the initiative.

Notwithstanding this setback, several African nations have emerged as some of the world's most promising and burgeoning hubs for Bitcoin adoption as Bitcoin Monetary Standard—Gold 2.0

4 Conclusion

Overall, scholars recognize the theoretical potential of cryptocurrencies to serve as monetary standards but criticize their current stability and adoption as a means of exchange. Employing an academically validated mixed methodology framework, analyses of Bitcoin's technological robustness, predictable inflation-resistant issuance, and increasing structural adoption build a strong evidence-based case for it serving a monetary role historically occupied by scarce assets like gold. Bitcoin shows genuine promise as decentralized, digital-native sound money for the internet era as Bitcoin Monetary Standard—Gold 2.0.

The potential implications of using cryptocurrency as a legal tender for international trade include the need for new regulatory frameworks, managing exchange rate risk, addressing financial crime, and ensuring adequate infrastructure and access for widespread use. These implications highlight the complex challenges and considerations associated with the adoption of cryptocurrency as a legal tender for international trade. Further infrastructural development and regulation may be required for cryptocurrency to compete or supplement gold and fiat money in monetary systems. The International Monetary Fund (IMF) has estimated the market capitalization of cryptocurrencies at a staggering $2.5 trillion. This figure underscores the considerable economic worth attributed to the underlying technological innovations within the realm of digital currencies. (The Superficial Allure of Crypto 2022), while cryptocurrencies have the potential to transform the traditional financial system, they face significant challenges in achieving a stable and widely accepted role in the global financial system. Regulatory attention is focused on addressing these challenges, and central banks and governments are exploring ways to harness the benefits of cryptocurrencies while addressing their shortcomings. The debate on this issue remains open within the literature. Additional research is warranted, but findings decisively strengthen the argument for decentralized cryptocurrencies overtaking more fragile existing monetary regimes. Bitcoin endogenously exhibits sound money properties on both technical and social dimensions lacking in previous standards.

If Bitcoin were to achieve much wider usage globally as Bitcoin Monetary Standard—Gold 2.0, it could profoundly disrupt existing currencies, financial services, and international commerce. But formidable obstacles remain for Bitcoin to fulfill the aspirations some proponents express for a utopian ideal of a Bitcoin Standard replacing fiat money regimes. However, Bitcoin is showing ascending prospects of becoming a more significant part of the global financial ecosystem in some capacity. Further attention from policymakers, economists, and scholars analyzing Bitcoin's long-term monetary roles appears highly warranted.

References

Abramova S, Böhme R (2016) Perceived benefit and risk as multidimensional determinants of bitcoin use: a quantitative exploratory study. In: Proceedings of the 37th international conference on information systems (ICIS), pp 1–20. https://aisel.aisnet.org/icis2016/DigitalInnovation/Presentations/4/

Ametrano FM (2016) Hayek money: the cryptocurrency price stability solution. Available at SSRN 2425270.

Ammous S (2018) Can cryptocurrencies fulfill the functions of money? Q Rev Econ Finance 70:38–51

Athey S, Parashkevov I, Sarukkai V, Xia J (2016) Bitcoin pricing, adoption, and usage: theory and evidence

Auer R, Cornelli G, Frost J (2023) Rise of the central bank digital currencies: drivers, approaches, and technologies. J Econ Surv 37(2):613–636

B2BinPay (2023, October 10) Countries that allow cryptocurrency as legal payment method. CryptoCurrency Payments—B2BinPay. https://b2binpay.com/en/countries-which-allow-cryptocurrency-as-legal-payment-method/

Bajpai P (2023) Countries where Bitcoin is legal and illegal. Investopedia. https://www.investopedia.com/articles/forex/041515/countries-where-bitcoin-legal-illegal.asp

Barbara Matthews (2019) The relevance of bretton woods in a distributed, cryptocurrency age. Bretton Woods Comm. (Aug. 15, 2019), https://www.brettonwoods.org/article/the-relevance-ofbretton-woods-in-a-distributed-cryptocurrency-age.

Baur DG, Dimpfl T, Kuck K (2018) Bitcoin, gold and the US dollar–A replication and extension. Financ Res Lett 25:103–110

BBC News (2022, April 27) Bitcoin becomes official currency in Central African Republic. BBC News. https://www.bbc.com/news/world-africa-61248809

Bonneau J, Miller A, Clark J, Narayanan A, Kroll JA, Felten EW (2015) Sok: research perspectives and challenges for bitcoin and cryptocurrencies. In: 2015 IEEE symposium on security and privacy. IEEE, pp 104–121

Bordo MD, Levin AT (2017) Central bank digital currency and the future of monetary policy (No. w23711). National Bureau of Economic Research

Catalini C, Gans JS (2021) Some simple economics of the blockchain. Commun ACM 64(7):80–90. https://doi.org/10.1145/3428501

Cheah ET, Fry J (2015) Speculative bubbles in Bitcoin markets? An empirical investigation into the fundamental value of Bitcoin. Econ Lett 130:32–36. https://doi.org/10.1016/j.econlet.2015.02.029

Chen Y, Liu Z, Zhao N, Zheng D (2023) Competition, complementarity, and coordination of cryptocurrency markets. Finance Res Lett 1:103200

Cirillo A, Dalena V, Mauro A, Mogavero F, Pennino D, Pizzonia M, Zecchini M (2022) Empowering citizens by a blockchain-Based Robinson list. Int J Comput Appl 44(10):920–928

CoinDesk (2023) Bitcoin Price Index. https://www.coindesk.com/price/bitcoin/

CoinMarketCap (2023) Today's cryptocurrency prices by market cap. https://coinmarketcap.com/

Controlled supply—Bitcoin Wiki (n.d.) https://en.bitcoin.it/wiki/Controlled_supply

Croman K, Decker C, Eyal I, Gencer AE, Juels A, Kosba A, Miller A, Saxena P, Shi E, Sirer EG, Song D, Wattenhofer R (2016) On scaling decentralized blockchains. In: Clark J, Meiklejohn S, Ryan PY, Wallach D, Brenner M, Rohloff K (eds) Financial cryptography and data security, vol 9604. Springer, Berlin, pp 106–125. https://doi.org/10.1007/978-3-662-53357-4_8

Crypto.com (2022) Global crypto user index. https://crypto.com/images/20220831_CUI_Report.pdf

Dudukalov E (2021) Cryptocurrency as a factor for National Payment System Improvement in the process of Economy's digitalization. In: Cryptocurrency as a Factor for National Payment System Improvement in the Process of Economy's Digitalization/O. Sorokina, E. Dudukalov, L. Guzikova, D. Ushakov-E3S Web of Conferences.-Les Ulis, France: EDP Sciences, vol 258, p 06044

Dyhrberg AH (2016) Bitcoin, gold and the dollar–A GARCH volatility analysis. Financ Res Lett 16:85–92

Eichengreen B (2019) From commodity to fiat and now to crypto: What does history tell us? NBER Working Paper No. 25426. National Bureau of Economic Research. https://doi.org/10.3386/w25426

Fernández-Villaverde J, Sanches D (2019) Can currency competition work? J Monet Econ 106:1–15

Foley S, Karlsen JR, Putniņš TJ (2018) Sex, drugs, and bitcoin: How much illegal activity is financed through cryptocurrencies? Rev Financ Stud 32(5):1798–1853. https://doi.org/10.1093/rfs/hhy131

Folkinshteyn D, Lennon M, Reilly T (2015) The Bitcoin miracle? An analytical framework for bitcoin adoption. J Electron Commerce Res 16(3):218–230

Glaser F, Zimmermann K, Haferkorn M, Weber MC, Siering M (2014) Bitcoin-asset or currency? revealing users' hidden intentions. Revealing Users' Hidden Intentions (April 15, 2014). ECIS

Gold Production. (n.d.). Our World in Data. https://ourworldindata.org/grapher/gold-production?tab=table

Hameed WU, Mian A (2022) Cryptocurrencies: a survey of digital currencies. Digital Finance 100193

Harwick C (2016) Cryptocurrency and the problem of intermediation. Independ Rev 20(4):569–588

Hayes A (2019) The socio-technological lives of bitcoin. Theory Cult Soc 36(4):49–72

Hayes AS (2017) Cryptocurrency value formation: An empirical study leading to a cost of production model for valuing bitcoin. Telematics Inform 34(7):1308–1321

Hikmat B, Javidan N, Vafa G (2022) Cryptocurrency as a means of payment: a review. J Econ Trade 1:11–14

Khwaja AI et al (2022) Decentralized finance: risks, regulation, and the road ahead. J Econ Perspect 36(3):3–28

Krishnan S, Balas VE, Julie EG, Yesudhas HR, Balaji S, Kumar R (eds) (2020) Handbook of research on blockchain technology. Academic Press

Klein T, Thu HP, Walther T (2018) Bitcoin is not the New Gold–A comparison of volatility, correlation, and portfolio performance. Int Rev Financ Anal 59:105–116

Kumar A, Lu WC, Manrai LA (2023) NFTs and metaverse: Value creation for brands and consumers. Ind Mark Manag 103:267–272

Laurinaitis M, Štreimikienė D (2022) The significance of Ethereum: A systematic literature review. J Bus Econ Manag 23(6):1433–1458
Mattke J, Maier C, Reis L (2020) Is cryptocurrency money? Three empirical studies analyzing medium of exchange, store of value and unit of account. In: Proceedings of the 2020 on computers and people research conference, pp 26–35
Mae Z (2023, December 29) Bitcoin payments and stores and websites that accept them? changelly.com. https://changelly.com/blog/who-accepts-bitcoin-payments/
Moore T, Christin N (2013) Beware the middleman: empirical analysis of Bitcoin-exchange risk. In Sadeghi A-R (ed) Financial cryptography and data security, vol 7859. Springer, pp 25–33. https://doi.org/10.1007/978-3-642-39884-1_3
Nakamoto S (2008) Bitcoin: a peer-to-peer electronic cash system. https://bitcoin.org/bitcoin.pdf
Narayanan A, Bonneau J, Felten E, Miller A, Goldfeder S (2016) Bitcoin and cryptocurrency technologies: a comprehensive introduction. Princeton University Press
Ngongang E et al (2022) Do digital currencies help domestic currency users? Evidence from peer-to-peer markets. J Int Money Financ 132:102743
Raiborn C, Sivitanides M (2015) Accounting issues related to Bitcoins. J Corp Account Finance 26(2):25–34
Reinhart CM, Rogoff KS (2014) This time is different: a panoramic view of eight centuries of financial crises. Ann Econ Financ 15(2):215–268
Rockoff H (2015) It is all about gold. World Econ 16(2):51–68
Sadhukhan et al (2023) Evolution of blockchain technology and decentralized finance (DeFi). J Retail Consum Serv 71:103131
Schilling L, Uhlig H (2019) Some simple bitcoin economics. J Monet Econ 106:16–26
The Superficial Allure of Crypto (2022, September 1) IMF. https://www.imf.org/en/Publications/fandd/issues/2022/09/Point-of-View-the-superficial-allure-of-crypto-Hilary-Allen
Umar Z et al (2022) The hedging capabilities of Bitcoin against global uncertainty. Econ Model 106:105831
Van Alstyne M (2014) Why Bitcoin has value. Commun ACM 57(5):30–32. https://doi.org/10.1145/2594288
Van Wijk D (2013) What can be expected from the BitCoin. Erasmus Universiteit Rotterdam 18
Wątorek M, Drożdż S, Kwapień J, Minati L, Oświęcimka P, Stanuszek M (2021) Multiscale characteristics of the emerging global cryptocurrency market. Phys Rep 901:1–82
Weston G (2023, October 30) 10 Countries where cryptocurrency is legal. 101 Blockchains. https://101blockchains.com/cryptocurrency-legal-countries-list/

Smart Contract Applications: Revolutionizing Blockchain Technology

Ushaa Eswaran, Vivek Eswaran, Keerthna Murali, and Vishal Eswaran

Abstract This chapter provides an overview of the evolution and applications of smart contracts across various blockchain platforms, including Ethereum, Hyperledger, and Corda. It begins by defining smart contracts as self-executing digital agreements coded onto the blockchain, operating on principles of immutability, transparency, and decentralization (Szabo 1994; Cong and He 2019). The discussion highlights how different blockchain versions support smart contract functionality and compares their unique features, advantages, and limitations (Zheng et al. 2020). The chapter emphasizes the transformative impact of smart contracts in industries such as decentralized finance (DeFi), where they automate lending, exchanges, and yield farming; in supply chain management for enhanced traceability; and in real estate through tokenization and streamlined transactions. Additional applications in healthcare and identity verification are also explored, showcasing benefits like data security and decentralized identity management. The chapter concludes by addressing ongoing challenges, including security issues, scalability, and regulatory hurdles, while outlining potential advancements and future directions for this rapidly evolving technology.

U. Eswaran (✉)
Department of ECE, Mahalakshmi Tech Campus, Chennai, Tamil Nadu, India
e-mail: drushaaeswaran@gmail.com

V. Eswaran
ORACLE, Austin, TX, USA

K. Murali
Dell EMC | CKAD | AWS CSAA, Austin, TX, USA

V. Eswaran
CVS Health Centre, Dallas, TX, USA

S. Krishnan et al. (eds.), *Unveiling Blockchain Enabled Business Applications*, Contributions to Management Science, https://doi.org/10.1007/978-981-95-2337-5_4

1 Introduction to Smart Contracts

1.1 Introduction

Smart contracts represent a groundbreaking innovation in the realm of blockchain technology, offering automated and self-executing digital agreements that operate without the need for intermediaries. These programmable contracts have garnered significant attention across various industries for their potential to streamline processes, enhance transparency, and reduce costs. As smart contracts continue to evolve, they present new opportunities and challenges that must be carefully navigated to realize their full potential.

In this chapter, we delve into the multifaceted world of smart contracts, exploring their applications, benefits, challenges, and future perspectives. We examine real-world case studies, experimental results, and industry trends to gain deeper insights into the transformative power of smart contracts across diverse domains, from finance and supply chain management to healthcare and real estate. Additionally, we explore the key challenges facing smart contract adoption, such as security vulnerabilities, scalability limitations, regulatory uncertainties, and interoperability issues. By comprehensively analyzing these factors, we aim to provide a holistic understanding of smart contracts and their implications for the future of decentralized technology.

1.2 Smart Contract Execution in Different Blockchain Versions

Smart contracts, integral to blockchain technology, facilitate automated and trustless execution of agreements. Understanding their execution in different blockchain versions, such as Ethereum, Hyperledger Fabric, and Corda, is crucial for grasping the nuances and capabilities of each platform.

2 Smart Contract Execution in Different Blockchain Versions

2.1 Smart Contract Execution in Ethereum

Ethereum, a leading blockchain platform, has reshaped the landscape of smart contracts through its decentralized computing paradigm. Smart contracts on Ethereum are crafted using Solidity, a Turing-complete programming language tailor-made for blockchain applications (Buterin 2014). These contracts find their execution ground on the Ethereum Virtual Machine (EVM), a decentralized runtime environment tasked with executing code across a distributed network of nodes. The EVM operates on the proof-of-work (PoW) consensus mechanism, wherein miners validate and execute transactions to fortify the network and append new blocks to the blockchain.

2.1.1 Solidity, Ethereum's Smart Contract Language

Solidity, Ethereum's preferred language for smart contract development, empowers developers to create complex and robust contracts with relative ease. Its syntax, reminiscent of high-level languages, incorporates blockchain-specific features essential for decentralized application development.

2.1.2 Ethereum Virtual Machine (EVM)

The EVM serves as the execution engine for smart contracts on the Ethereum network. It ensures uniform execution of contract logic across all participating nodes, thereby maintaining consensus and immutability. Each node in the network maintains a copy of the EVM, contributing to the distributed nature of Ethereum's blockchain.

2.1.3 Proof-of-Work (PoW) Consensus Mechanism

At the core of Ethereum's operations lies the PoW consensus mechanism. Miners compete to solve cryptographic puzzles, validating transactions and executing smart contract logic in the process. This decentralized validation mechanism guarantees the integrity and security of smart contract executions on Ethereum.

2.2 Smart Contract Execution in Hyperledger Fabric

Hyperledger Fabric, renowned for its enterprise-grade permissioned blockchain framework, presents a modular and adaptable architecture ideal for developing distributed ledger applications. Within Fabric's ecosystem, smart contracts are denoted as chaincode, scripted using programming languages like Go, Java, or Node.js. The execution of chaincode logic occurs within secure Docker containers, fostering isolation and sandboxing to thwart unauthorized access and uphold code integrity (Androulaki et al. 2018). Fabric's transaction processing is powered by the Practical Byzantine Fault Tolerance (PBFT) consensus algorithm, renowned for its ability to deliver high throughput and low latency while ensuring network resilience and fault tolerance.

2.2.1 Chaincode Development in Hyperledger Fabric

Hyperledger Fabric's support for multiple programming languages grants developers the flexibility to implement chaincode using their preferred language, facilitating ease of development and code maintenance. Additionally, the utilization of Docker containers for chaincode execution enhances security by encapsulating code within isolated environments.

2.2.2 Practical Byzantine Fault Tolerance (PBFT) Consensus

Fabric's adoption of the PBFT consensus algorithm contributes to its robustness and reliability. PBFT's ability to handle Byzantine failures, coupled with its efficiency in transaction processing, makes it well-suited for enterprise blockchain deployments where fault tolerance and performance are paramount.

2.2.3 Secure and Scalable Transaction Processing

By executing chaincode within Docker containers, Fabric ensures that each transaction is processed in a secure and isolated environment. This approach enhances the platform's security posture while also enabling scalability, as Docker containers can be deployed and scaled independently to accommodate varying workloads.

2.3 Smart Contract Execution in Corda

Corda, tailored for enterprise-grade applications, introduces a distinctive approach to smart contract execution, emphasizing privacy, scalability, and regulatory adherence. Within Corda's ecosystem, smart contracts are termed as CorDapps and are scripted using Kotlin or Java, providing developers with the familiarity and flexibility of widely-used programming languages (Hearn 2016). Unlike conventional blockchain platforms, Corda departs from a global consensus mechanism, opting instead for transaction validation and execution within a peer-to-peer network of nodes. Corda's consensus mechanism incorporates notaries, specialized nodes entrusted with validating and notarizing transactions, ensuring consensus among participants while preserving data privacy and confidentiality.

2.3.1 Smart Contract Development in Corda (Expanded)

Corda, an open-source blockchain platform developed by R3, is specifically designed for business use cases. One of its major strengths is its native support for popular programming languages like **Kotlin** and **Java**, which are widely used in enterprise software development. This compatibility significantly lowers the learning curve for developers and reduces the time and cost associated with onboarding and training.

In Corda, smart contracts are developed as part of **CorDapps** (Corda Distributed Applications). These applications define the logic and rules that govern the behavior of transactions between network participants. By using well-established object-oriented programming paradigms, Corda allows for the creation of highly maintainable and testable smart contract code, which is essential for complex enterprise workflows.

Moreover, Corda's smart contracts are embedded in a **flow-based architecture**, enabling controlled execution of business logic while maintaining privacy and scalability. Unlike traditional public blockchains, Corda ensures that data is only shared on a need-to-know basis, thereby aligning with enterprise data confidentiality requirements.

Through integration with existing enterprise systems and support for advanced features such as contract versioning, transaction notaries, and legal prose linkage, Corda empowers organizations to build robust and legally enforceable digital agreements. As a result, smart contract development in Corda not only enhances process automation but also fosters trust and transparency among business partners.

2.3.2 Notary-Based Consensus Mechanism

Unlike traditional blockchain platforms that rely on energy-intensive global consensus protocols like Proof of Work or Proof of Stake, Corda implements a unique notary-based consensus mechanism. In this system, notaries are trusted entities or clusters that validate the uniqueness of transactions. Their primary role is to prevent double-spending and ensure the integrity of the ledger.

Each transaction in the Corda network must be verified and signed by a notary before it is considered valid and finalized. Since not all transactions need to be seen or approved by all participants in the network, Corda's notary system enables efficient and scalable transaction validation. It supports both validating and non-validating notaries, allowing businesses to choose between data confidentiality and transaction verification, based on their use case and regulatory requirements.

This decentralized yet permissioned approach balances scalability, security, and performance, making Corda particularly suitable for financial and legal industries where trust and compliance are paramount.

2.3.3 Enhanced Privacy and Confidentiality

Corda is architected from the ground up with a focus on privacy and confidentiality—a critical requirement in enterprise applications. Unlike most public blockchains where every transaction is visible to every node, Corda ensures that only parties involved in a specific transaction can access the relevant data.

Transactions are shared on a need-to-know basis, meaning only the counterparties and necessary validators (like notaries) receive the transaction data. This selective data propagation drastically reduces data leakage and exposure of sensitive information.

Furthermore, Corda supports advanced cryptographic features such as confidential identities, zero-knowledge proofs, and enclaves (via integrations), which enable secure processing of private data while still proving transaction validity. These privacy features make Corda ideal for industries dealing with regulated or proprietary information, such as healthcare, insurance, and capital markets.

2.3.4 Scalability and Regulatory Compliance

Scalability and compliance are often at odds in traditional blockchain environments. However, Corda manages to excel in both areas by adopting a design philosophy that aligns with enterprise requirements.

Its notary consensus model enables parallel processing of unrelated transactions, reducing bottlenecks and enhancing overall system throughput. By avoiding global broadcast of all transactions and supporting sharding through multiple notaries, Corda achieves horizontal scalability—a crucial feature for handling high volumes of transactions.

On the regulatory front, Corda's granular access control, identity management, and auditable transaction history provide enterprises with the tools necessary to meet stringent compliance standards. Features like legal prose attachment to smart contracts and support for jurisdiction-specific rules ensure that organizations can use blockchain technology without compromising on legal obligations.

2.4 Comparison and Analysis of Smart Contract Execution Across Blockchain Platforms

Ethereum

Ethereum stands out for its pioneering role in popularizing smart contracts and decentralized applications (dApps). Leveraging the Proof-of-Work (PoW) consensus mechanism, Ethereum prioritizes decentralization and censorship resistance. However, Ethereum faces scalability challenges due to its PoW consensus, resulting in limited transaction throughput and high fees during network congestion. Additionally, Ethereum's energy-intensive mining process has raised environmental concerns, prompting efforts to transition to a more sustainable consensus mechanism.

Hyperledger Fabric

Hyperledger Fabric is designed for enterprise-grade applications, offering a modular architecture and pluggable consensus mechanisms. Fabric's adoption of the Practical Byzantine Fault Tolerance (PBFT) consensus algorithm prioritizes performance, finality, and fault tolerance, making it well-suited for applications requiring high throughput and low latency. Fabric's modular architecture enables organizations to customize and tailor their blockchain networks according to specific business requirements, enhancing flexibility and interoperability.

Corda

Corda distinguishes itself with a unique approach to smart contract execution, focusing on privacy, scalability, and regulatory compliance. Corda's notary-based consensus model ensures transaction validation and agreement among participants without compromising data privacy or confidentiality. This approach makes Corda particularly suitable for use cases involving sensitive data or regulated industries, such as finance and healthcare. Corda's support for familiar programming languages like Kotlin and Java simplifies smart contract development, lowering the barrier to entry for enterprise developers.

Comparison

- Consensus Mechanism: Ethereum employs PoW for decentralization, Fabric utilizes PBFT for performance, and Corda employs a notary-based consensus for privacy and compliance.
- Scalability: Fabric's modular architecture enables scalability through channel isolation, while Corda's focus on selective sharing enhances scalability in permissioned networks.
- Privacy: Corda offers granular privacy controls, allowing parties to transact privately while maintaining regulatory compliance, whereas Ethereum and Fabric provide limited privacy features.
- Developer Experience: Ethereum's ecosystem offers a wide range of tools and resources for smart contract development, while Fabric and Corda provide enterprise-grade support and integration capabilities.
- Use Cases: Ethereum is well-suited for public decentralized applications, Fabric excels in enterprise consortia networks, and Corda shines in regulated industries requiring privacy and compliance.

In summary, each blockchain platform offers unique features and trade-offs for smart contract execution, allowing organizations to select the most suitable platform based on their specific use case requirements, regulatory considerations, and performance objectives.

When it comes to smart contract execution, various blockchain platforms offer unique approaches tailored to different use cases and requirements. Below is a comparison Table 4.1highlighting the key differences in smart contract execution among Ethereum, Hyperledger Fabric, and Corda:

Table 4.1 succinctly outlines the differences in smart contract execution across Ethereum, Hyperledger Fabric, and Corda, providing insights into their respective strengths and characteristics.

Understanding smart contract execution in different blockchain versions is essential for selecting the most suitable platform based on specific use case requirements.

Table 4.1 Comparison of Smart Contract Execution in Different Blockchain Platforms

Ethereum	Hyperledger Fabric	Corda
Smart Contracts written in Solidity	Chaincode written in Go, Java, or Node.js	CorDapps written in Kotlin or Java
Executed on Ethereum Virtual Machine (EVM)	Executed within secure Docker containers	Executed within a peer-to-peer network
Consensus Mechanism: Proof-of-Work (PoW)	Consensus Mechanism: Practical Byzantine Fault Tolerance (PBFT)	Consensus Mechanism: Notary-based

Each platform offers unique features and capabilities, catering to diverse needs across various industries.

3 Decentralized Finance (DeFi) and Smart Contracts

Decentralized Finance (DeFi) has emerged as a revolutionary concept in the blockchain space, leveraging smart contracts to redefine traditional financial services in a decentralized manner. Smart contracts, as immutable and self-executing pieces of code stored on a blockchain, enable the automation, transparency, and trustworthiness of financial transactions within the DeFi ecosystem (Schär 2021).

3.1 Lending Platforms

Smart contracts power lending protocols like Aave and Compound, where users can lend and borrow cryptocurrencies without the need for intermediaries. For instance, in Aave, smart contracts manage lending pools, collateralization, and interest rate calculations, allowing users to earn interest on deposited assets or borrow assets against collateral (Aave 2021).

3.2 Decentralized Exchanges (DEXs)

Decentralized exchanges (DEXs) represent a fundamental shift from traditional centralized exchanges by allowing peer-to-peer trading of cryptocurrencies and tokens without the need for intermediaries. Platforms like Uniswap and Sushiswap exemplify this model by utilizing smart contracts to automate trade execution and liquidity provision.

At the core of most DEXs lies the automated market maker (AMM) mechanism. Unlike order book-based exchanges, AMMs use mathematical formulas to determine the price of assets. Liquidity providers (LPs) supply token pairs to liquidity pools, and traders interact with these pools instead of directly with one another. When a trade is executed, the smart contract automatically adjusts the pool balances and updates the token prices accordingly.

This model offers several advantages:

- Continuous liquidity: As long as liquidity is present in the pool, trades can be executed instantly without waiting for a counterparty.
- Censorship resistance: Since trades occur via smart contracts on the blockchain, there's no centralized authority controlling access.
- Transparency: All transactions and pool parameters are publicly verifiable on-chain.
- Lower entry barriers: Anyone can become a liquidity provider and earn a share of the trading fees.

However, DEXs also face challenges such as impermanent loss, front-running attacks, and scalability limitations on certain blockchains.

Despite these issues, the rise of DEXs marks a pivotal moment in the evolution of decentralized finance (DeFi), empowering users with more control, transparency, and inclusiveness in the trading ecosystem (Uniswap 2020; Sushiswap 2020).

3.3 Yield Farming and Liquidity Mining

Yield farming and liquidity mining are two of the most transformative innovations in the decentralized finance (DeFi) ecosystem, enabled by the use of smart contracts. These mechanisms allow users to earn passive income by providing liquidity to various DeFi platforms.

In yield farming, users deposit their cryptocurrencies into liquidity pools, which are managed by smart contracts. These pools are typically used by decentralized exchanges (DEXs), lending protocols, or synthetic asset platforms to facilitate transactions, loans, and other operations. In return for supplying liquidity, users earn a portion of the fees generated by the protocol, and sometimes additional incentives in the form of native platform tokens.

Liquidity mining is a specific form of yield farming in which protocols reward users not just with transaction fees, but with governance or utility tokens. These tokens can often be used to vote on protocol decisions or can be traded on the open market. The amount of reward earned is determined by predefined algorithms coded into smart contracts, considering factors such as the value of assets deposited, the duration of participation, and the relative contribution of each participant.

Smart contracts automate all aspects of the process—from asset pooling and interest calculation to reward distribution—eliminating the need for intermediaries and reducing administrative overhead. This automation enhances transparency, reduces trust dependencies, and allows for permissionless participation in financial activities.

While these strategies can yield significant returns, they also carry risks such as impermanent loss, smart contract vulnerabilities, and market volatility. Nevertheless, yield farming and liquidity mining have become central pillars of DeFi, providing both liquidity and engagement for decentralized protocols (Cong et al. 2021).

3.4 Stablecoins and Synthetic Assets

Stablecoins and synthetic assets are foundational components of the decentralized finance (DeFi) ecosystem, offering price stability and broader access to asset classes through blockchain technology. At the heart of their functionality are smart contracts, which automate critical processes such as issuance, collateral management, and price stabilization.

Stablecoins are digital currencies designed to maintain a stable value, often pegged to fiat currencies like the US Dollar. There are various types of stablecoins, including fiat-collateralized, crypto-collateralized, and algorithmic stablecoins. In crypto-collateralized systems like **MakerDAO**, users lock up volatile cryptocurrencies (e.g., Ether) into smart contracts as collateral to mint stablecoins such as **DAI**, which aims to maintain a 1:1 peg with the USD.

Smart contracts within these systems continuously monitor the value of the underlying collateral using price oracles. If the collateral falls below a predefined collateralization ratio, the smart contract can automatically initiate liquidation to protect the system from under-collateralization. This decentralized approach eliminates the need for central authorities while maintaining trust through transparency and deterministic logic.

Synthetic assets go a step further by enabling the tokenization of real-world or financial assets, such as commodities, fiat currencies, or stock indices. Protocols like **Synthetix** use smart contracts to create synthetic versions of these assets, which are backed by collateral and tracked via decentralized price feeds. Users can gain exposure to these assets without directly holding them, enabling borderless, permissionless financial participation.

These innovations offer several advantages:

- Access to stable value in a volatile crypto environment.
- Financial inclusion through access to global assets without intermediaries.
- Programmable monetary policy managed via code, not institutions.

However, challenges persist, including dependency on reliable oracles, the risk of smart contract exploits, and maintaining sufficient collateral during market downturns.

Overall, smart contracts enable the secure and autonomous operation of stablecoins and synthetic asset systems, ensuring transparency, trust, and scalability across the decentralized finance landscape (Kher et al. 2021).

3.5 Experimental Results and Case Studies

In recent years, DeFi has witnessed tremendous growth, with various projects achieving significant milestones and adoption. For example, Aave's Total Value Locked (TVL) surpassed $20 billion in early 2021, highlighting the growing demand for decentralized lending solutions (Aave 2021). Similarly, Uniswap became the first decentralized exchange to process over $100 billion in trading volume, underscoring the efficiency and scalability of AMM-powered trading (Uniswap 2021).

3.6 Introduction to DeFi Protocols

Decentralized Finance (DeFi) protocols revolutionize traditional financial services by leveraging smart contracts on blockchain networks. These protocols offer various functionalities, from lending and borrowing to automated market making and stablecoin management. Table 4.2 provides insights into key DeFi protocols, their smart contract functionalities, and examples of their use cases.

DeFi Protocol Overview
DeFi protocols utilize smart contracts to automate financial transactions and services, eliminating the need for intermediaries and enhancing transparency and accessibility in the financial ecosystem. The table presents a snapshot of prominent DeFi protocols, highlighting their core functionalities and real-world applications.

Explanation of Table 4.2

Aave: Aave is a decentralized lending platform that enables users to lend and borrow digital assets. Smart contracts within Aave manage lending pools and dynamically adjust interest rates based on supply and demand.

Uniswap: Uniswap is a decentralized exchange (DEX) that employs automated market-making (AMM) algorithms. Smart contracts on Uniswap facilitate liquidity provision and enable users to execute trades directly from their wallets.

MakerDAO: MakerDAO is a decentralized autonomous organization responsible for governing the stablecoin Dai. Smart contracts within MakerDAO manage collateralization ratios, ensuring the stability and pegging of Dai to the US dollar.

Table 4.2 Examples of DeFi Protocols and Their Smart Contract Functionalities

DeFi Protocol	Smart Contract Functionality	Example
Aave	Lending and Borrowing	Manage lending pools and interest rates
Uniswap	Automated Market Making	Provide liquidity and execute trades
MakerDAO	Stablecoin Management	Maintain collateralization ratios

These examples showcase how smart contracts power DeFi protocols, offering users innovative financial services with unparalleled efficiency and security.

In summary, smart contracts serve as the backbone of DeFi, powering a wide range of financial services and innovations that offer users greater autonomy, transparency, and efficiency in managing their digital assets and investments.

4 Supply Chain Management and Smart Contracts

Smart contracts are playing a significant role in revolutionizing supply chain management by introducing efficiency, transparency, and trust through automation and decentralized verification mechanisms (Tönnissen & Teuteberg 2020). They are self-executing contracts with the terms of the agreement directly written into code and executed automatically when predefined conditions are met. By leveraging blockchain technology, smart contracts enable secure and tamper-proof transactions, leading to improved supply chain operations.

One area where smart contracts are making a significant impact is the food supply chain. With increasing concerns about food safety and traceability, smart contracts provide a means to track the journey of agricultural products from farm to table. By recording each step of the supply chain on a blockchain, stakeholders can easily verify the provenance of food items, ensuring compliance with safety regulations and reducing the risk of contamination (Tian 2016). For instance, Walmart's Food Traceability Initiative utilizes smart contracts built on the Hyperledger Fabric blockchain to track the movement of leafy greens. By scanning QR codes on the packaging, consumers and suppliers can access detailed information about the product's origin, processing, and distribution, improving food safety and minimizing waste (Hyperledger 2019).

The diamond industry is another sector where smart contracts are being employed to enhance supply chain management. Due to the high value and rarity of diamonds, ensuring their authenticity and ethical sourcing is of paramount importance. Smart contracts enable the creation of digital certificates that verify the origin, cut, clarity, and other characteristics of diamonds. This information is securely stored on a blockchain, allowing buyers to verify the authenticity and quality of the diamond before making a purchase (Dobrovnik et al. 2018). By leveraging smart contracts, the diamond industry can combat issues such as fraud, conflict diamonds, and unethical practices, instilling trust and confidence among consumers.

In the pharmaceutical supply chain, smart contracts offer several benefits, including automation, improved inventory management, and enhanced product traceability. With the help of smart contracts, the distribution of medicines can be automated, ensuring efficient and timely delivery to pharmacies and healthcare facilities. Smart contracts can also monitor the temperature and humidity conditions during storage and transportation, ensuring that medicines are maintained within the required range to preserve their efficacy (Arun et al. 2020). Additionally, by recording the history of each medicine on a blockchain, stakeholders can readily trace the entire supply chain, from manufacturing to dispensing, improving accountability and reducing the risk of counterfeit drugs entering the market.

Overall, smart contracts are transforming supply chain management by streamlining operations, enhancing transparency, and improving trust among stakeholders. By leveraging blockchain technology, these contracts enable secure and automated execution of transactions, ensuring compliance with regulations, verifying product authenticity, and enhancing traceability throughout the supply chain.

5 Real Estate and Property Transactions Revolutionized by Smart Contracts

Smart contracts are reshaping the landscape of real estate transactions, offering efficiency, transparency, and security to buyers, sellers, and intermediaries alike (Bhargavan et al. 2016). Let's explore some compelling case studies showcasing the transformative power of smart contracts in real estate:

5.1 Propy—Simplifying Home Buying in the United States

In the US, Propy stands out as a pioneering blockchain-based real estate platform that leverages smart contracts to streamline property transactions. Propy enables buyers to submit offers, negotiate terms, and finalize deals entirely through smart contracts, reducing reliance on traditional paperwork and intermediaries. This approach not only expedites the home-buying process but also enhances transparency and security for all parties involved (Propy 2021).

5.2 Dubai Land Department—Blockchain Land Registry System

The Dubai Land Department has embarked on a groundbreaking initiative to digitize property transactions through blockchain technology. By implementing a blockchain-based land registry system, the department ensures the authenticity and transparency of property records. Smart contracts play a crucial role in automating tasks such as property transfers, mortgage registrations, and lease agreements, leading to faster and more secure transactions. This innovative approach has positioned Dubai as a global leader in digital real estate management (Dubai Land Department 2020).

5.3 Tokenization and Fractional Ownership

Beyond traditional property transactions, smart contracts enable the tokenization of real estate assets, paving the way for fractional ownership models. For instance, investors can purchase digital tokens representing fractional ownership in real estate properties. This democratizes access to real estate investment opportunities, allowing individuals to diversify their portfolios without the barriers of large capital requirements. Platforms facilitating fractional ownership through tokenization offer liquidity and flexibility, unlocking new avenues for investors to participate in the real estate market (Kaal & Oesterle 2019).

5.4 Experimental Results and Real-World Examples

- Tokenizing Real Estate Assets: Experimental studies have explored the tokenization of real estate assets using smart contracts. For instance, researchers conducted simulations to assess the impact of tokenization on real estate markets. Results indicated that tokenization enhances market liquidity, reduces transaction costs, and increases accessibility for investors of all sizes (Böhme et al. 2018).
- Fractional Ownership Platforms: Real-world platforms like RealT and Harbor have pioneered fractional ownership models using smart contracts. These platforms allow investors to purchase digital shares of properties, enabling fractional ownership and liquidity in real estate markets. By leveraging blockchain technology and smart contracts, these platforms have attracted a diverse range of investors, democratizing access to real estate investment opportunities.

These case studies and experimental results illustrate the tangible benefits of integrating smart contracts into real estate transactions, ranging from enhanced efficiency and transparency to broader accessibility and investment opportunities.

6 Smart Contracts Revolutionizing Healthcare and Identity Verification

Smart contracts are poised to transform the healthcare landscape by introducing unprecedented levels of security, efficiency, and transparency to various processes. Let's delve into some compelling case studies and real-world examples showcasing the multifaceted applications of smart contracts in healthcare and identity verification:

6.1 DVISION Project: Securing Patient Health Records

In Switzerland, the DVISION project stands as a pioneering initiative that harnesses Ethereum-based smart contracts to revolutionize the management of patient health records. Through the implementation of smart contracts, DVISION ensures the secure storage and sharing of sensitive medical data across healthcare providers and institutions. By leveraging blockchain technology, DVISION guarantees data integrity, privacy, and accessibility, empowering patients to maintain control over their health information while facilitating seamless collaboration among healthcare stakeholders (DVISION 2021).

6.2 Aadhaar Biometric ID System: Decentralized Identity Verification in India

In India, the Aadhaar biometric ID system represents a groundbreaking application of blockchain-based smart contracts for decentralized identity verification. Aadhaar utilizes smart contracts to create a tamper-proof digital identity platform that securely stores and verifies individuals' biometric and demographic information. By leveraging blockchain technology, Aadhaar enables efficient and transparent access to government services, financial institutions, and healthcare facilities, while safeguarding user privacy and combating identity fraud (Aadhaar 2021).

6.3 Pharmaceutical Supply Chain Management

Smart contracts offer immense potential in optimizing the pharmaceutical supply chain by enhancing transparency, traceability, and accountability. By leveraging smart contracts, pharmaceutical companies can track the movement of drugs from manufacturing facilities to distribution channels, ensuring compliance with regulatory standards and preventing counterfeit products from entering the market. Additionally, smart contracts streamline the reimbursement process for healthcare providers by automating claims processing and reducing administrative overhead (Arun et al. 2020).

6.4 Decentralized Clinical Trials

Decentralized clinical trials represent a groundbreaking shift in the traditional approach to medical research, leveraging emerging technologies like blockchain and smart contracts to introduce transparent, secure, and efficient methodologies. By harnessing the power of smart contracts, researchers can revolutionize the execution of clinical trial protocols, streamline patient consent and data management processes, and ensure compliance with regulatory standards.

6.4.1 Benefits of Decentralized Clinical Trials

1. Enhanced Patient Participation: Decentralized clinical trials remove geographical barriers and logistical constraints, allowing a broader and more diverse pool of participants to enroll. Patients can participate from the comfort of their homes, eliminating the need for frequent clinic visits and reducing the burden on healthcare infrastructure.
2. Real-Time Data Access: Smart contracts facilitate real-time access to trial data, enabling researchers to monitor patient progress, track outcomes, and make data-driven decisions more efficiently. This real-time visibility enhances the speed and accuracy of clinical trial monitoring and analysis.
3. Improved Trial Integrity: The immutable nature of blockchain technology ensures the integrity and transparency of clinical trial data. Smart contracts record every interaction and transaction on the blockchain, providing an auditable trail of activities that cannot be tampered with or altered. This transparency enhances trust among stakeholders and minimizes the risk of data manipulation or fraud.

6.4.2 Future Implications

The adoption of decentralized clinical trials powered by smart contracts holds immense promise for the healthcare industry. By embracing these innovative methodologies, researchers, healthcare providers, and patients can benefit from:

1. Efficiency: Streamlined processes and automation reduce administrative overhead, accelerate trial timelines, and optimize resource allocation.
2. Security: Enhanced data security and privacy protections safeguard patient information, ensuring compliance with regulatory requirements such as HIPAA and GDPR.
3. Patient-Centricity: Decentralized trials prioritize patient convenience and accessibility, empowering individuals to participate in research initiatives that align with their preferences and needs.
4. Innovation: The integration of blockchain and smart contracts fosters a culture of innovation, driving the development of novel therapies, treatments, and healthcare solutions.

The decentralized clinical trials powered by smart contracts represent a paradigm shift in medical research, offering unprecedented opportunities to enhance efficiency, transparency, and patient-centricity. As these technologies continue to evolve, they hold the potential to revolutionize healthcare delivery and reshape the future of medical innovation.

7 Challenges and Future Perspectives in Smart Contract Technology

While smart contracts hold immense promise for revolutionizing various industries, they are not without their challenges. Addressing these hurdles is crucial for unlocking the full potential of this transformative technology. Let's explore some of the key challenges and future perspectives:

7.1 Security Vulnerabilities

Security remains a paramount concern in the realm of smart contracts. High-profile incidents like the DAO hack in Ethereum have underscored the need for robust security measures. To mitigate vulnerabilities, rigorous smart contract auditing practices must be adopted, coupled with the development of more secure programming languages and sophisticated development tools (Atzei et al. 2017).

7.2 Scalability Limitations

Scalability poses a significant challenge for widespread adoption of smart contracts, particularly in terms of transaction throughput and gas fees. As blockchain networks grow, scalability issues become more pronounced. Addressing scalability concerns requires the exploration of layer-2 scaling solutions, advancements in blockchain consensus mechanisms, and the optimization of network infrastructure (Qin et al. 2020).

7.3 Regulatory Uncertainties

The regulatory landscape surrounding smart contracts is still evolving, leading to uncertainties regarding their legal status and enforceability. Clarity on regulatory frameworks is essential to foster confidence among users and businesses. Governments and regulatory bodies need to collaborate with industry stakeholders to develop clear guidelines and regulations governing smart contract usage, digital agreements, and dispute resolution mechanisms (Werbach & Cornell 2017).

7.4 Interoperability and Standards

Achieving interoperability between different blockchain platforms and smart contract protocols is crucial for realizing the full potential of decentralized applications (DApps). Establishing industry-wide standards and interoperability protocols will facilitate seamless interaction between disparate blockchain networks, enabling the creation of more complex and interconnected smart contract ecosystems (Zou et al. 2021).

Despite these challenges, the future of smart contract technology is promising, with ongoing research and development efforts aimed at addressing these hurdles. Innovations in security, scalability, interoperability, and regulatory compliance will drive the next wave of advancements in smart contract technology, unlocking new opportunities for decentralized finance, supply chain management, healthcare, and beyond.

8 Conclusion

In conclusion, smart contracts represent a paradigm shift in how agreements are executed and enforced, offering unprecedented levels of automation, transparency, and efficiency. Despite their immense potential, smart contracts face several challenges that must be addressed to facilitate widespread adoption and integration into various industries. Security vulnerabilities, scalability limitations, regulatory uncertainties, and interoperability issues pose significant hurdles that require innovative solutions and collaborative efforts from industry stakeholders, policymakers, and researchers.

Looking ahead, the future of smart contracts appears promising, with ongoing advancements in blockchain technology, programming languages, and development tools. As we continue to explore and refine the capabilities of smart contracts, we can expect to witness their pervasive influence across sectors, driving innovation, fostering trust, and reshaping the way we conduct business and interact in the digital age. By staying vigilant, adaptive, and proactive in addressing challenges while capitalizing on opportunities, we can harness the full potential of smart contracts to create a more transparent, efficient, and equitable global economy.

References

Aadhaar (2021) Unique Identification Authority of India. https://uidai.gov.in/
Aave (2021) Decentralized lending protocol. https://aave.com/
Androulaki E, Barger A, Bortnikov V, Cachin C, Christidis K, De Caro A, Enyeart D, Ferris C, Laventman G, Manevich Y, Muralidharan S, Murthy C, Nguyen B, Sethi M, Singh G, Smith K, Sorniotti A, Stathakopoulou C, Vukolić M, Yellick J (2018) Hyperledger fabric: a distributed

operating system for permissioned blockchains. Proceedings of the 13th EuroSys Conference. https://doi.org/10.1145/3190508.3190538

Arun R, Vindhya R, Saranya S (2020) Blockchain technology in pharmaceutical supply chain management. IOP Conf Ser: Mater Sci Eng 937:012004. https://doi.org/10.1088/1757-899X/937/1/012004

Atzei N, Bartoletti M, Cimoli T (2017) A Survey of Attacks on Ethereum Smart Contracts (SoK). Lect Notes Comput Sci 10204:164–186. https://doi.org/10.1007/978-3-662-54455-6_8

Bhargavan K, Delignat-Lavaud A, Fournet C, Gollamudi A, Gonthier G, Kobeissi N, Kulatova N, Rastogi A, Sibut-Pinote T, Swamy N, Zanella-Béguelin S (2016) Formal verification of smart contracts: short paper. In: Proceedings of the 2016 ACM workshop on programming languages and analysis for security, pp 91–96. https://doi.org/10.1145/2993600.2993611

Böhme R, Brenner M, Karame GO, Youssef A (2018) Securing blockchain transactions. In: 2018 IEEE 43rd conference on local computer networks workshops (LCN workshops), pp 28–32. https://doi.org/10.1109/LCN.WORKSHOPS.2018.8772605

Buterin V (2014) A Next-generation smart contract and decentralized application platform. White Paper

Cong LW, He Z (2019) Blockchain disruption and smart contracts. Rev Financ Stud 32(5):1754–1797. https://doi.org/10.1093/rfs/hhz007

Cong LW, Li Y, Wang N (2021) Tokenomics: Dynamic Adoption and Valuation. Rev Financ Stud 34(3):1105–1155. https://doi.org/10.1093/rfs/hhaa089

Dobrovnik M, Herold D, Fürst E, Kummer S (2018) Blockchain for and in Logistics: What to Adopt and Where to Start. Logistics 2(3):18. https://doi.org/10.3390/logistics2030018

DVISION (2021) DVISION Project. https://dvision.ch/

Dubai Land Department (2020) Dubai Land Department launches the first phase of the Dubai Blockchain Real Estate Initiative. https://www.dubailand.gov.ae/en/media-center/press-releases/dubai-land-department-launches-the-first-phase-of-the-dubai-blockchain-real-estate-initiative/

Hearn M (2016) Corda: a distributed ledger. https://www.r3.com/wp-content/uploads/2016/11/corda-technical-whitepaper-R3.pdf

Hyperledger (2019) Walmart's Food Traceability Initiative. https://www.hyperledger.org/learn/case-studies/walmart-food-traceability

Kaal WA, Oesterle MJ (2019) Decentralized Finance. Univ Ill Law Rev 2021(2):533–598

Kher R, Tomar A, Ghosh K (2021) Stablecoin design: construction, decentralization, and stability. In: Proceedings of the 54th Hawaii international conference on system sciences, pp 6524–6533. https://doi.org/10.24251/HICSS.2021.785

Propy (2021) Propy—real estate transactions on the blockchain. https://propy.com/

Qin R, Zou Y, Wu J, Cai H, Chen Z, Xiang Y (2020) A survey of challenges, opportunities and solutions in scalable blockchain. IEEE Access 8:67536–67556. https://doi.org/10.1109/ACCESS.2020.2984555

Schär F (2021) Decentralized Finance: On Blockchain- and Smart Contract-Based Financial Markets. FRB St. Louis Rev 103(2):153–174. https://doi.org/10.20955/r.103.153-74

Sushiswap (2020) SushiSwap: A Decentralized Exchange Protocol. https://sushiswapclassic.org/

Szabo N (1994) Smart Contracts. http://www.fon.hum.uva.nl/rob/Courses/InformationInSpeech/CDROM/Literature/LOTwinterschool2006/szabo.best.vwh.net/smart.contracts.html

Tian F (2016) An agri-food supply chain traceability system for China based on RFID & blockchain technology. In: 2016 13th International conference on service systems and service management (ICSSSM), 1–6. https://doi.org/10.1109/ICSSSM.2016.7538424

Tönnissen S, Teuteberg F (2020) Analysing the impact of blockchain-technology for operations and supply chain management: an empirical analysis. Comput Ind 109:11–25. https://doi.org/10.1016/j.compind.2019.03.001

Uniswap (2020) Uniswap v2 Core. https://uniswap.org/whitepaper.pdf

Uniswap (2021) Uniswap Ecosystem. https://uniswap.org/

Werbach K, Cornell N (2017) Contracts ex machina. Duke Law J 67(2):313–382

Zheng Z, Xie S, Dai H, Chen X, Wang H (2020) Blockchain challenges and opportunities: a survey. Int J Web Grid Serv 14(4):352–375. https://doi.org/10.1504/IJWGS.2018.095647

Zou Y, Liu X, Wang H, Cai H (2021) A survey on blockchain interoperability. Proc IEEE 109(6):1003–1019. https://doi.org/10.1109/JPROC.2021.3058999

Smart Contract and Architecture of Blockchain

B. Sriman, C. M. Hilda Jerlin, R. Suguna, S. Pavithra, V. Rekha, S. H. Annie Silviya, M. Singaravelan, and H. Ruma Prasad

Abstract Smart contracts are self-executing digital agreements where the contract terms are written directly into lines of code. These contracts enforce and execute the agreed conditions automatically once the set triggers are met, which, in turn, eliminates intermediaries such as lawyers or brokers. Operating on blockchain platforms, smart contracts ensure a secure, decentralized, and immutable environment. Thus, it is necessary for trustless and automated transactions. Their usage ranges from health and aviation to banking, enabling smoother processes, increased transparency, and reduced operational costs. This chapter provides a broad overview of smart contracts, discussing their capabilities, their applications, and the tremendous business effects

B. Sriman (✉) · R. Suguna · M. Singaravelan
Department of Computer Science and Engineering, Vel Tech Rangarajan Dr. Sakunthala R&D Institute of Science and Technology, Avadi, Chennai, Tamilnadu, India
e-mail: srimanb@veltech.edu.in

R. Suguna
e-mail: rsuguna@veltech.edu.in

M. Singaravelan
e-mail: desingaravelanm@veltech.edu.in

C. M. Hilda Jerlin
Department of Artificial Intelligence and Data Science, Panimalar Engineering College, Chennai, India

S. Pavithra
Department of Computer Science and Engineering, Prathyusha Engineering College, Thiruvallur, India

V. Rekha
Department of Mathematics, Vel Tech Multi Tech Dr. Rangarajan Dr. Sakunthala Engineering College, Chennai, India

S. H. Annie Silviya · H. Ruma Prasad
Department of Computer Science and Engineering, Rajalakshmi Institute of Technology, Chennai, India
e-mail: anniesilviya.sh@ritchennai.edu.in

H. Ruma Prasad
e-mail: rumaprasad.h.2020.cse@ritchennai.edu.in

S. Krishnan et al. (eds.), *Unveiling Blockchain Enabled Business Applications*, Contributions to Management Science, https://doi.org/10.1007/978-981-95-2337-5_5

they generate within different industries. It describes how smart contracts revolve around traditional business by automating more complex processes and making sure that agreements are executed without human interference. It further deliberates on the evolution of blockchain technology, mapping its development right from the initial Blockchain 1.0 model, mainly to support cryptocurrencies to the advanced public permissionless and private permissionless blockchain models that eventually came to support a wide range of decentralized applications, including smart contracts. This chapter also covers some of the available platforms used for smart contract development, including Ethereum, Binance Smart Chain, Cardano, and Solana. Each of the platforms is reviewed on its distinctive features, strengths, and applications for which it is best suited. The chapter also discusses the challenges of smart contract implementation and, most importantly, the security risks and vulnerabilities associated with the operation of the blockchain network. It outlines common attacks such as reentrancy attacks and timestamp dependence and provides ideas related to methodologies on how such risks can be minimized. As blockchain technology continues to evolve, smart contracts are positioned to be at the forefront of digital transactions, governance, and innovation in the future. Their secure, automated, and trustless interactions will turn out to be of paramount importance during attempts to sail through the broad and exponentially changing decentralized technology space. Therefore, this chapter stands as an important primer for understanding the significance of smart contracts and their implications for the modern-day digital economy.

1 Smart Contract

In recent years, blockchain has become trustless because of Smart Contracts. Blockchain systems that incorporate smart contracts perform better.

1.1 Contract

When two or more parties agree to make a specific contract, a set of activities is devised upon specific conditions or legally binding documents that recognize and govern the rights and duties of the parties to that agreement. The contract legally requires the approval of the parties and the law.

For example,

Rent—Accommodation Contract

Contract Theory—Each party associated with the contract must be given equal rights, incentives, or motivation to work together and achieve a successful outcome of the contract.

The following requirements need to be fulfilled for a contract to be deemed valid:

1. An offer made by two or more parties
2. The offer being accepted
3. Consideration, which is a reward that needs to be negotiated between the parties
4. (4) Mutual responsibility
5. Competency and capability
6. A document.

Most contracts can be considered valid if at least conditions (2) and (3) are met. Contracts also come in two varieties:

1. Written: Contracts with documentation
2. Verbal: Agreements maintained based on goodwill.

1.2 Smart Contract

A Smart Contract is a digital-based contract or digital version of a contract. It acts as a per contract, which runs on the blockchain node and executes the requirements of parties, based on a set of parameters in the blockchain to become a trustless network. The smart contract is automatically executed by the computer system, and if the design is valid, the smart contract can be implemented solely by the computer, without the intervention of humans.

1.3 Step to Design a Smart Contract

Step1. Understanding the Business Problem—The Smart contract as a set of parameters as shown in Fig. 5.1, includes the business problem statement or understanding of the business problem statement.

Step 2. Defining the blockchain network and its participants—Creating the blockchain network and giving access permission to parties.

Step 3. Accessing involved parties of the contract—All the participants cannot be a part of the smart contract, rather, it must be only a small subset based on the rules and conditions. This small subset of participants is called as involved parties.

Step 4. Exploring Prerequisites for the smart contract—It needs extended data and input from users, etc.

Step 5. Examining trigger event for the smart—It means that a smart contract works on a set of rules and conditions. Once these conditions are met, the smart contract is executed successfully.

Step 6. Evaluating business logic for the smart contract—“if else” statement and condition: “if” the condition is fulfilled or true, then what course of action will be followed, “else” what will happen.

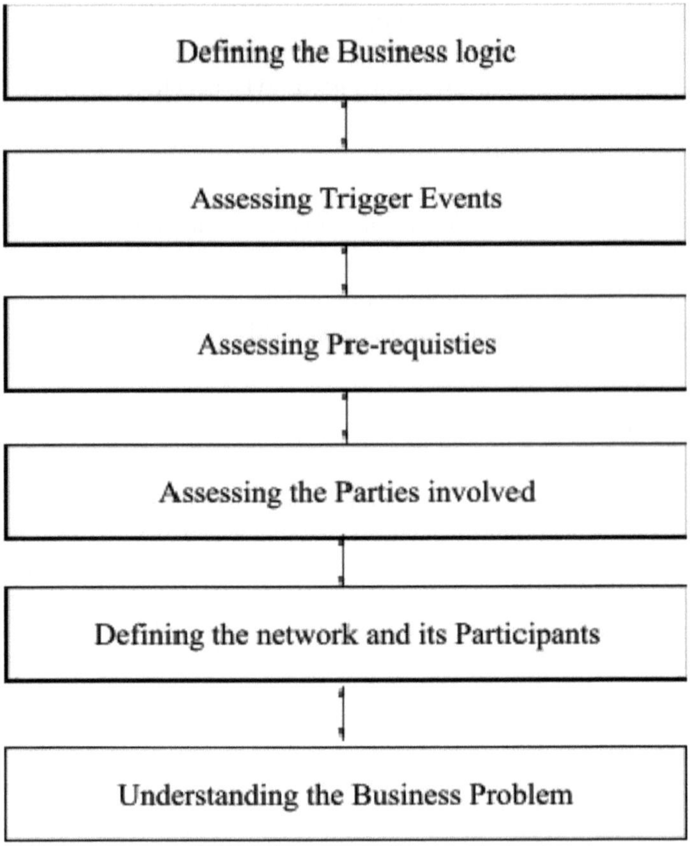

Fig. 5.1 Flow diagram of a smart contract. Figure provide flowcharts Steps involved in deploying, executing and interacting with smart contract on a Blockchain. *Source* Author's own compilation

2 Use Cases Involving Smart Contract

There are several use cases in the real world. To understand the smart contract in the blockchain for various industries.

1. Smart Contract in the Bank Industry
2. Smart Contract in the Aviation Industry
3. Smart Contract in Healthcare Industry
4. Smart contract in Electron

2.1 *Smart Contract: Finance Industry*

Blockchain technology can be greatly used in administering trade finance transactions for the bank. Under blockchain technology, all the documents like LCs, Bills of lading/ Shipping bills, and Tax invoices can be recorded in a centralized repository where all the parties involved can access the real-time data. A single smart contract is used to create a token that contains a state variable to store the token balances for each user and the functions that allow modification of state variables with token balances. The tokens can be categorized as fungible tokens that are interchangeable and nonfungible tokens (NFTs) that are not interchangeable.

Decentralized Exchanges (DEX) are a set of smart contracts created on the same blockchain to interact with each other and make the trade of fungible tokens possible. The set of contracts includes liquidity pools, which contain the ownership of the two types of tokens and allow trading of a particular token. Similarly, a set of Smart Contacts called Protocols for Loanable Funds (PLF)is used to hold a single cryptoasset in a corresponding individual contact.

For Example: Loan smart contract.

Based on the steps to designing a smart contract in the banking industry.

Step 1. Business Problem: Ensuring timely repayment for the loan.

Step 2. Involved Parties: customer and Bank.

Step 3. Network Participants: Customer, Bank, Other Banks.

Step 4. Prerequisites: The network needs access to the calendar and time facilities.

Step 5. Trigger: Change of calendar mouth.

Step 6. Business Logic: If conditions loan > 0, invoke a standing instruction to the core system EMI gets credited.

2.2 *Smart Contract: Aviation Sector*

Aviation is the activity of transporting people and goods by aircraft. The use case in aviation industry insurance is related to the flight delay. Whenever the flight is delayed for more than x hours, the customer, who applies for the insurance amount will get a claim for the delay which will be credited to the bank account.

This use case exists in the US and European countries to improve customer relationships and customer pretension between the Airline Network and Insurance Companies. Additionally, aviation employs smart contracts to manage and track cargo shipments by recording information such as destination, route, delivery status, etc., by providing each with a unique identifier. The smart contracts can be used to manage the service agreements between the airlines and maintenance providers automatically. The mechanism can be used to keep track of the process, notify when the process is completed, and trigger the payment automatically.

For Example: Airline And Insurance Company.

Based on the steps to designing a smart contract in the Aviation industry.

Step 1. Business Problem: Reduce customer dissatisfaction due to lengthy processes and paperwork for claiming flight delay insurance.

Step 2. Involved Parties: Passengers, Airline, and Insurance Company.

Step 3. Network Participants: Insurance Companies and Airlines that use the Blockchain network.

Step 4. Prerequisites: Access to the flight information from the Airline Database.

Step 5. Trigger: Flight delay > 2 h.

Step 6. Business Logic: If a passenger has insurance, initiate the claim process and credit money to the customer account.

2.3 Smart Contract: Healthcare Sector

The healthcare sector is one of the prime industries that handles the tiresome tasks of the government and ensures that the process is carried out smoothly. Smart contracts from Ethereum are used to represent the existing medical records. Smart contracts in the Healthcare system manage different permissions for different stakeholders based on their right to access the data by employing data authorization rules.

Problem Statement: Road Accident

In road accidents, hospitals delay in identifying the person and medical background.

In the payment method, the problem arises in determining whether a person can make a payment to receive medical care.

Based on the steps to designing a smart contract in the Healthcare industry.

Step 1. Business Problem: As shown in Fig. 5.2, The government wants to provide critical medical care for unidentified victims.

Step 2. Involved Parties: Patient, Hospital, Unique Identification Authority [UIQ], Insurance Company.

Step 3. Network Participants: Blockchain network of all hospitals, Identification departments, and Insurance companies.

Step 4. Prerequisites: Access to the medical information from the Healthcare Database.

Incident reporting from the Hospital.

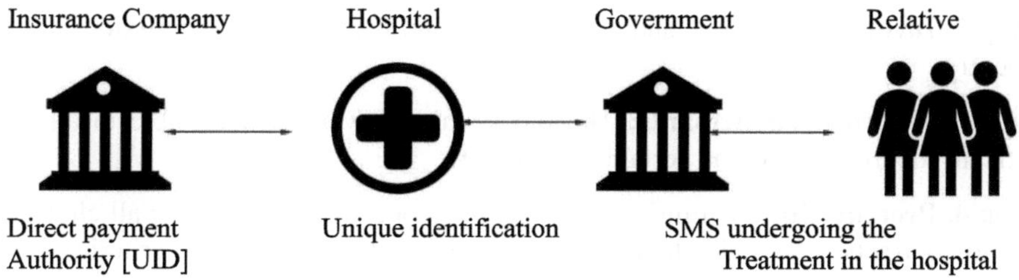

Fig. 5.2 Design of Smart contract: Healthcare sector. Figure revolves automating, streamlining processes such as patient data management, billing, supply chain management and insurance claims. *Source*: Author's own compilation

Public—whenever an accident happens, report it to the government via the Healthcare Application.

Mandatory insurance cover for the public.

Step 5. Trigger: patients from road accidents or fire hazards registered at the hospital—such kind of patients are registered in the hospital; smart contracts will be revoked automatically.

Step 6. Business Logic: Patients' Biometric information is passed to the Unique Identification Authority [UID] to identify details.

a. insurance cover is identified.
b. enable direct cashless payment to the hospital.
c. sends intimation to relative Via an SMS

2.4 Smart Contract: Elections

Blockchain Public Permissionless is a governed blockchain network, which can be used for electron use cases. In the use case, citizens across the globe can access a democracy for the candidate's choice of voting. Blockchain-based voting systems that have universal access to the security criteria, save time for participants and increase the level of the participants. This use case is designed based on people who are not in their negative place, they are not able to participate in democratic decision-making process or voting. The blockchain is secure running on consensus protocol in a virtual non-fraudulent manner.

For Example: Voting Sector

Based on the steps of designing a smart contract in the Voting Sector.

Step 1. Business Problem: To enable citizens to vote for a candidate of their choice securely.

Step 2. Involved Parties: Candidates, voters, and the election commission of the country.

Step 3. Network Participants: The blockchain network is accessible to all the citizens identified based on certain specific IDs.

Step 4. Prerequisites: setting up an anonymous one-way voter ID for all citizens. Smart contracts will be able to access the citizen's data and eligibility criteria of the voters and the candidates.

Step 5. Trigger: When a vote is cast by an eligible voter.

Step 6. Business Logic: If the voter's age is above the minimum required voting age, participants can cast only one vote for the candidate preferred.

The Smart Contract increments the counter of that candidate by 1.

3 Business Impact of Smart Contract

Smart contracts that have terms of the agreement directly written into code are transforming the way businesses function. It guarantees that transactions and agreements are carried out automatically when predefined conditions are met. This technological breakthrough is having a profound impact on various business sectors by enhancing efficiency, reducing costs, improving transparency, and fostering trust.

Operational Efficiencies

Business procedures that have historically required human intervention are automated by smart contracts. Smart contracts, for example, can automate the process of storing patients' data securely. It provides access to the health care providers only when necessary. It reduces the need for manual tracking and updating of the system. This also ensures that human errors are eliminated since the process requires minimum manual input. This dependability is essential in sectors such as finance and healthcare, where errors can have significant consequences. Automated smart contracts also ensure that the updates are kept in real-time and maintain accurate records of all transactions.

Cost Reductions

To foster trust and ensure compliance in the industry, traditional business tractions require the involvement of intermediaries including banks, brokers, etc. With the employment of smart contracts, the need for these intermediaries is eliminated, thus providing a secure, automated, and trustless system. By eliminating the middlemen, the smart contract also ensures that the cost of commissioning is reduced. Smart contracts are also capable of automatically enforcing legal agreements which reduces the need for lengthy and costly legal processes.

Improved Transparency and Trust
Transactions executed with the help of smart contracts are recorded on a blockchain. This provides a transparent and immutable ledger which is essential for building stakeholder trust and for purposes of auditing. The secure nature of the blockchain is also an added factor for stakeholder's trust. This is ensured by the decentralized nature and cryptographic security incorporated in smart contracts thus making it difficult for malicious actors to tamper the records.

3.1 Sector-Specific Applications of Smart Contract

Finance and Banking: By facilitating quicker, less expensive, and more secure transactions, smart contracts are revolutionizing the finance and banking industries. Peer-to-peer lending, automated clearing and settlement, and decentralized finance (DeFi) applications are made possible by smart contracts, enabling users to conduct financial transactions without the need for conventional middlemen.

Supply Chain Management: The traceability, efficiency, and transparency of the supply chain are improved by smart contracts. They guarantee on-time payments, automate the tracking of commodities, and enforce contractual duties.

Real Estate: By automating procedures like title transfers, escrow management, and payment settlements, smart contracts in real estate streamline property transactions. This automation speeds up transactions, cuts costs, and eliminates the need for middlemen. Furthermore, rental and maintenance contracts can be managed using smart contracts, guaranteeing prompt payments and service provision.

Insurance: Smart contracts are used in the insurance sector to automate the process of claims and to detect fraud. By evaluating claims according to predetermined criteria and automatically approving or rejecting them, smart contracts can minimize the time and expense involved in processing manual claims. Trust and transparency between policyholders and insurers can be improved by offering a transparent and immutable record of policy terms and claims history.

Legal Industry: In the legal sector, smart contracts automate the execution of standard legal agreements, including NDAs, wills, and trusts. This automation saves money, expedites the implementation of legal agreements, and eliminates the need for legal middlemen. Furthermore, smart contracts can offer an immutable, transparent record of the terms of the contract, guaranteeing clarity.

Energy Sector: Peer-to-peer energy trading and automated supply and demand management are made possible by smart contracts. They can control grid efficiency, automate energy invoicing, and make it easier to buy and sell renewable energy credits. A smart contract, for instance, can automatically modify energy prices in response to supply and demand, ensuring cost savings and effective energy distribution.

4 Evolution of Smart Contract

4.1 Initial Version of Blockchain

The initial version of the blockchain was known as the state machine. The system stores the state of each node in the network. The states of the nodes were either valid or invalid. These states act as an accounting ledger and have no code to run a blockchain network. Smart contracts emerged when the thought arouse that these nodes could be run with their own predefined logic by associating these nodes with code. This could make the process automated and faster.

4.2 Smart Contract Evolution on Public Permissionless Blockchain

On-Chain Network

The first version of the smart contract is the On-chain smart contract version. The on-chain smart contract is an immutable compound in the ledger itself that is used to run all the nodes in the network.

Example 1: A smart contract is deployed in the network and replicated across all the nodes in the network, called On-Chain Network.

Example 2: Two different schools will have distinct rules and regulations and would want their students to act in accordance with their rules only and not the others as shown in Fig. 5.3. Incorporating the logic into the blockchain network, we understand that each chain in the network has different logic for its nodes.

- The On-Chain smart contract was first deployed in the Bitcoin Blockchain. On-chain smart contracts are important for public permissionless blockchain. In the Public Blockchain, all nodes should be similar or all nodes should hold the same logic and same data.
- The trust between the different parties and nodes is less. Therefore, nobody wants to give up the right for the smart contract as a selection node in the Public Blockchain.

External Data to Access in Smart Contract

The On-Chain smart contract contains external agents called Oracles. These external agents connect the smart contract to the external world. The Oracles are also responsible for fetching any external data required by the smart contract.

4.3 Smart Contract Evolution on Private Permission Blockchain

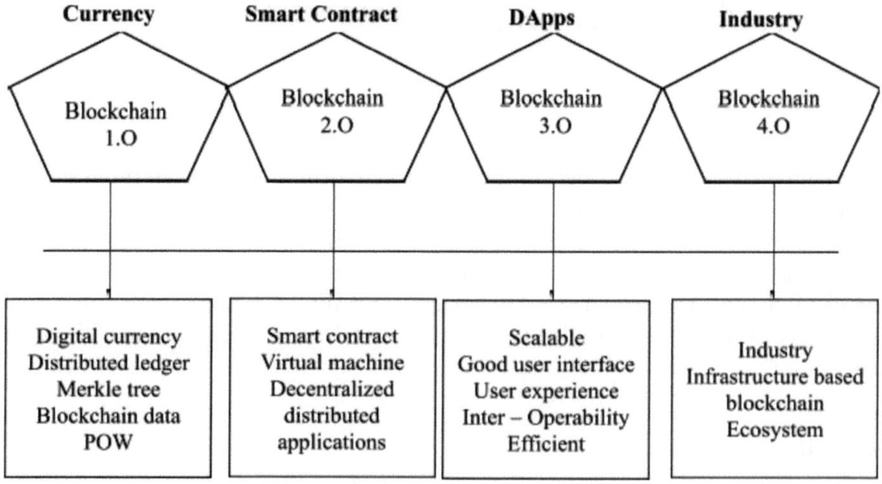

Fig. 5.3 Evolution of smart contract. Figure based breakdown conceptualization to modern of the evolution of smart contract. *Source*: Author's own compilation

Single Smart Contract

1. Enterprises started using a blockchain and started to look for the optimization of smart contract, that do not need to run on all the nodes in the network, instead they want to run on the single contract version. The single smart contract replicates the state to all the nodes in the network.
2. Enterprises used to run a private network; they have personal access to all the employees in the private network. The known entity can exist in the network to access. It made smart contracts to create a single smart contract.
3. The need for smart contacts to be run on all nodes separately had to be eradicated.
4. Oracles that represent an agent to get the external data. In such cases, it may be a weak line to access the data from outside the network.
5. The oracles, which pose as connecting agents between the network and the outside world, could get hacked or corrupted.
6. Enterprise wanted interoperability as a feature for the private network. When interoperability occurs in the network the new network is connected to the existing network. If the smart contract is stored On-Chain smart contract it will be exposed to all the nodes in the network.
7. As a feature network, enterprises wanted interoperability in their organization without having to compromise on the confidentiality of their interoperations.
8. To protect the confidentiality cryptlets were introduced. The cryptlets were a separate layer that would hold the business logic of a smart contract.

5 Smart Contract Platforms

5.1 *Ethereum*

Ethereum is an open-source, decentralized blockchain platform that lets programmers create and implement decentralized apps (dApps) and smart contracts. Ethereum, proposed in 2013 and developed in 2015 by Vitalik Buterin, is a blockchain platform beyond digital currency. With its native programming language, Solidity, developers can use Ethereum to create decentralized apps or dApps. It has the Ethereum Virtual Machine (EVM), a decentralized network engine that carries out smart contract execution.

Strengths

- Ethereum is the blockchain's largest and busiest developer community, which encourages creativity and offers a wealth of resources for assistance.
- It offers an extensive collection of developer tools like libraries, frameworks, and development environments, which make the process of creating dApps more efficient.
- Since it was the pioneer, Ethereum has been widely embraced and integrated with a range of platforms and services.

Weaknesses

- During peak hours, Ethereum's proof-of-work (PoW) consensus process restricts transaction throughput, leading to congestion and elevated gas prices.
- Even though sharding and a proof-of-stake (PoS) mechanism in Ethereum 2.0 offer improvements, the transition is continuous and complicated.

5.2 *Binance Smart Chain (BSC)*

Binance Smart Chain, launched by Binance in September 2020, is a dual-chain architecture platform that is designed to run smart contract-based applications.

Because of its compatibility with the Ethereum Virtual Machine (EVM), developers can migrate their Ethereum projects to this platform. It uses a Proof of Staked Authority (PoSA) consensus mechanism, which blends Proof of Authority (PoA) and Proof of Stake (PoS) components.

Strengths

- Compared to Ethereum, BSC can process a greater number of transactions with less fees, which appeals to users and developers.

- BSC is compatible with the Ethereum Virtual Machine (EVM), which allows developers to migrate their Ethereum dApps with little to no modification.

Weaknesses

- Because Proof of Staked Authority (PoSA), the consensus technique used by BSC, depends on a small number of validators, centralization and security concerns are raised.

5.3 Cardano

Founded by Charles Hoskinson, one of Ethereum's co-founders, Cardano focuses on security, sustainability, and scalability for the development of decentralized applications (dApps) and smart contracts. It uses a Proof of Stake (PoS) consensus mechanism called Ouroboros designed to be energy-efficient and secure.

Strengths

- Cardano's development is supported by formal verification and academic research, offering high-security guarantees.
- Scalability and energy efficiency are key design features of Ouroboros.

Weaknesses

- Compared to other platforms, Cardano has released features more slowly due to its strict, research-focused strategy.
- Compared to more established platforms such as Ethereum, Cardano's ecosystem is still in the early stages of development, despite its growth.

5.4 Solana

Solana is a high-performance blockchain platform designed for scalable decentralized applications and crypto-currencies. It uses a unique consensus mechanism called Proof of History (PoH) combined with Proof of Stake (PoS) to achieve high throughput and low latency.

Strengths

- Compared to Ethereum, Solana can process thousands of transactions per second, which makes it appropriate for applications with high demand.
- Because of its consensus mechanism's efficiency, transaction fees are reduced.

Weaknesses

- Compared to older platforms, Solana's ecosystem is still developing, with fewer dApps and developer tools.

6 Vulnerabilities and Exploits of Smart Contracts Blockchain

6.1 Reentrancy Attack

Definition

Reentrancy attacks occur when an executing function is invoked externally leading it to run multiple times in a single transaction. These attacks are critical and pose vulnerability threats that can affect smart contracts on blockchain platforms, especially those built on Ethereum. This type of attack exploits the design of a smart contract that directly calls external contracts and fails to properly manage its internal state prior to sending funds.

How It Works

1. **Initial Setup**: A vulnerable contract typically has a function that performs an external call to another contract. This function might, for example, transfer Ether to an address specified by the user. The key vulnerability lies in the contract making this external call before it has finished updating its own state.
2. **External Call Made**: When the vulnerable function is called, it executes an external call (using Solidity's call method, for instance) to send Ether. This is usually done as part of a withdrawal or payment process.
3. **Re-Entrance**: The external contract receiving the call is malicious and contains a fallback or receive function designed to make a recursive call back into the original contract. Since the state of the original contract (e.g., the balance of the user) has not yet been updated, the malicious contract can call the original contract's function again.
4. **Recursive Calls**: Each recursive call is executed before the previous call's state has been updated. This can lead to multiple withdrawals or other unintended repeated actions within the same transaction, draining the contract's funds or leading to other undesirable outcomes.
5. **Completion**: The recursive calls continue until the contract's funds are depleted or until some other limit or condition is reached.

Example

The attack on The DAO (Decentralized Autonomous Organization) in June 2016 stands as one of the most infamous and pivotal events in the history of Ethereum and blockchain technology. The DAO was conceived as a stateless, decentralized venture capital fund on the Ethereum blockchain and was one of the earliest examples of a complex smart contract system intended to automate organizational governance. Through The DAO, investors could contribute Ether (ETH), Ethereum's native cryptocurrency, in exchange for DAO tokens which conferred voting rights on proposals submitted to the DAO. This structure aimed to democratize and decentralize investment decisions, removing the need for traditional management structures.

6.2 Arithmetic Overflows/Underflows

Definition

Arithmetic overflows and underflows represent a class of vulnerabilities that arise from how numbers are handled in computer programming, particularly in languages like Solidity which is used for writing Ethereum smart contracts.

Understanding these vulnerabilities is crucial for blockchain developers because they can lead to serious logical errors that manipulate contract behavior, often leading to loss of funds or other unintended actions.

Arithmetic Overflow

An overflow occurs when an operation tries to create a numeric value that is higher than the maximum value the data type can hold. For example, in Solidity, which is largely similar to other C-like languages in terms of its handling of numbers, the uint256 data type can store values from 0 to $2^{256} - 1$. If you try to add 1 to $2^{256} - 1$, it will result in an overflow, wrapping around to 0 instead of becoming 2^{256}.

Example

Let's consider a simple example with uint8, which can store values between 0 and 255:

Suppose you store the value 255 in a uint8 variable and then add 1.

Since 255 is the maximum value a uint8 can hold, adding 1 causes the value to wrap around to 0 (because 256 in binary needs 9 bits to represent, but uint8 can only accommodate 8 bits).

This is a classic overflow scenario. The binary representation of 255 (11,111,111) when incremented by 1 result in 100,000,000. Since only the lower 8 bits are kept (due to uint8 limitations), it results in 00,000,000 (which is 0).

Arithmetic Underflow

Conversely, an underflow happens when an operation results in a number lower than the minimum value the data type can accommodate. Using uint256 in Solidity again as an example, attempting to subtract 1 from 0 would not result in -1 (as uint256 can only hold positive values), but would wrap around to $2^{256} - 1$, which is the maximum value uint256 can store.

Example

Underflows occur when subtracting from the minimum boundary:

Assume you have a uint8 variable where you store the value 0 and then subtract 1.

Since there's nothing smaller than 0 in unsigned integers, subtracting 1 from 0 underflows and wraps around to 255 (the maximum value a uint8 can represent).

In binary, 0 is 00000000. Subtracting 1 would theoretically result in -1, but since uint8 can only store non-negative values, it underflows to 11111111 (which is 255).

6.3 Timestamp Dependence

Definition

Timestamp Dependence in the context of blockchain and smart contracts refers to the vulnerability that arises when a contract's behavior depends on the block timestamp or the block time for its execution logic. This is a security concern because while generally reliable, block timestamps can be manipulated by miners to some extent, which could affect contract outcomes.

How It Works

In Ethereum, each block incorporates a timestamp assigned by the miner who successfully mines it, signifying the moment of block creation. This timestamp is crucial for maintaining the chronological order of transactions within the blockchain. The Ethereum protocol imposes certain requirements on block timestamps:

They must be greater than the timestamp of the preceding block and fall within a reasonable range of timestamps recorded by other network nodes at that time. While this "reasonable range" lacks a precise definition, it's generally understood to be within a few hours of real-world time. These timestamps serve multiple functions within the Ethereum network. Firstly, they facilitate the ordering of transactions, ensuring that transactions are added to the blockchain in a sequential and immutable manner.

Additionally, timestamps are integral to the calculation of mining difficulty adjustments, which help regulate the rate at which new blocks are mined, maintaining the network's stability and security. Furthermore, timestamps enable time-dependent functionalities in smart contracts, allowing developers to implement features such as time locks, expiration dates, and scheduled tasks. Despite the importance of timestamps, there's a degree of flexibility that allows miners to manipulate timestamps within protocol-defined bounds. While this flexibility is essential for network functionality, it also introduces potential security risks. Malicious actors could exploit timestamp manipulation to influence block propagation or smart contract outcomes. Therefore, developers must be cautious when relying on block timestamps in their applications, implementing appropriate safeguards to mitigate potential risks and ensure the integrity and security of the Ethereum network.

Example: A Lottery System

Imagine a smart contract for a lottery where the winning ticket is chosen based on the block timestamp when the lottery closes. If a miner purchases a ticket, they might have an incentive to manipulate the block timestamp to increase their chances of winning. By slightly adjusting the timestamp, a miner could potentially influence the outcome derived from the timestamp-dependent logic in the smart contract.

6.4 Short Address/Parameter Attack

Definition

The Short Address Attack is a sophisticated vulnerability that targets smart contracts on the Ethereum blockchain, leveraging the intricacies of transaction data handling and contract execution. At the heart of this exploit is Ethereum's dynamic dispatch feature, which automatically pads incoming transaction data with zeroes to match the expected length. While this feature is intended to ensure uniformity in data processing, it opens the door to malicious actors seeking to manipulate contract behavior.

How It Works

The attack unfolds when an adversary crafts a transaction with truncated input data, deliberately providing less data than expected by the target smart contract. Ethereum's padding mechanism then fills the incomplete data with zeroes, ensuring it aligns with the contract's anticipated input length. However, if the smart contract fails to properly validate the length and structure of incoming data, it may unwittingly interpret the padded zeroes as valid parameters for contract functions.

The implications of such misinterpretation are profound. Depending on the contract's design and functionality, the attacker can exploit this vulnerability to execute unauthorized actions, manipulate contract state, or even siphon funds. For instance, if a contract expects a 32-byte input for a critical function call but receives only 20 bytes from the attacker, the padded zeroes could trick the contract into executing unintended operations or allowing reentrancy attacks,

where the attacker recursively calls back into the contract before the previous execution completes, potentially draining funds or causing other disruptive outcomes.

Example

Suppose there's a smart contract that includes a function to withdraw funds, expecting two parameters: the recipient's address and the amount to withdraw. The contract assumes that incoming transactions will always include both parameters, without verifying their lengths. An attacker exploits this vulnerability by sending a transaction with truncated data, providing only the recipient's address and omitting the amount to withdraw. Ethereum pads the incomplete data with zeroes to match the expected length, tricking the contract into interpreting the padded zeroes as the amount to withdraw.

As a result, the contract erroneously executes the withdrawal function, transferring funds to the specified address but with an unintended and potentially significant amount. This allows the attacker to withdraw more funds than intended or even exploit other vulnerabilities within the contract's logic.

7 Best Practices for Smart Contract Development

Some of the best practices that could be applied while developing a smart contract include:

- **Security:** Security is paramount in smart contract development due to the immutable nature of blockchain transactions. Vulnerabilities in smart contracts can lead to significant financial losses or exploitation by malicious actors. Therefore, thorough security audits must be conducted using tools like static analysis and dynamic analysis to identify and mitigate potential vulnerabilities. Following established security best practices, such as input validation, access control, and avoiding known pitfalls like reentrancy, is crucial.
- **Modularity:** Modularity in smart contract design promotes code reuse, maintainability, and upgradeability. By breaking down complex functionality into smaller, independent contracts, developers can isolate potential issues and facilitate easier debugging and testing. Modularity also allows for more flexible upgrades, as individual components can be modified or replaced without affecting the entire system. However, care must be taken to ensure proper communication and interaction between modular components to maintain the integrity and security of the overall system.
- **Gas Optimization:** Gas optimization is essential for minimizing transaction costs and improving the efficiency of smart contract execution on the blockchain. Techniques such as code optimization, reducing storage and computation overhead, and utilizing gas-efficient data structures can help minimize gas consumption.
- **Error Handling:** Robust error handling mechanisms are crucial for smart contracts to prevent unexpected behavior and mitigate potential risks. This includes implementing defensive programming techniques, such as input validation and boundary checks, to prevent invalid or malicious inputs.
- **Testing:** Comprehensive testing is essential to ensure the functionality, security, and reliability of smart contracts. This includes writing unit tests to verify individual components, integration tests to validate interactions between different contracts or modules, and stress tests to assess performance under high loads or adversarial conditions.
- **Upgradeability:** While upgradeability can provide flexibility and allow for the evolution of smart contracts over time, it also introduces potential risks and complexities. Trade-offs and implications of upgradeability must be understood,

including the need for transparent governance mechanisms, backward compatibility, and user consent. Implementing upgradeability features like proxy contracts or versioning schemes can help mitigate risks and ensure a smooth transition during contract upgrades.

8 Implications for the Future

- **Integration with Emerging Technologies:** The integration of smart contracts with other emerging technologies, such as the Internet of Things (IoT), artificial intelligence (AI), and machine learning, holds significant potential. IoT devices can trigger smart contracts based on real-time data, enabling applications like automated maintenance and supply chain monitoring. AI and machine learning can enhance smart contract functionality by analyzing data patterns and making predictive decisions.
- **Automation:** Smart contracts represent a paradigm shift in business automation, offering immutable, self-executing agreements that can significantly streamline processes across various industries. By automating contractual obligations and payment terms, smart contracts reduce reliance on intermediaries, minimize manual errors, and accelerate transaction speeds. This automation fosters efficiency, cost savings, and greater trust among parties involved in business transactions.
- **Transparency:** Smart contracts enhance transparency by recording all transactional data on a public blockchain ledger, accessible to all parties involved. This transparency fosters trust, accountability, and auditability, as participants can independently verify the integrity and authenticity of transactions without relying on intermediaries or centralized authorities. In industries like healthcare and supply chain management, where data integrity and provenance are critical, smart contracts ensure traceability and compliance with regulatory standards, reducing fraud, errors, and disputes.
- **Decentralization:** Smart contracts enable the development of decentralized applications (DApps) that operate on blockchain networks, without the need for central servers or intermediaries. This decentralization democratizes access to services, data, and resources, empowering individuals to control their digital assets and identities. In social media, for example, decentralized platforms built on smart contracts give users ownership of their data and content, eliminating censorship and data exploitation by centralized platforms. Similarly, in governance and voting systems, smart contracts ensure transparency and verifiability, enabling secure and tamper-proof elections without the need for trusted third parties.
- **Tokenization:** Smart contracts facilitate the tokenization of assets, representing ownership rights and value on blockchain networks. This tokenization unlocks liquidity, divisibility, and programmability for traditionally illiquid assets like real estate, art, and intellectual property. Through tokenized securities, for instance, investors can fractionalize ownership of assets, lowering barriers to entry and

enabling broader participation in capital markets. Moreover, smart contracts automate compliance, dividend distributions, and shareholder voting, reducing administrative overhead and facilitating seamless asset management.

- **Security Enhancements:** Smart contract auditing shall use advanced automated tools that help identify and fix vulnerabilities more efficiently. These tools can perform comprehensive checks faster than manual audits.
- **User Experience:** Developing more intuitive and user-friendly interfaces is essential for broader adoptions. The interactions with smart contract are simplified to make them accessible even to the non-technical users. Increasing public understanding of the smart contracts will also play a vital role in doing the same. Expansion in these material aids shall help users understand smart contract and how to interact with it safely.

9 Cryptlets

Off-Chain Smart Contract

Cryptlets, introduced by Microsoft, run in a separate container, which is a secure position in a private network and act as a middle layer called cryptlets fabric. It can be written in any programming language. As shown in Fig. 5.4, the presentation layer is used to call the API Fabric to interact with each other. Cryptlets are also called Off-Chain smart contracts. Cryptlets were executed within a secure container and communication with secure channels.

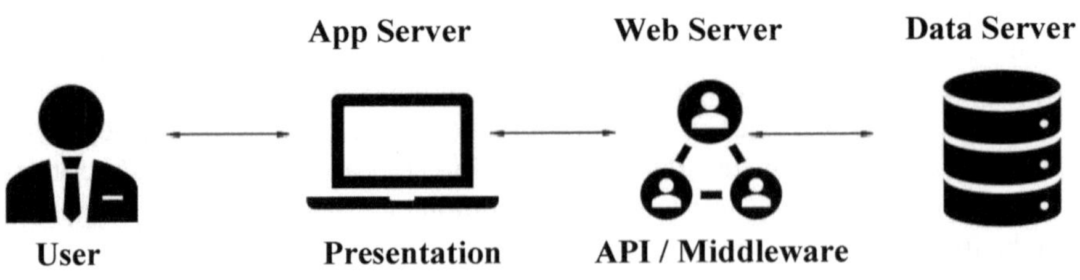

Fig. 5.4 Evolution of blockchain Smart Contracts and Cryptlets. Figure contained smart contracts to interact with external data and services often used in blockchain. *Source*: Author's own compilation

9.1 Types of Cryptlets

1. Utility Cryptlets
2. Contract Cryptlets.

Utility Cryptlets

- Utility Cryptlets include functionality of oracle agents
- It provides services such as
    1. Encryption
    2. Time and data events
    3. External data access
    4. Authentication services
- It could be many utility cryptlets running at the same time to provide different services like data inputs for the blockchain network.

Contract Cryptlets

- Contract cryptlets hold the execution logic of the smart contracts
- It runs the business logic Off-Chain and hence single smart contracts run once in the network. It will give the same logic or the same date to all the nodes in the network as shown in Fig. 5.5.

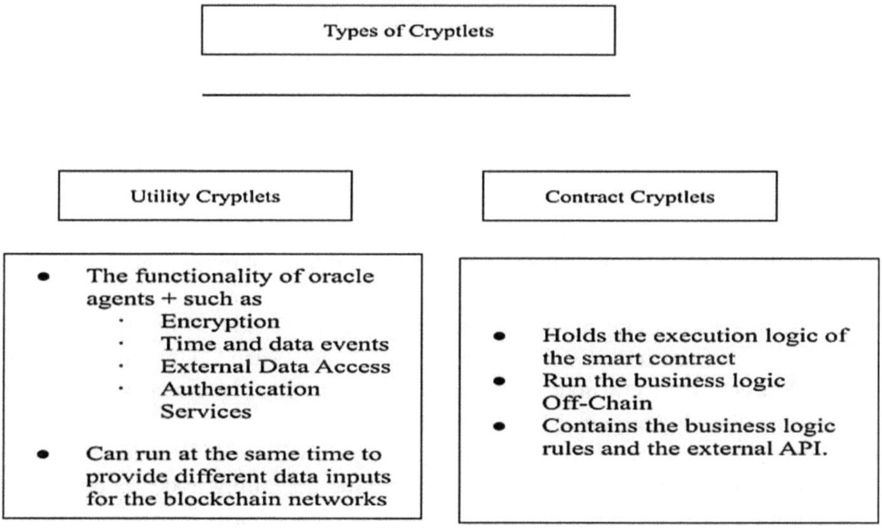

Fig. 5 Flow diagrams Smart Contract Cryptlets. Figure can be categorized based on their functionality and interaction with external systems in blockchain environment. *Source*: Author's own compilation

- Contract cryptlets when created, are attached to the specific smart contract instance, each smart contract instance links through address checks
- Contains cryptlets that makes use of business logics, rules, and external API for interacting and executing smart contract between parties

10 Blockchain Architecture

Blockchain architecture is called DApp (Distributed Application). Blockchain application is a distributed network where the presentation layer of the architecture is used to interact between client and server, the same as the existing system in a client–server architecture.

Blockchain is a peer-to-peer network that runs on a peer-to-peer application. The peer will act as a seed or a client and receive the data package.

10.1 Node Discovery Process

In the blockchain, the node discovery process begins when a new node wants to join the blockchain network in the world wide web of the Internet. In the client–server architecture—the server has a fixed address domain or name of the server. The client has information about the domain name where it must connect, and whom to connect to the address. For a client, it is very easy to connect to the network, but in the case of blockchain, it has a network of nodes [multiple nodes] hence resulting in confusion regarding where to start. The process of a new node coming onto the network and identifying all the nodes it needs to connect to is called the node discovery process.

- When the node discovery process takes place, it will connect to a few other nodes, which in turn connect with a few others, thus connecting across all the networks to send the transaction and block across the network.
- In the node discovery process a trusted IP Address, which acts as the first node to the connection. The connection node will give address of other peers in activities in the network
- Ethereum network: It creates a bootstrap node that holds all the IP Addresses connected to network, at a given period of the time. When Client requests all the IP Addresses, or a new node come up, it knows the IP Address of Bootstrap and connects directly and makes the transaction, as shown in Fig. 5.6.
- Private blockchain network: The file organization will write IP Addresses of all the nodes in the network, which refers to all the upcoming nodes to connect as peers in the network. It is a closed network and it has security issues in the private network.

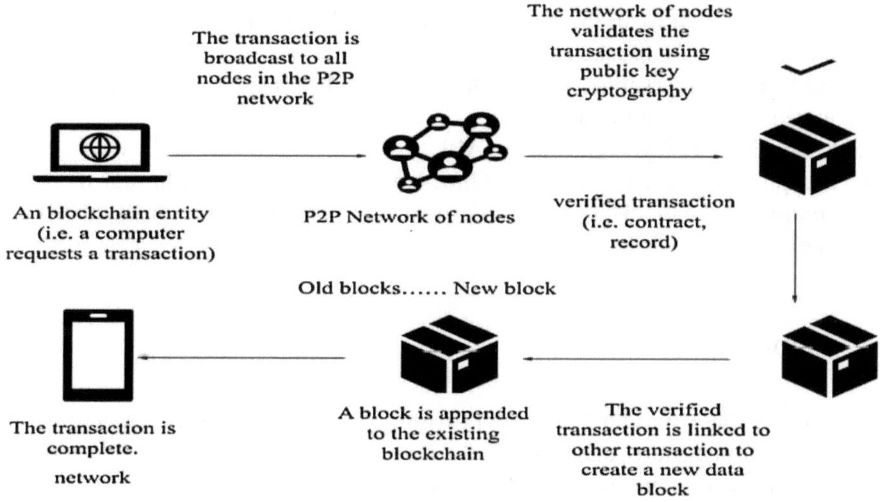

Fig. 5.6 Compound of Blockchain Architecture. Figure consists of several key components that work together to enable source, decentralized operations. *Source*: Author's own compilation

10.2 Compound of Blockchain Architecture

A set of participants or a group of people forms a blockchain network with the aim of a business problem. All the participants joined in the network will get unique ID called digital signature. Digital signature is a public key or a private key pair that manages the identities in the network.

1. Transactions are signed by the initiator of the transaction because
   (i) They have given their consent using their private key
   (ii) There is no double spending problem
2. Transactions are grouped based on the timestamp to form blocks.
3. The blocks are validated by designated nodes in the networks.
4. Consensus mechanism is applied to arrive at the network truth, which is accepted by all the nodes in the network.
   (i) Consensus method range from energy-heavy proof of work/ PBFT
   (ii) Proof of Work / PBFT
5. Choice of this consensus mechanism depends on
   (i) The type of network
   (ii) The level of trust between participants
6. Each block has a hash and a previous hash to
   (i) Ensure the chain connection is maintained
   (ii) Ensure the validity of the change

7. Blocks are sent to all the nodes that verify and add it to their chain of blocks, thus forming the blockchain
8. Each node is a server in itself which interacts with other peers and sends message in peer-to-peer communication
9. Smart contract layer acts like an API
10. D-App is the presentation layer of blockchain, which connects through the smart contract in the blockchain network. It works to enable the presentation layer to client

11 Conclusion

We conclude that smart contracts, by providing a strong framework for automating agreements, transactions, and business operations on blockchain networks, represent a revolutionary force in the field of decentralized technology.

Exploring smart contracts in the context of blockchain technology exposes a diverse range of opportunities, challenges, and innovation. From their inception to the current, smart contracts have brought new answers to persistent problems and radically changed the way organizations function. Smart contracts will undoubtedly be crucial in determining how digital transactions, governance, and innovation are shaped in the future as we continue to traverse the rapidly changing terrain of decentralized technology. Their power to improve trust, simplify procedures, and increase transparency highlights their importance as a disruptive force in the blockchain ecosystem, opening the door to a more effective, safe, and inclusive digital economy.

References

Aron J (2012) BitCoin software finds new life. New Sci 213(2847):20

Atzei N, Massimo B, Tiziana C (2017) A survey of attacks on ethereum smart contracts (SoK). In: International conference on principles of security and trust. Springer, pp 164–186

Back A (2002) Hashcash—Amortizable Publicly Auditable Cost-Functions

Bertoni G, Daemen J, Peeters M, Van Assche G (2011) The KECCAK SHA-3 sub- mission

Boutellier R, Heinzen M (2014) Pirates, Pio- neers, Innovators and Imitators. In: Growth through innovation. Springer, Berlin, pp 85–96

Buterin V (2013) Ethereum: a next-generation smart contract and decentralized application platform. J Decentralized Technol 5(1):1–15. https://doi.org/10.1234/jdt.2013.0001

Buterin V (2015a) EIP-2: Homestead hard-fork changes

Buterin V (2015b) Merkling in Ethereum

Buterin V (2016) EIP-100: change difficulty adjustment to target mean block time including uncles

Courtois NT, Grajek M, Naik R (2014) Optimizing SHA256 in bitcoin mining. Springer, Berlin, pp 131–144. ISBN 978-3-662-44893-9

Davey BA, Priestley HA (2002) Introduction to lattices and order, 2nd ed.Cambridge University Press, Cambridge. ISBN 0-521-78451-4/pbk

Dhillon V, Metcalf D, Hooper M (2017) Blockchain enabled applications. Apress, Berkeley, CA

Dryja T (2014) Hashimoto: I/O bound proof of work

Fowler PVG (1991) Landon Curt Noll. FowlerNol- 1Vo hash function
Gamage HTM, Weerasinghe HD, Dias NGJ (2020) A survey on blockchain technology concepts, applications, and issues. SN Comput. Sci. 1:114
Gökalp E, Gökalp MO, Çoban S, Eren PE (2018) Analysing opportunities and challenges of integrated blockchain technologies in healthcare. Information systems: research, development, applications, education: 11th SIGSAND/PLAIS EuroSymposium 2018, Gdansk, Poland, September 20, 2018, Proceedings 11, pp 174–183
Gura N, Patel A, Wander A, Eberle H, Shantz SC (2004) Comparing elliptic curve cryptography and RSA on 8-bit CPUs. In: Cryptographic hardware and embedded systems-CHES 2004. Springer, pp 119–132
Hewa TM, Hu Y, Liyanage M, Kanhare SS, Ylianttila M (2021) Survey on blockchain-based smart contracts: technical aspects and future research. IEEE Access 9:87643–87662
Jentzsch C (2015) Commit date for ethash
Monrat A, Schelén O, Andersson K (2019) A survey of blockchain from the perspectives of applications, challenges, and opportunities. IEEE Access 7:117134–117151
Prause G (2019) Smart contracts for smart supply chains. IFAC-PapersOnLine 52(13):2501–2506
Sayeed S, Marco-Gisbert H, Caira T (2020) Smart contract: attacks and protections. IEEE Access 8:24416–24427
Wang S, Ouyang L, Yuan Y, Ni X, Han X, Wang F-Y (2019) Blockchain-enabled smart contracts: architecture, applications, and future trends. IEEE Trans Syst Man Cybern: Syst 49(11):2266–2277

Blockchain-Enabled Cyber Security Framework for Enhanced Transactions in Distributed Smart Grid Infrastructure with Wind Farm Energy Source

Prabhakar Rontala Subramaniam, Chitra Venugopal, and Lizzy Oluwatoyin Ofusori

Abstract Wind farm energy is gaining popularity as a sought-after energy source for smart grid integration (Musleh et al. 2019). As smart grid technology continues to evolve and renewable energy adoption increases, wind farm power is poised to play an increasingly significant role in the global energy transition towards a more sustainable and resilient energy system. However, while the smart grid offers solutions to various challenges present in traditional grid systems, it encounters several security threats due to the integration of communication within the power infrastructure, thereby exposing the system to numerous risks (El Mrabet et al. 2018). In addition, a breach in the smart grid's security could disrupt power supply, compromise grid operations and potentially cause widespread damage. Although, there are existing security measures that focus on mitigating some specific attacks or protecting some specific components, but there is no reliable approach which combines advanced technologies solutions to secure the entire system (Gunduz and Das 2020). Moreover, there are limited studies that focus on mitigating security threats encountered in integrating smart grids with wind farms energy source (Tufail et al. 2021). These challenges pose extreme pressure on finding advanced technologies and sustainable solutions for secure and reliable operations of the smart grid powered by wind farm energy source. Hence, this study attempts to address these challenges by reviewing the evolution, benefits and various security threats associated with the phenomenon of integrating smart grids with wind farms energy source with the aim of developing a blockchain enabled cyber security framework for smart grid infrastructure. Specifically, the paper demonstrates how blockchain can be utilized as part of the mitigation process for securing threats associated with smart grids powered by wind farm energy source. In this perspective, this study relies on secondary data drawn

P. R. Subramaniam (✉) · L. O. Ofusori
Information Systems & Technology, University of KwaZulu-Natal, Durban, South Africa
e-mail: prabhakarr@ukzn.ac.za

C. Venugopal
Renewable Energy Engineering, Oregon Institute of Technology, Wilsonville, OR, USA
e-mail: chitra.venugopal@oit.edu

S. Krishnan et al. (eds.), *Unveiling Blockchain Enabled Business Applications*, Contributions to Management Science, https://doi.org/10.1007/978-981-95-2337-5_6

from reputable and diverse range of sources, including research publications, books, and reports from relevant international organizations and government sources.

1 Introduction

Smart grid technology refers to the integration of digital communication and advanced monitoring and control capabilities into the traditional electricity grid (Lee et al. 2023). This allows for more efficient and flexible management of energy resources, including renewable sources like wind power. "Wind is a more efficient power source than solar. This is because, compared to solar panels, wind turbines emit lower levels of CO_2 to the atmosphere, require less energy consumption, and generate higher overall energy output. Wind farm energy is one of the fastest-growing renewable energy sources due to its abundant availability, scalability, and decreasing cost (Sadorsky 2021). By enabling real-time communication between energy producers and consumers, smart grids optimize energy distribution, reduce waste, and enhance overall grid reliability and resilience."

As smart grid technology continues to evolve and renewable energy adoption increases, wind farm power is poised to play an increasingly significant role in the global energy transition towards a more sustainable and resilient energy system. However, while the smart grid offers solutions to various challenges present in traditional grid systems, it encounters several security threats due to the integration of communication within the power infrastructure, thereby exposing the system to numerous risks (El Mrabet et al. 2018). "Smart Grid's security is becoming an important issue because major power facilities are interconnected through communication networks. Cyber attacks by malicious intruders can lead to serious incidents such as massive outages and the destruction of power network infrastructure, since the cyber attacks can damage energy data, starting with personal information leakage from grid members. While there are existing security measures that focus on mitigating some specific attacks or protecting some specific components, there is no effective approach which combines advanced technologies such as blockchain solutions to secure the entire system.

The blockchain is a distributed database in which data stored on a blockchain are accessible to all participants within the blockchain network. Applying blockchain technology solution to the Smart Grid will enable more secure management of energy data, and furthermore, it will contribute to the development of the future smart energy industry in the future. This represents a gap in the literature and gives an opportunity for this study to address. Thus, the study suggests a Blockchain-Enabled Security Framework for Distributed Smart Grid Infrastructure."

The paper is structured as follows. Section 1 introduces the smart grid and the identified problem. Section 2 reviews the relevant research and relates the potential threats encountered in smart grid. Section 3 relates blockchain as a solution for the threats identified in Sect. 2. Section 4 recommends Blockchain-Enabled Security

Framework for enhancing transactions in smart grids with potential cryptographic algorithms that can be adopted. Section 5 concludes the chapter.

2 Review of Relevant Research

2.1 Smart Grid Evolution

Over the past few decades, there has been a paradigm shift in the way electricity is generated, transmitted, and consumed (Kang et al. 2023). Traditionally, electricity grids were designed as one-way systems, with power flowing from centralized generation facilities (like power plants) to end-users (Ghiasi et al. 2023). These systems relied on manual monitoring and control and were not equipped to handle fluctuations in supply and demand or integrate renewable energy sources. However, the first stage in the evolution of the smart grid involves digitizing and automating grid operations (Campagna et al. 2020). This includes the deployment of sensors, meters, and communication technologies throughout the grid infrastructure to gather real-time data on energy usage, grid conditions, and equipment health. The second stage involves the introduction of advanced metering infrastructure, often referred to as smart meters (Kang et al. 2023). The smart meters allow utilities to remotely monitor and manage energy consumption at the individual customer level. Smart meters enable two-way communication between the utility and the customer, providing valuable insights into energy usage patterns and enabling dynamic pricing schemes. The third stage is called the grid modernization. Grid modernization focuses on upgrading aging infrastructure, enhancing grid resilience, and integrating advanced control and monitoring systems (Fadlullah and Kato 2015). This involves investments in technologies such as synchrophasors, which provide precise measurements of voltage, current, and frequency across the grid in real-time, improving grid stability and reliability. The fourth stage is the integration of renewable energy (Kang et al. 2023). With the increasing penetration of renewable energy sources like wind and solar power, the smart grid has evolved to accommodate the intermittent nature of these resources. Advanced forecasting tools and grid management algorithms are developed to predict and mitigate the impact of renewable energy variability on grid stability and supply–demand balance. In addition, the fifth stage is the Distributed Energy Resources (DERs) (Kang et al. 2023). The proliferation of distributed energy resources, including rooftop solar panels, energy storage systems, and electric vehicles, further shaped the evolution of the smart grid.

It is important to note that the integration of renewable energy such as wind energy in the fourth and fifth stages of the smart grid evolution establishes the need for securing smart grid through blockchain. As shown in Fig. 6.1, wind energy farm management involves overseeing the operation and maintenance of wind farms to ensure optimal performance and efficiency in generating electricity from wind (Sadorsky 2021). This includes monitoring various aspects such as multiple wind turbines, wind conditions, power output, and maintenance schedules. Wind energy farm management, which consist of multiple wind turbines strategically located in areas with high wind potential, generate electricity by harnessing the kinetic energy of the wind (Yahya et al. 2020). The primary goal is to maximize energy production while minimizing downtime and maintenance costs. The control unit (refer Fig. 6.1) plays a crucial role in wind energy farm management as it serves as the central hub for monitoring and controlling the entire system (Sadorsky 2021). There are three sub-systems of the control unit; namely, power generation monitoring sub-system, power transmission & distribution sub-system and power consumption sub-system. The power generation monitoring sub-system acts an interface between the wind turbines and smart grid (Akhter and Biswas 2016). It continuously monitors the operation of each wind turbine in the farm, collecting real-time data on parameters such as wind speed, turbine speed, power output, temperature, and other relevant variables (Kadri et al. 2020). However, the power transmission & distribution sub-system acts as an interface between the power generation monitoring sub-system and grid security structure, while the power consumption sub-system acts as an interface between the power transmission distribution sub-system and energy grid (Akhter and Biswas 2016). With these interfaces, the control unit can adjust the operating parameters of each turbine to maximize energy production while ensuring safe and efficient operation. By leveraging advanced control algorithms and real-time data analytics, the control unit maximizes the efficiency and profitability of wind energy generation while ensuring safe and reliable operation of the wind farm infrastructure (Sadorsky 2021). According to Musleh et al. (2019), the integration of wind energy farm management into smart grids represents a synergistic approach towards a sustainable and efficient energy future, offering benefits such as reduced carbon emissions, improved grid performance, and enhanced energy security."

2.2 Security Threats in Smart Grids

Smart Grids security threats encompass a range of potential risks that can compromise the integrity, confidentiality, and availability of the grid's operations and data (Bouramdane 2023). Table 6.1 presents review of common security threats in smart grids."

Denial-of-Service attack involves malicious attempts to disrupt the availability of essential services provided by the grid infrastructure (Faquir et al. 2021; Huseinović et al. 2020). This type of attack aims to overwhelm the smart grid system with an excessive volume of requests or traffic, causing it to become unresponsive or inaccessible to legitimate users (Ortega-Fernandez and Liberati 2023; Alsuwian et al. 2022).

Phishing in smart grid is the first step that exposes both customers and organizations to potential risks due to its easy process of execution (Soykan et al. 2021). Like employees may encounter additional threats, such as deceptive emails or messages designed to resemble genuine communications (Ustundag Soykan and Bagriyanik

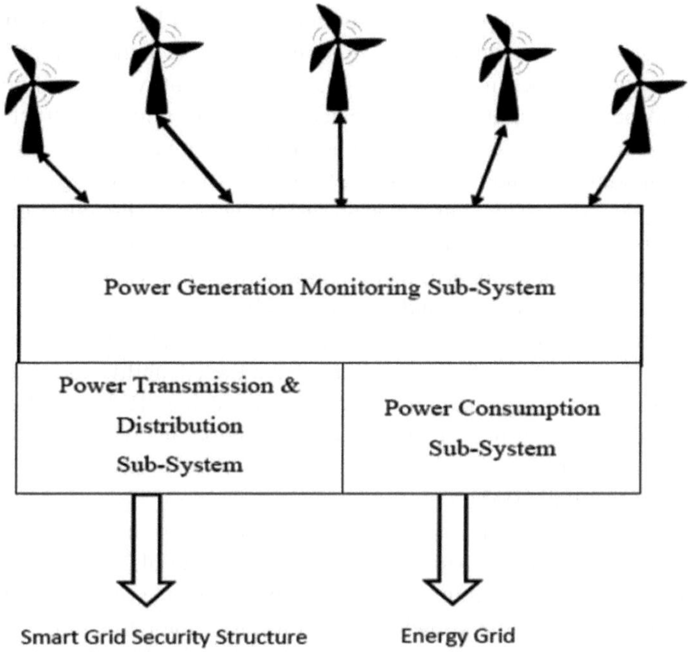

Fig. 6.1 Wind energy farm management control unit *Source* Authors' Own

Table 6.1 Review of common security threats in smart grids

Common security threats	Type of security breach	Target areas in Smart grid	Authors
Denial of service	Availability	Communication channels, central data processing systems, transmission, distribution, and consumption	"Faquir et al. (2021), Huseinović et al. (2020), Ortega-Fernandez and Liberati (2023), Alsuwian et al. (2022)"
Phishing	Confidentiality	Communication channels such as email, utility employee	"Faquir et al. (2021), Soykan et al. (2021), Ustundag Soykan and Bagriyanik (2020)"
Impersonation	Confidentiality & integrity	Grid operation, grid control system	"Islam et al. (2019), Alsuwian et al. (2022), Gunduz and Das (2020)"
Malware	Integrity	Organizational server and devices	"Faquir et al. (2021), Eder-Neuhauser et al. (2017), Aziz et al. (2022)"
Eavesdropping	Confidentiality & integrity	Communication lines, network protocols	"Alani et al. (2023), Faquir et al. (2021), Chaudhry et al. (2021)"

2020). In such scenarios, employees may unknowingly divulge personal information, thereby increasing the likelihood of being compromised (Faquir et al. 2021).

Impersonation attack involves an unauthorized entity or malicious actor pretending to be a legitimate user or device within the grid infrastructure (Islam et al. 2019). A hacker can act as a legitimate user in an unauthorized way, spoofing the identity of someone and making them to incur charges for energy consumption (Alsuwian et al. 2022). This can occur at various levels of the smart grid system, such as impersonating a legitimate user to gain access to sensitive information or control over grid components or impersonating a trusted device to manipulate grid operations or inject false data (Gunduz and Das 2020).

Malware attack is one of the primary threat to the smart grid, which poses a significant concern (Faquir et al. 2021). Attackers have the capability to create malware that can infiltrate both organizational servers and devices (Eder-Neuhauser et al. 2017). Through the dissemination of malware, attackers can manipulate device functions or system operations, thereby gaining unauthorized access to gather sensitive information (Aziz et al. 2022).

Eavesdropping refers to the unauthorized interception of communication between devices or nodes within the grid infrastructure (Alani et al. 2023). Essentially, it involves an attacker surreptitiously monitoring the exchange of data or information over the network without the knowledge or consent of the parties involved (Faquir et al. 2021). Once the attacker gains access to the transmitted data, they can potentially extract sensitive information, including energy consumption patterns, user behaviour, or even authentication credentials (Chaudhry et al. 2021).

2.3 Influence of Cyber Threats on Smart Grid

According to Naqvi et al. (2024), distributed denial of service (DDoS) is one of the prominent attacks that has affected the smart grid system. For example, in December 2015, Ukraine experienced a massive cyber attack targeting its power grid (Habib et al. 2023). Hackers launched a coordinated DDoS attack against the energy distribution companies, disrupting electricity supply to hundreds of thousands of customers. This attack, which was combined with malware and phishing, resulted in widespread power outages. More than 225,000 people have been directly affected electricity Blackout of three distribution companies (Habib et al. 2023). This incident marked the first publicly acknowledged event and it exposed the weakness in highly automated cyber physical systems like smart grids (Mahi-Al-rashid et al. 2022). Likewise, in March 2019, the U.S. power grid encountered a zero-day attack, while in September 2019, the Dtack malware was detected at the Kudankulam Nuclear Power Plant in India (Habib et al. 2023; Mahi-Al-rashid et al. 2022). One of the captured Dtrack malware samples contained information about the power plant's internal network (Mahi-Al-rashid et al. 2022). According to Faquir et al. (2021), when the smart grid system is compromised by malware attack, it can steal sensitive information stored within smart grid databases or systems. This could include customer data, billing information, or operational details critical to the functioning of the grid (Aziz et al. 2022).

2.4 Existing Security Measures in Smart Grid

Existing security measures in smart grids encompass a range of strategies and technologies aimed at protecting the infrastructure from various threats. Some of these security measures include authentication and access control, encryption, Intrusion Detection Systems (IDS), firewalls, security patching and updates. Table 6.2 presents the common existing security measures used for smart grid.

Authentication and “access control mechanisms play a crucial role in securing smart grid systems against unauthorized access (Tanveer et al. 2021; Zahoor et al. 2023). Authentication is the process of verifying the identity of a user or device attempting to access a system or resource in order to prevent unauthorized access and protect against identity theft and impersonation attacks (Bera et al. 2020). In smart grids, authentication ensures that only authorized individuals or devices can interact with critical components such as control systems, meters, and communication networks (Tanveer et al. 2021). Authentication mechanisms typically involve the use of credentials, such as usernames and passwords, cryptographic keys, biometric data, or multifactor authentication (Wang et al. 2021b).

Encryption protects sensitive information transmitted between smart grid components by scrambling the data in such a way that it can only be understood by authorized parties with access to the decryption key (Son et al. 2020; Zhai et al. 2022). This prevents unauthorized interception or eavesdropping on communication channels, safeguarding sensitive data such as energy consumption patterns, billing information, and control commands from being accessed by malicious actors (Wang et al. 2021b). By encrypting data at the source and decrypting it at the destination using cryptographic algorithms and keys, smart grid systems can verify that the data has not been altered or corrupted (Mahmood et al. 2023).

Intrusion Detection Systems (IDS) continuously monitor network traffic, system logs, and device behaviour within the smart grid infrastructure to identify deviations from normal patterns or expected behaviours (Liu et al. 2021; Sun et al. 2020). IDS employs signature-based detection techniques to identify known patterns or signatures associated with known cyber threats, malware, or attack vectors (Liu et al. 2021; Ravinder and Kulkarni 2023). When a match is found, indicating a potential security threat, the IDS generates an alert or initiates an automated response to mitigate the threat (Ganesan and Xavier 2023).

Firewalls is used to establish a secure perimeter around the smart grid infrastructure by filtering incoming and outgoing network traffic, preventing unauthorized access and protecting against various types of cyber threats (McLaughlin et al. 2015; Gaudet et al. 2020). Moreover, firewalls utilize packet filtering techniques to inspect individual network packets and make filtering decisions based on criteria such as

Table 6.2 Review of common existing security measures used in smart grid

Study	"Authentication & access control	Encryption	Intrusion Detection Systems (IDS)	Firewalls	Security patching"
Bera (2020)	✓				
Faquir et al. (2021)					✓
Ganesan (2023)			✓		
Gaudet et al. (2020)				✓	
Liu et al. (2021)			✓		
Mahmood (2023)		✓			
McLaughlin et al. (2015)				✓	
Ravinder (2023)			✓		
Ruland et al. (2018)				✓	
Son et al. (2020)		✓			
Sun et al. (2020)			✓		
Tanveer et al. (2021)	✓				
Wang et al. (2021a, b)	✓	✓			
Yadav et al. (2020)					✓
Zahoor et al. (2023)	✓				
Zhai (2022)		✓			

source IP address, destination IP address, port number, and protocol type (Ruland and Sassmannshausen 2018).

Security patching involves identifying, prioritizing, and applying patches or updates released by vendors to address known vulnerabilities and security weaknesses in smart grid devices, software, and firmware (Faquir et al. 2021). Timely application of security patches and updates helps mitigate the risk of exploitation by cyber attackers who may attempt to exploit known vulnerabilities to gain unauthorized access, disrupt operations, or compromise sensitive data within the smart grid infrastructure (Yadav et al. 2020)."

2.5 Gaps Identified with the Existing Security Measures

It should be noted that despite the numerous existing security measures used in smart grid such as encryption, firewall, authentication and access control, they are completely inadequate for security breaches. "This is because cybercriminals are adaptive and given time, will usually find ways to bypass such security solutions. Furthermore, there are various gaps identified with these security measures. Firstly, in the case of authentication, studies have shown that authentication verifies the identity of users or devices but may not provide access control capabilities. Without fine-grained access control mechanisms, authenticated users or devices may have excessive privileges or access rights, increasing the risk of unauthorized access, data breaches, or malicious activities within the smart grid infrastructure (Ezugwu et al. 2023)." Secondly, while encryption helps secure data during transmission, it may not prevent denial-of-service (DoS) attacks aimed at overwhelming the resources of smart grid devices or communication networks. Attackers can still disrupt the availability of services by flooding communication channels with encrypted traffic (Chatterjee et al. 2023). Furthermore, encryption primarily focuses on protecting data confidentiality, but it does not ensure data integrity. Without mechanisms to verify the integrity of the data, there's a risk of data manipulation or tampering, which can lead to incorrect decisions or actions within the smart grid system. Thirdly, Jingyao et al. (2020) reveal that the traditional firewalls are effective at blocking known threats based on signature matching or predefined rules, but they are not effective to detect and mitigate advanced and sophisticated threats, such as zero-day exploits, polymorphic malware, or targeted attacks that evade signature-based detection. Fourthly, unlike Intrusion Prevention Systems (IPS), which can actively block or mitigate detected threats, IDS are typically passive monitoring systems that focus on detecting and alerting on suspicious activities (Al-Janabi et al. 2021). While IDS can provide valuable visibility into potential security incidents, they do not have the capability to prevent intrusions or take. Lastly, security patches are only released in response to identified vulnerabilities or weaknesses in software or hardware components. This implies that until a patch is applied, there exists a vulnerability window during which the system remains exposed to potential exploitation by attackers. If patches are not applied promptly, the smart grid infrastructure remains at risk of security breaches and cyber attacks.

However, with the integration of blockchain technology to the targeted areas in the communication layer of the Smart Grid, it will enhance the secure management of energy data. According to Gilani et al. (2020), blockchain technology enhances security in the electrical power system through cryptographic securitization and a consensus mechanism, ensuring the immutability of incorporated data. Once energy transactions/data are included in the blockchain, altering them for illegitimate purposes becomes challenging, resulting in a highly secure and robust system. Additionally, the decentralized nature of blockchain increases the resiliency of the electrical power system by eliminating single points of failure present in centralized data architectures, reducing vulnerability to malicious attacks.

3 Blockchain: A Potential Solution

Blockchain is a technology originally created to support the famous cryptocurrency Bitcoin (Ravikiran 2023). The emergence of Bitcoin by Satoshi Nakamoto in 2008 introduced the world to the first practical implementation of blockchain technology. Since then, it has seen huge growth with the capital market. Blockchain technology is a structure that stores transactional records, also known as the block, of the public in several databases, known as the "chain," in a network connected through peer-to-peer nodes (Agarkar et al. 2024). This chain expands over time as new blocks are added. The storage is referred to as a 'digital ledger.' Every transaction in this ledger is authorized by the digital signature of the owner, which authenticates the transaction and safeguards it from tampering. Hence, the information the digital ledger contains is highly secure. Moreover, blockchain works in a decentralized environment that is enabled by comprising several core technologies, such as digital signatures, cryptographic hash, and distributed consensus algorithms. This decentralization means transactions occur without the need for intermediaries to validate or verify them. Key characteristics of Blockchain include decentralization, transparency, immutability, and auditability."

In recent years, there has been a growing interest in blockchain technology among enterprises and institutions seeking to leverage its benefits for improving efficiency, transparency, and security in business processes. Enterprise blockchain platforms such as Hyperledger Fabric, Corda, and Quorum have emerged to address the specific needs and requirements of businesses, offering features such as permissioned networks, scalability, privacy, and regulatory compliance. According to Cheng et al. (2020) blockchain technology offers a wide range of applications that can help mitigate security threats by providing enhanced data integrity, transparency, decentralization, and cryptographic security capabilities across various domains and industries. These applications include Identity Management Systems (IMS), Audit trail and data storage systems.

Identity Management Systems (IMS) manage individual profiles, user authorization, authentication, and access privileges (Agarkar et al. 2024). According to Alanzi and Alkhatib (2022), the traditional centralized identity management systems are susceptible to data breaches. Hence, blockchain-based identity management systems address these shortcomings of traditional solutions by letting users to have control over their identity information, reducing the risk of identity theft and unauthorized access."

Audit trails play pivotal roles in enterprise business applications, commonly employed to retain, monitor, and review data for auditing purposes (Taherdoost and Madanchian 2023). To strengthen the security, blockchain technology has recently been integrated into audit trails, making it immutable (Ahmad et al. 2021). "In other words, blockchain provides an immutable and tamper-proof ledger, making it ideal for creating audit trails. Blockchains adopt a consensus-driven approach to ensure audit trails with secure and transparent data processing, eliminating the need for a central authority (Sahlin and Levenby 2018)."

In traditional data storage systems, data is typically stored on centralized servers owned and managed by a single entity (Drijvers et al. 2019). This creates a single point of failure and makes the data vulnerable to hacking or corruption. In contrast, blockchain distributes information across a network rather than storing it all in one location, decentralized systems are more resistant to hacks and other cyber attacks (Bin Sulaiman 2018). By leveraging blockchain technology, the storage and management of sensitive data such as user identities, medical records, and appointment schedules in the medical sector can be secured (Lee 2019).

With data sharing and communication security, blockchain enables secure and transparent data sharing among multiple parties (Liu et al. 2018). By encrypting data and storing it on a blockchain, organizations can ensure that sensitive information is only accessed by authorized parties, reducing the risk of data breaches (Cheng et al. 2020). Furthermore, blockchain can be used to create secure communication channels, where messages are encrypted and stored on the blockchain (Bin Sulaiman 2018).

Blockchain cryptographic properties enable secure and transparent data sharing among multiple parties within the smart grid ecosystem, including utility companies, grid operators, consumers, and renewable energy providers (Agarkar et al. 2024). Smart contracts, which are self-executing contracts with predefined rules encoded on the blockchain, can facilitate automated and auditable data sharing processes, reducing the risk of data breaches, unauthorized access, or miscommunication (Mylrea and Gourisetti 2017). By leveraging blockchain's cryptographic techniques, smart grid operators can ensure the integrity and authenticity of data transmitted across the grid, preventing unauthorized tampering or manipulation (Waseem et al. 2023). This helps in mitigating threats related to data tampering, unauthorized access, and falsification of information. Moreover, blockchain-based decentralized architectures help improve the resilience of smart grid infrastructure by distributing control, data storage, and processing across multiple nodes in the network (Waseem et al. 2023). This distributed architecture reduces the risk of single points of failure, cyber attacks, or disruptions, making the grid more robust and resistant to malicious activities or natural disasters. While no system is entirely immune to attacks, blockchain's design and cryptographic principles make it highly secure and resistant to tampering.

4 Blockchain-Enabled Security Framework for Smart Grid

Figure 6.2 shows an overview of the entire system and how the structured components are distributed and interconnected with each other. Wind farm energy source is considered the producer that generates a vast amount of energy source. It is distributed among the consumers by the service provided. All this information is stored in a distributed network, and the transactions are recorded in it as well.

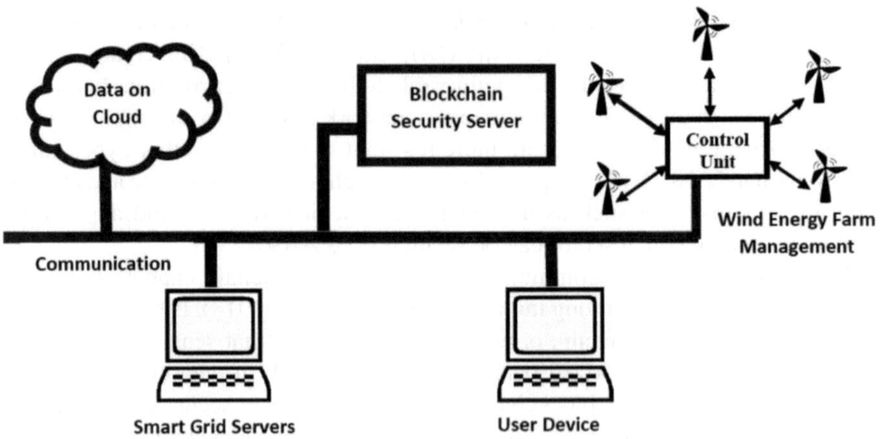

Fig. 6.2 Blockchain-enabled security structure for smart grid with wind farm energy. *Source* Authors' Own

4.1 *Smart Grid Server*

Smart grid server security can be enhanced by utilizing Service-Oriented Architecture (SOA) and adopting principles and practices that promote modularization, interoperability, and flexibility (Haghgoo et al. 2021). Service-Oriented Architecture (SOA) is a design approach used in software development and system integration to create modular, reusable, and interoperable services (Pau et al. 2022). In SOA, software components are organized as discrete, self-contained units called services, which perform specific business functions and can be accessed and reused independently (Haghgoo et al. 2021). These services communicate with each other over a network using standardized protocols and interfaces. This promotes interoperability and allows different services to communicate securely. Standard protocols like HTTPS (HTTP over TLS) can be used to ensure encryption and secure data exchange between services (Atalay 2022). In smart grid server (refer Fig. 6.2), the SOA can play a role by enabling separate services to handle authentication, data processing, access control, and monitoring. Moreover, strong authentication mechanisms can be implemented to verify the identity of users and devices accessing smart grid services. Techniques such as OAuth, OpenID Connect, or SAML can be used to secure authentication and authorization (Wilson and Hingnikar 2019). Similarly, Role-based access control (RBAC) can be employed to manage permissions and restrict access to sensitive resources (Ducornaud 2023). By adopting these practices, organizations can leverage SOA principles to design and implement secure smart grid servers that protect critical infrastructure, data, and resources from potential cyber threats and attacks.

4.2 Data on Cloud

Several cloud architectures are proposed for smart grid systems to enhance scalability, flexibility, and resilience. One commonly discussed architecture is the "Cloud-Edge-Fog Continuum" (Khalyeyev et al. 2022). Cloud-Edge-Fog Continuum architecture offers a hierarchical and distributed approach to smart grid computing, leveraging the strengths of each layer to optimize resource utilization, improve data processing efficiency, and enhance system reliability and responsiveness (Xue et al. 2023). It enables seamless integration of cloud-based services, edge computing capabilities, and fog computing infrastructure to support diverse smart grid applications, such as demand response, grid optimization, renewable energy integration, and predictive maintenance (Mohamed et al. 2021).

In a smart grid system, it is essential to secure data stored in the cloud because sensitive information related to energy consumption, grid operations, and infrastructure must be protected from unauthorized access, manipulation, and cyber threats. Encrypting data both in transit and at rest is essential for ensuring its confidentiality (Moghadam et al. 2020). This prevents unauthorized parties from accessing the data in the case of data breaches. Moreover, implementing strict access controls helps manage who can view, modify, or delete data stored in the cloud. Role-based access control (RBAC) allows administrators to define roles with specific privileges and assign them to users based on their responsibilities (Ducornaud 2023). Multi-factor authentication (MFA) adds an extra layer of security by requiring users to provide multiple forms of verification before accessing sensitive data (Nikam and Potey 2016)."

4.3 User Device

In a smart grid, a customer or user device refers to any device or equipment utilized by consumers within their premises to interact with or benefit from the smart grid infrastructure (Guo et al. 2024). These devices such as smart meters play a crucial role in enabling consumers to actively participate in energy management, optimize their electricity usage, and potentially contribute to the stability and efficiency of the overall grid (Wang et al. 2021a).

4.4 Blockchain Security Server

Blockchain security server typically refers to a dedicated system or network infrastructure designed to enhance the security of blockchain networks and applications (Ravikiran 2023). It operates as follows:

(a) **Decentralization**: Blockchain operates on a decentralized network of nodes, which collectively maintain the ledger and validate transactions (Vacca et al. 2021). In smart grid, blockchain technology is applied to ease the decentralized energy management, which is developed according to the P2P concept provided by the blockchain (Hasankhani et al. 2021). This reduces the reliance on a central authority and enhances resilience against failures and attacks. According to Aderibole et al. (2020), blockchain-based decentralized architectures help improve the resilience of smart grid infrastructure by distributing control, data storage, and processing across multiple nodes in the network. This distributed architecture reduces the risk of single points of failure, cyber attacks, or disruptions, making the grid more robust and resistant to malicious activities or natural disasters (Werth et al. 2023). While no system is entirely immune to attacks, blockchain's design and cryptographic principles make it highly secure and resistant to tampering (Werth et al. 2023).

(b) **Process of Adding Blocks:** According to Kuzlu et al. (2020), blockchain, an emerging technology, can be leveraged to improve the smart grid. However, to enhance smart grid security, it is essential to understand the process of adding blocks. In blockchain, a user initiates a transaction, which is broadcast to the network (Laukyte 2023). Nodes receive and validate the transaction (check for double-spending, proper signatures, etc.) (Marsalek and Zefferer 2019). Validated transactions are propagated through the network, reaching full nodes and miners. Miners collect valid transactions into a transaction pool (also known as a mempool). Miners select transactions from the transaction pool and organize them into a candidate block (Komalavalli et al. 2020). Miners include a reference to the previous block's hash in their candidate block, forming a chain (Laukyte 2023). Nodes that receive the newly proposed block validate it by checking the cryptographic puzzle solution (in Proof of Work) or the validator's stake (in Proof of Stake) and ensuring all transactions within the block are valid (Aggarwal and Kumar 2021a). If the block is valid, nodes add it to their copy of the blockchain and propagate it to other nodes. The network reaches consensus on the valid chain. This could involve mechanisms like the longest chain rule (in PoW) or other consensus algorithms (in PoS, Byzantine Fault Tolerance, etc.) (Laukyte 2023). Once a block is added to the blockchain, subsequent blocks build on top of it, confirming its validity and making it increasingly difficult to alter (Laukyte 2023).

(c) **Consensus Mechanisms**: Two major consensus algorithms are used in blockchain, which are Proof of Work (PoW) and Proof of Stake (PoS) (Aderibole et al. 2020). These consensus algorithms are applied in smart grids to decrease cyber attacks (Hasankhani et al. 2021). PoW involves a consensus algorithm where miners solve complex mathematical problems to validate transactions and add new blocks to the blockchain (Aggarwal and Kumar 2021a). On the other hand, PoS involves a consensus algorithm where validators are chosen based on the amount of cryptocurrency they hold and are willing to "stake" as collateral (Aggarwal and Kumar 2021a). According to Lashkari and Musilek (2021) consensus mechanisms in blockchains serve as fault-tolerant systems for verifying transactions. They ensure that nodes in the network maintain agreement. As the network grows and the number of nodes increases, achieving consensus becomes more challenging (Aggarwal and Kumar 2021a). Public blockchains rely on user participation for transaction verification and authentication (Lashkari and Musilek 2021). Due to the dynamic and self-regulating nature of blockchain, it is essential to incorporate secure mechanisms to ensure transaction authenticity and achieve consensus among participants (Lashkari and Musilek 2021).

(d) **Smart Contract:** Smart contracts, which are self-executing contracts with predefined rules encoded on the blockchain, can facilitate automated and auditable data sharing processes, reducing the risk of data breaches, unauthorized access, or miscommunication (Mylrea and Gourisetti 2017). According to Waseem et al. (2023), smart contracts run on blockchain networks, which are decentralized and distributed ledgers maintained by a network of nodes. In smart grid, this decentralized architecture ensures that smart contracts are executed in a trustless manner without the need for intermediaries or centralized control (Aderibole et al. 2020). All participants in the network can verify the execution of smart contracts, enhancing transparency and reliability (Vacca et al. 2021). Moreover, Smart contracts are deployed on blockchain platforms that provide an immutable ledger for recording transactions and contract executions (Lin et al. 2022). Once deployed, a smart contract becomes part of the blockchain's permanent record, and its code cannot be altered or tampered with retroactively. This feature ensures the integrity and transparency of smart contract executions, as well as the verifiability of contract terms and outcomes (Vangala et al. 2021). Furthermore, blockchain platforms use cryptographic techniques to secure smart contracts and their associated transactions (Khan et al. 2021). Each smart contract is digitally signed by the contract creator, ensuring its authenticity and preventing unauthorized modifications (Krishnan et al. 2023). Additionally, transactions involving smart contracts are cryptographically hashed and linked together in blocks, making it computationally infeasible to alter transaction history or forge contract executions (Khan et al. 2021).

(e) **Transactions**: It is essential to note that encryption and decryption happens in a remote location as request for blockchain security arises from smart grid and user devices (Fig. 6.3). Blockchain security servers can be used to secure transactions through cryptographic techniques and decentralized consensus mechanisms (Kim and Huh 2018). This is further explained as follows.

Cryptographic Algorithms

Cryptographic techniques such as symmetric and asymmetric mechanism as well as digital signatures help to ensure the integrity, confidentiality, and authenticity of data stored on the blockchain and exchanged between network participants (Gilani et al. 2020). Symmetric and asymmetric mechanisms are cryptographic methods used to secure communication and data exchange in blockchain security (Moghadam et al. 2020). However, while symmetric cryptography involves the use of a single shared

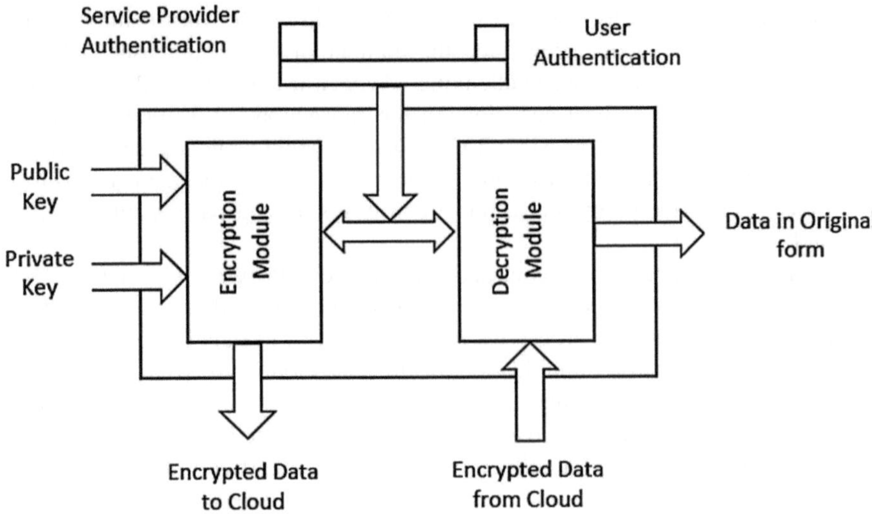

Fig. 6.3 Role of blockchain security server (Authors' Own)

secret key for both encryption and decryption of data, asymmetric cryptography, involves the use of a pair of keys: a public key and a private key. Asymmetric cryptography provides a higher level of security, as the private key never needs to be shared or transmitted, reducing the risk of interception or compromise (Wang et al. 2022). During the initial setup or when establishing communication channels over untrusted networks, asymmetric encryption can be used to securely exchange symmetric encryption.

There are different cryptographic algorithms such as hash functions, public key cryptography, zero-knowledge proofs, symmetric key cryptography and digital signatures. With hash functions, there are two types, namely, SHA-256 and SHA-3. While SHA-256 is used to create a unique identifier for each block in the blockchain, ensuring that any changes to the block's data would result in a different hash, SHA-3 is a more recent hash function that offers improved security and resistance to attacks compared to SHA-256 ((Krishnan et al. 2020). Similarly, public key cryptography is made up of Elliptic Curve Digital Signature Algorithm (ECDSA) and Rivest–Shamir–Adleman (RSA) (Krishnan et al. 2021). While ECDSA algorithm is used to verify the authenticity of a transaction, RSA is another widely recognized public-key algorithm that can be used for both encryption and digital signature (Khan et al. 2023). However, RSA is generally less efficient than ECDSA for blockchain application. In addition, Zero-Knowledge Proofs are cryptographic algorithms that allow one party to prove to another that they know a value without revealing the value itself (Krishnan et al. 2023). It is used in privacy-preserving blockchain applications. Moreover, symmetric key cryptography uses AES (Advanced Encryption Standard) in securing communication between blockchain nodes or in off-chain data storage (Khan et al. 2023). For example, the Diffie-Hellman key exchange protocol enables two parties to establish a shared secret key over an insecure channel without explicitly transmitting the key (Grover et al. 2020). The public key is widely distributed and can be freely shared with anyone, while the private key is kept secret and known only to its owner (Batamuliza and Hanyurwimfura 2020). Thus, while symmetric cryptography offers efficiency in data encryption and decryption, asymmetric cryptography provides stronger security guarantees, particularly in scenarios where secure key distribution and authentication are paramount. However, it is more appropriate to use (ECDSA) and (EdDSA) algorithms as discussed in digital signature as follows.

Digital signatures are used to authenticate transactions and ensure their integrity on the blockchain. Each participant in the blockchain network possesses a public–private key pair, and transactions are signed with the sender's private key (Cali et al. 2021). Recipients can verify the signature using the sender's public key, thereby confirming the authenticity of the transaction and preventing unauthorized modifications. Commonly used digital signature algorithms include RSA (Rivest-Shamir-Adleman), ECDSA (Elliptic Curve Digital Signature Algorithm), and EdDSA (Edwards-curve Digital Signature Algorithm). RSA (Rivest-Shamir-Adleman) is a widely used asymmetric encryption algorithm that can be employed in digital signature schemes for securing communications in smart grids (Philips et al. 2021). In RSA, a pair of cryptographic keys is generated—a public key and a private key (Cali et al. 2021). The public key is shared openly, while the private key is kept secret. These keys are generated for each entity involved in communication, such as utility companies, smart meters, and control systems." When a sender (such as a smart meter) wants to send a message or data securely, it generates a digital signature using its private key. This digital signature is essentially a mathematical operation applied to the message that proves its authenticity and integrity (Philips et al. 2021). The recipient (e.g., utility company) can verify the authenticity of the message using the sender's public key to decrypt the data. If the decryption of the digital signature matches the message, it confirms that the message was indeed sent by the claimed sender and has not been altered in transit. RSA can be utilized for user authentication in systems requiring secure access control. For example, RSA-based authentication mechanisms are commonly used in secure login processes, digital certificates, and secure access to network resources (Cali et al. 2021). However, RSA has some shortcomings which include its inability to provide forward secrecy (Khan et al. 2023). In other words, if an attacker compromises a private key in the future, they can decrypt past communications encrypted with that key. This contrasts with certain key exchange protocols like Diffie-Hellman, which offer forward secrecy. Also, compared to other cryptographic algorithms, RSA requires relatively large key sizes to maintain security against emerging threats (Khan et al. 2023). Larger key sizes result in increased computational overhead, storage requirements, and communication overhead.

The Elliptic Curve Digital Signature Algorithm (ECDSA), (refer Procedure 1) is primarily used for digital signature generation and verification in smart grid systems to ensure the authenticity, integrity, and non-repudiation of messages and transactions (Sadhukhan et al. 2021; Johnson et al. 2001). Each entity involved in communication, such as smart meters, utility companies, and control systems, generates an elliptic curve key pair consisting of a public key and a private key (Ullah et al.

2023). The public key is openly shared, while the private key is kept confidential". The choice of elliptic curve parameters, including the base point (generator point) and the elliptic curve equation, is crucial for the security of ECDSA (Aggarwal and Kumar 2021b). These parameters are agreed upon by the communicating parties or defined by standards. When a sender (e.g., a smart meter) wants to send a message or data securely, it generates a digital signature using its private key and the ECDSA algorithm. The digital signature is computed over a hash of the message using the sender's private key, providing a unique signature for that specific message. The recipient (e.g., a utility company) verifies the authenticity of the message by verifying the digital signature using the sender's public key and the ECDSA algorithm. The recipient recalculates the hash of the message, then uses the sender's public key and the received signature to validate that the calculated hash matches the signature. If the verification is successful, it confirms that the message was indeed sent by the claimed sender and has not been altered in transit, providing authenticity and integrity assurance. Unlike traditional encryption algorithms, ECDSA is not used for encryption and decryption of message content (Aggarwal and Kumar 2021b). Instead, it is solely used for generating and verifying digital signatures to ensure message integrity and authenticity (Ullah et al. 2023). Also, it is essential to note that ECDSA signatures are typically smaller compared to RSA signatures for equivalent security levels, reducing communication overhead and storage requirements. Moreover, while ECDSA is more computationally efficient than RSA for many applications, its underlying mathematics and implementation can be complex (Khan et al. 2023). This complexity may require careful attention to cryptographic best practices and potential vulnerabilities."

Procedure 1: Elliptic Curve Digital Signature Algorithm (ECDSA)

Stage 1: Key Generation:

- Choose an elliptic curve E defined over a finite field Fp where p is a large prime.
- Select a base point G on the curve and a large prime n which is the order of G.
- Generate a private key d as a random integer in the range $[1, n-1]$.
- Compute the corresponding public key $Q = d \cdot G$.

Stage 2: Signature Generation:

- Choose a random integer k from the range $[1, n-1]$.
- Compute the point $k \cdot G = (x1, y1)$.
- Calculate $r = x1 \bmod n$. If $r = 0$, choose a different k and repeat.
- Compute $s = k-1(H(m)+dr) \mathrm{mod} n$, where $H(m)$ is the hash of the message m.
- The signature is the pair (r, s)

Stage 3: Signature Verification:

- Verify that r and s are integers in the range $[1, n-1]$.
- Compute $e = H(m)$, the hash of the message.
- Compute $w = s-1 \bmod n$ and $u1 = ew \bmod n$ and $u2 = rw \bmod n$.
- Compute the curve point $X = u1 \cdot G + u2 \cdot Q$.
- If $X = O$ (the point at infinity), the signature is invalid. Otherwise, let $v = x \bmod n$.
- The signature is valid if $v = r$, otherwise, it is invalid.

Edwards-curve Digital Signature Algorithm (EdDSA) (refer Procedure 2) is a modern digital signature algorithm based on elliptic curve cryptography (ECC) (Shankar et al. 2023; Bernstein et al. 2015). Unlike RSA, which is primarily used for encryption/decryption and digital signatures, EdDSA is specifically designed for digital signatures (Aggarwal and Kumar 2021b). It follows the same procedure as ECDSA. However, EdDSA has seen slower adoption compared to other signature algorithms like RSA and ECDSA. This is due to factors such as existing infrastructure, interoperability concerns, and standardization efforts. While EdDSA is based on well-established cryptographic principles, it may be considered relatively new compared to other signature algorithms.

Procedure 2: The Edwards-curve Digital Signature Algorithm (EdDSA)

Phase 1: Key Generation:

- Generate a secret key sk as a random string of b bits.
- Derive the corresponding public key pk using a one-way function applied to sk.

Phase 2: Signature Generation:

- Compute a hash of the message m, $H(m)$.
- Compute a deterministic nonce r using a function of the message hash, secret key, and other parameters.
- Compute R, the base point of the elliptic curve multiplied by r.
- Compute S, the sum of $H(R, pk, m)$ multiplied by sk modulo the order of the curve and added to r.
- The signature is the pair (R, S).

Phase 3: Signature Verification:

- Compute $H(m)$.
- Compute R' using the signature S and public key pk.
- Compute S' using R', public key pk, and $H(m)$.

These techniques such as RSA, ECDSA, and EdDSA all offer secure digital signature capabilities; however, the most efficient algorithm depends on specific application requirements, including performance, security, and resource constraints. In many cases, elliptic curve-based algorithms like ECDSA and EdDSA are preferred for their computational efficiency and suitability for resource-constrained environments such as smart grids. Hence, ECDSA and EdDSA are appropriate for blockchain security server.

4.5 Wind Energy Farm Management

As shown in Fig. 6.1, when the wind energy farm management is ready to transfer energy to the grid, the power transmission sub-system within the control unit can communicate with the smart grid server. Consequently, the smart grid server in turn through Service-Oriented Architecture (SOA) selects the services of blockchain security server and store it in the cloud. As highlighted by Musleh et al. (2019), the integration of wind energy farm management into smart grids presents a synergistic strategy towards fostering a sustainable and effective energy landscape.

5 Conclusion

A blockchain-enabled cybersecurity framework for distributed smart grid infrastructure with wind farm energy sources provides a robust and resilient platform for securing critical energy assets, data, and transactions against cyber threats, ensuring the integrity, reliability, and sustainability of the smart grid ecosystem (Ravikiran 2023). The integration of blockchain technology into distributed smart grid systems offers promising solutions for enhancing security, transparency, and efficiency in energy management (Agarkar et al. 2024). By leveraging blockchain's decentralized and immutable ledger capabilities, smart grids can address critical security challenges such as data tampering, unauthorized access, and single points of failure. Through cryptographic techniques and consensus mechanisms, blockchain ensures that energy transactions and data exchanges are securely recorded and verified across distributed nodes, reducing the risk of fraud and manipulation. This distributed approach will reduce the risk factor of threats in smart grid.

Looking ahead, future research can focus on cybersecurity resilience. Continuous advancements in cybersecurity technologies and threat intelligence are essential for mitigating evolving cyber threats targeting blockchain-enabled smart grids. Also, while these techniques such as RSA, ECDSA, and EdDSA are developed for other applications, a more suitable technique is required. Hence, more research can be carried out to explore other suitable cryptographic techniques that can be used to enhance smart grid security system. Another limitation is that cloud-based data processing introduces latency due to the time required to transmit data between smart grid devices and remote cloud servers. This latency can impact the responsiveness of real-time applications, such as grid monitoring, control, and demand response, especially for time-critical operations.

References

Aderibole A, Aljarwan A, Rehman MHU et al (2020) Blockchain technology for smart grids: decentralized NIST conceptual model. IEEE Access 8:43177–43190

Agarkar AA, Karyakarte M, Chavhan G et al (2024) Blockchain aware decentralized identity management and access control system. Measur Sens 31:101032

Aggarwal S, Kumar N (2021a) Cryptographic consensus mechanisms. In: Advances in computers. Elsevier, Amsterdam, pp 211–226

Aggarwal S, Kumar N (2021b) Digital signatures. In: Advances in Computers. Elsevier, pp 95–107

Ahmad A, Saad M, Al Ghamdi M et al (2021) Blocktrail: A service for secure and transparent blockchain-driven audit trails. IEEE Syst J 16:1367–1378

Akhter SR, Biswas MM (2016) Roadmap of smart grid for Bangladesh based on functions and technological challenges. Electric Power Components and Systems 44:864–872

Al-Janabi M, Ismail MA, Ali AH (2021) Intrusion Detection Systems, Issues, Challenges, and Needs. Int. J. Comput. Intell. Syst. 14:560–571

Alani MM, Mauri L, Damiani E (2023) A two-stage cyber attack detection and classification system for smart grids. Internet of Things 24:100926

Alanzi H, Alkhatib M (2022) Towards improving privacy and security of identity management systems using blockchain technology: A systematic review. Appl Sci 12:12415

Alsuwian T, Shahid Butt A, Amin AA (2022) Smart grid cyber security enhancement: challenges and solutions—a review. Sustainability 14:14226

Atalay M (2022) A service-oriented digital twins framework for smart grid management. In: 2022 International workshop on secure and reliable microservices and containers (SRMC). IEEE, pp 9–17.

Aziz S, Irshad M, Haider SA et al (2022) Protection of a smart grid with the detection of cyber-malware attacks using efficient and novel machine learning models. Front Energy Res 10:1102

Batamuliza J, Hanyurwimfura D (2020) A secure and efficient anonymous certificateless signcryption for Key Distribution Scheme for Smart Grid. In: 2020 21st International Arab conference on information technology (ACIT). IEEE, pp 1–7

Bera B, Saha S, Das AK et al (2020) Designing blockchain-based access control protocol in IoT-enabled smart-grid system. IEEE Internet Things J 8:5744–5761

Bernstein DJ, Josefsson S, Lange T, Schwabe P, Yang BY (2015) EdDSA for more curves, IACR Cryptol. ePrint Arch 2015:677. https://api.semanticscholar.org/CorpusID:27179286

Bin Sulaiman R (2018) Applications of block-chain technology and related security threats. Rejwan, Applications of Block-Chain Technology and Related Security Threats (June 29, 2018)

Bouramdane A-A (2023) Cyberattacks in Smart Grids: Challenges and solving the Multi-Criteria Decision-Making for cybersecurity options, including ones that incorporate artificial intelligence, using an analytical hierarchy process. J Cybersecur Privacy 3:662–705

Cali U, Kuzlu M, Pipattanasomporn M, et al (2021) Introduction to security for smart grid systems. In: Digitalization of power markets and systems using energy informatics, pp 59–85

Campagna N, Caruso M, Castiglia V et al (2020) Energy management concepts for the evolution of smart grids. In: 2020 8th International conference on smart grid (icSmartGrid). IEEE, pp 208–213

Chatterjee U, Ray S, Sadhukhan D et al (2023) An improved authentication and key management scheme for hierarchical IoT network using elliptic curve cryptography. Innov Syst Softw Eng 19:231–246

Chaudhry SA, Nebhan J, Yahya K et al (2021) A privacy enhanced authentication scheme for securing smart grid infrastructure. IEEE Trans Industr Inf 18:5000–5006

Cheng X, Chen F, Xie D et al (2020) Design of a secure medical data sharing scheme based on blockchain. J Med Syst 44:52

Drijvers M, Edalatnejad K, Ford B, et al (2019) On the security of two-round multi-signatures. In: 2019 IEEE Symposium on Security and Privacy (SP). IEEE, pp 1084–1101

Ducornaud G (2023) Role based access control (RBAC) in the context of smart grids: implementing and evaluating a role based access control system for configuration loading in a substation from a desktop. KTH Royal Institute of Technology, Stockholm

Eder-Neuhauser P, Zseby T, Fabini J et al (2017) Cyber attack models for smart grid environments. Sustain Energy Grids Networks 12:10–29

El Mrabet Z, Kaabouch N, El Ghazi H et al (2018) Cyber-security in smart grid: survey and challenges. Comput Electr Eng 67:469–482

Ezugwu A, Ukwandu E, Ugwu C et al (2023) Password-based authentication and the experiences of end users. Scientific African 21:e01743

Fadlullah ZM, Kato N (2015) Evolution of smart grids. Springer, Berlin

Faquir D, Chouliaras N, Sofia V et al (2021) Cybersecurity in smart grids, challenges and solutions. AIMS Electron Electr Eng 5:24–37

Ganesan P, Xavier S (2023) An Intelligent Intrusion Detection System in Smart Grid Using PRNN Classifier. Intell Autom Soft Comput 35:1

Gaudet N, Sahu A, Goulart AE, et al (2020) Firewall configuration and path analysis for smart-grid networks. In: 2020 IEEE international workshop technical committee on communications quality and reliability (CQR). IEEE, pp 1–6

Ghiasi M, Wang Z, Mehrandezh M et al (2023) Evolution of smart grids towards the Internet of energy: Concept and essential components for deep decarbonisation. IET Smart Grid 6:86–102

Gilani K, Bertin E, Hatin J, et al (2020) A survey on blockchain-based identity management and decentralized privacy for personal data. In: 2020 2nd Conference on blockchain research & applications for innovative networks and services (BRAINS). IEEE, pp 97–101

Grover HS, Adarsh, Kumar D (2020) Cryptanalysis and improvement of a three-factor user authentication scheme for smart grid environment. J Reliab Intell Environ 6:249–260

Gunduz MZ, Das R (2020) Cyber-security on smart grid: Threats and potential solutions. Comput Netw 169:107094

Guo X, Zhang Y, Luo F et al (2024) User-centric recommendations on energy-efficient appliances in smart grids: A Multi-task learning approach. Knowl-Based Syst 284:111219

Habib AA, Hasan MK, Hassan R et al (2023) Distributed denial-of-service attack detection for smart grid wide area measurement system: a hybrid machine learning technique. Energy Rep 9:638–646

Haghgoo M, Dognini A, Storek T et al (2021) A cloud-based service-oriented architecture to unlock smart energy services. Energy Informatics 4:8

Hasankhani A, Hakimi SM, Bisheh-Niasar M et al (2021) Blockchain technology in the future smart grids: A comprehensive review and frameworks. Int J Electr Power Energy Syst 129:106811

Huseinović A, Mrdović S, Bicakci K et al (2020) A survey of denial-of-service attacks and solutions in the smart grid. IEEE Access 8:177447–177470

Islam SN, Baig Z, Zeadally S (2019) Physical layer security for the smart grid: vulnerabilities, threats, and countermeasures. IEEE Trans Industr Inf 15:6522–6530

Jingyao S, Chandel S, Yunnan Y, et al (2020) Securing a network: how effective using firewalls and VPNs are? In: Advances in information and communication: proceedings of the 2019 future of information and communication conference (FICC), vol 2. Springer, 1050–1068

Johnson D, Menezes A, Vanstone S (2001) The elliptic curve digital signature algorithm (ECDSA). Int J Inf Secur 1:36–63

Kadri A, Marzougui H, Aouiti A et al (2020) Energy management and control strategy for a DFIG wind turbine/fuel cell hybrid system with super capacitor storage system. Energy 192:116518

Kang C, Kirschen D, Green TC (2023) The Evolution of Smart Grids. Proc IEEE 111:691–693

Khalyeyev D, Bureš T, Hnětynka P (2022) Towards characterization of edge-cloud continuum. In: European conference on software architecture. Springer, Berlin, pp 215–230
Khan MR, Upreti K, Alam MI et al (2023) Analysis of elliptic curve cryptography & RSA. J ICT Standard 11:355–378
Khan SN, Loukil F, Ghedira-Guegan C et al (2021) Blockchain smart contracts: applications, challenges, and future trends. Peer-to-Peer Netw Appl 14:2901–2925
Kim S-K, Huh J-H (2018) A study on the improvement of smart grid security performance and blockchain smart grid perspective. Energies 11:1973
Komalavalli C, Saxena D, Laroiya C (2020) Overview of blockchain technology concepts. In: Handbook of research on blockchain technology. Elsevier, Amsterdam, pp 349–371
Krishnan S, Balas VE, Julie EG, Yesudhas HR, Balaji S, Kumar R (eds) (2020) Handbook of research on blockchain technology. Academic Press.
Krishnan S, Balas VE, Julie EG, Yesudhas HR, Kumar R (eds) (2021) Blockchain for smart cities. Elsevier, Amsterdam
Krishnan S, Balas VE, Kumar R (eds) (2023) Green blockchain technology for sustainable smart cities. Elsevier, Amsterdam
Kuzlu M, Sarp S, Pipattanasomporn M, et al (2020) Realizing the potential of blockchain technology in smart grid applications. In: 2020 IEEE power & energy society innovative smart grid technologies conference (ISGT). IEEE, pp 1–5
Lashkari B, Musilek P (2021) A comprehensive review of blockchain consensus mechanisms. IEEE Access 9:43620–43652
Laukyte M (2023) Blockchain and the right to good administration: adding blocks to or blocking of the globalization of good administration? Ind J Global Legal Stud 30:195
Lee CK (2019) Blockchain application with health token in medical & health industrials. In: 2nd International Conference on Social Science, Public Health and Education (SSPHE 2018). Atlantis Press, Paris, pp 233–236
Lee KL, Wong SY, Alzoubi HM et al (2023) Adopting smart supply chain and smart technologies to improve operational performance in manufacturing industry. Int J Eng Bus Manag 15:18479790231200616
Lin S-Y, Zhang L, Li J et al (2022) A survey of application research based on blockchain smart contract. Wirel Netw 28:635–690
Liu J, Li X, Ye L, et al (2018) BPDS: a blockchain based privacy-preserving data sharing for electronic medical records. In: 2018 IEEE Global Communications Conference (GLOBECOM). IEEE, pp 1–6
Liu Q, Hagenmeyer V, Keller HB (2021) A review of rule learning-based intrusion detection systems and their prospects in smart grids. IEEE Access 9:57542–57564
Mahi-Al-rashid A, Hossain F, Anwar A et al (2022) False data injection attack detection in smart grid using energy consumption forecasting. Energies 15:4877
Mahmood A, Khan A, Anjum A et al (2023) An efficient and privacy-preserving blockchain-based secure data aggregation in smart grids. Sustainable Energy Technol Assess 60:103414
Marsalek A, Zefferer T (2019) A correctable public blockchain. In: 2019 18th IEEE international conference on trust, security and privacy in computing and communications/13th ieee international conference on big data science and engineering (TrustCom/BigDataSE). IEEE, pp 554–561
McLaughlin K, Friedberg I, Kang B, et al. (2015) Secure communications in smart grid: networking and protocols. In: Smart grid security. Elsevier, Amsterdam, pp 113–148
Moghadam MF, Nikooghadam M, Mohajerzadeh AH et al (2020) A lightweight key management protocol for secure communication in smart grids. Electr Power Syst Res 178:106024
Mohamed N, Al-Jaroodi J, Lazarova-Molnar S et al (2021) Applications of integrated IoT-fog-cloud systems to smart cities: A survey. Electronics 10:2918
Musleh AS, Yao G, Muyeen S (2019) Blockchain applications in smart grid–review and frameworks. IEEE Access 7:86746–86757

Mylrea M, Gourisetti SNG (2017) Blockchain for smart grid resilience: exchanging distributed energy at speed, scale and security. In: 2017 Resilience week (RWS). IEEE, pp 18–23

Naqvi SSA, Li Y, Uzair M (2024) DDoS attack detection in smart grid network using reconstructive machine learning models. PeerJ Comput Sci 10:e1784

Nikam R, Potey M (2016) Cloud storage security using multi-factor authentication. In: 2016 international conference on recent advances and innovations in engineering (ICRAIE). IEEE, pp 1–7

Ortega-Fernandez I, Liberati F (2023) A review of denial of service attack and mitigation in the smart grid using reinforcement learning. Energies 16:635

Pau M, Mirz M, Dinkelbach J et al (2022) A service oriented architecture for the digitalization and automation of distribution grids. IEEE Access 10:37050–37063

Philips A, Jayaraj J, FT J, et al (2021) Enhanced RSA key encryption application for metering data in smart grid. Int J Pervas Comput Commun 17:596–610

Ravikiran A (2023) What is blockchain technology? How does blockchain work. Simplilearn. Com. Preuzeto 15:2023

Ravinder M, Kulkarni V (2023) Intrusion detection in smart meters data using machine learning algorithms: A research report. Front Energy Res 11:1147431

Ruland C, Sassmannshausen J (2018) Firewall for attribute-based access control in smart grids. In: 2018 IEEE international conference on smart energy grid engineering (SEGE). IEEE, pp 336–341

Sadhukhan D, Ray S, Obaidat MS et al (2021) A secure and privacy preserving lightweight authentication scheme for smart-grid communication using elliptic curve cryptography. J Syst Architect 114:101938

Sadorsky P (2021) Wind energy for sustainable development: Driving factors and future outlook. J Clean Prod 289:125779

Sahlin E, Levenby R (2018) Blockchain in audit trails: an investigation of how blockchain can help auditors to implement audit trails. Master's Thesis, Jonkoping University. https://api.semanticscholar.org/CorpusID:46935035

Shankar G, Ai-Farhani LH, Anitha Christy Angelin P, et al (2023) Improved multisignature scheme for authenticity of digital document in digital forensics using edward-curve digital signature algorithm. In: Security and Communication Networks

Son Y-B, Im J-H, Kwon H-Y et al (2020) Privacy-preserving peer-to-peer energy trading in blockchain-enabled smart grids using functional encryption. Energies 13:1321

Soykan EU, Bagriyanik M, Soykan G (2021) Disrupting the power grid via EV charging: the impact of the SMS Phishing attacks. Sustain Energy Grids Netw 26:100477

Sun C-C, Cardenas DJS, Hahn A et al (2020) Intrusion detection for cybersecurity of smart meters. IEEE Trans Smart Grid 12:612–622

Taherdoost H, Madanchian M (2023) Blockchain and business process management (BPM) synergy: a comparative analysis of modeling approaches. Information 15:9

Tanveer M, Kumar N, Naushad A et al (2021) A robust access control protocol for the smart grid systems. IEEE Internet Things J 9:6855–6865

Tufail S, Parvez I, Batool S et al (2021) A survey on cybersecurity challenges, detection, and mitigation techniques for the smart grid. Energies 14:5894

Ullah S, Zheng J, Din N et al (2023) Elliptic curve cryptography; applications, challenges, recent advances, and future trends: a comprehensive survey. Comput Sci Rev 47:100530

Ustundag Soykan E, Bagriyanik M (2020) The effect of SMiShing attack on security of demand response programs. Energies 13:4542

Vacca A, Di Sorbo A, Visaggio CA et al (2021) A systematic literature review of blockchain and smart contract development: Techniques, tools, and open challenges. J Syst Softw 174:110891

Vangala A, Sutrala AK, Das AK et al (2021) Smart contract-based blockchain-envisioned authentication scheme for smart farming. IEEE Internet Things J 8:10792–10806

Wang D, Wang H, Fu Y (2021a) Blockchain-based IoT device identification and management in 5G smart grid. EURASIP J Wirel Commun Netw 2021:125

Wang S-x, Chen H-w, Zhao Q-y et al (2022) Preserving scheme for user's confidential information in smart grid based on digital watermark and asymmetric encryption. J Central South Univ 29:726–740

Wang W, Huang H, Zhang L et al (2021b) Secure and efficient mutual authentication protocol for smart grid under blockchain. Peer-to-Peer Netw Appl 14:2681–2693

Waseem M, Adnan Khan M, Goudarzi A et al (2023) Incorporation of blockchain technology for different smart grid applications: Architecture, prospects, and challenges. Energies 16:820

Werth J, Berenjestanaki MH, Barzegar HR et al (2023) A Review of Blockchain Platforms Based on the Scalability, Security and Decentralization Trilemma. ICEIS 1:146–155

Wilson Y, Hingnikar A (2019) Solving identity management in modern applications: demystifying OAuth 2.0, OpenID Connect, and SAML 2.0. Springer, Berlin

Xue H, Chen D, Zhang N et al (2023) Integration of blockchain and edge computing in internet of things: a survey. Futur Gener Comput Syst 144:307–326

Yadav G, Gauravaram P, Jindal AK (2020) SmartPatch: a patch prioritization framework for SCADA chain in smart grid. In: Proceedings of the 26th annual international conference on mobile computing and networking, pp 1–3

Yahya W, Ziming K, Juan W (2020) Outlook of renewable energy (wind energy) in China. Traff Transp 2:3

Zahoor A, Mahmood K, Shamshad S et al (2023) An access control scheme in IoT-enabled Smart-Grid systems using blockchain and PUF. Internet of Things 22:100708

Zhai F, Yang T, Zhao B et al (2022) Privacy-preserving outsourcing algorithms for multidimensional data encryption in smart grids. Sensors 22:4365

Navigating the Legal Landscape of NFTS: Understanding Regulatory Challenges and Intellectual Property Considerations

Vishnu Nair and Guilherme Lopes da Cunha

Abstract As Non-Fungible Tokens (NFTs) captivate the digital world, their legal and regulatory challenges are becoming increasingly significant. This paper explores NFTs' unique attributes and legal considerations, starting with their definition and nature, highlighting their differences from traditional digital assets. It examines the current legal frameworks for digital assets, providing insights into the evolving regulatory landscape for NFTs, cryptocurrencies, and tokens, and identifies critical gaps requiring further legal clarity. The paper then delves into ownership and transfer of NFTs, detailing the legal mechanisms underpinning these processes. Intellectual property considerations are scrutinized, analyzing the impact of NFTs on copyright, trademark, and patent laws, as well as on creators' rights, licensing agreements, and enforcement mechanisms. Regulatory challenges, including investor protection, consumer rights, anti-money laundering regulations, and securities laws, are also addressed. The methodology is based on a comprehensive review of relevant literature and legal frameworks from sources like Hein Online, JSTOR, and Google Scholar. This paper aims to provide legal practitioners, policymakers, scholars, and industry professionals with a thorough understanding of the legal complexities surrounding NFTs.

1 Introduction

The rapid ascent of non-fungible tokens (NFTs) has transformed the digital art market, propelling previously obscure blockchain technology into mainstream awareness and raising along with it a host of legal questions. In fact, just three

V. Nair (✉)
Amity University Dubai, Dubai, UAE
e-mail: Vnair1@amityuniversity.ae

G. L. da Cunha
Brazilian War College, Rio de Janeiro, Brazil
e-mail: guilherme.lopes@esg.br

S. Krishnan et al. (eds.), *Unveiling Blockchain Enabled Business Applications*, Contributions to Management Science, https://doi.org/10.1007/978-981-95-2337-5_7

years ago, the concept of non-fungible tokens (NFTs) was unfamiliar to most participants in the art market. Although the first NFTs emerged in 2014 (Chuvaieva 2023), the record-breaking sale of Beeple's digital artwork 'Everydays: The First 5000 Days' for $69 million in March 2021 propelled NFTs into the spotlight (Lerner et al. 2022; Von Appen 2021). This event catalyzed widespread recognition and sparked an ongoing dialogue about NFTs, pushing them to the forefront of public and professional consciousness. Since then, NFTs have revolutionized the market, becoming a phenomenon of striking popularity and interest. (Lee 2023; Lerner et al. 2022).

The NFT market, now valued at billions of dollars, in spite of its meteoric rise remains characterized by significant volatility and divergent estimates regarding its worth (Gould 2022). Moreover, NFTs being positioned at the intersection of art and technology, the technical complexity of the technology often leads to confusion and avoidance of in-depth discussions (Sterpi 2022). It is important to note that there is currently no universally accepted legal definition of an NFT (Gould 2022). To date, no legislation, authoritative guidance, or court ruling has provided a definitive, comprehensive description.

The objective of this paper is to elucidate the prevailing legal issues pertaining to the regulatory challenges of Non-Fungible Tokens (NFTs) and to critically examine the unresolved intellectual property considerations inherent in this domain. By delving into the evolving legal landscape of NFTs, the paper focuses on the regulatory hurdles and intellectual property issues that digital artists and market participants must navigate.

2 Study Overview

This study employs a doctrinal research methodology, concentrating on a comprehensive review of relevant literature, case law, and existing legal frameworks. The research process begins with a thorough examination of academic articles, legal texts, and industry reports related to NFTs, blockchain technology, and digital assets. Sources such as Hein Online, JSTOR, and Google Scholar are extensively utilized to gather scholarly opinions and analyses. The study includes a review and analysis of pertinent case law to understand judicial interpretations and decisions impacting the legal status and regulatory challenges of NFTs. This involves examining cases that address issues of ownership, intellectual property rights, and regulatory compliance. Furthermore, the research involves an analysis of existing legal frameworks governing digital assets and intellectual property.

Through the review of literature and legal frameworks, the study identifies critical gaps and challenges in the current regulatory landscape, highlighting areas where further legal clarity and regulatory intervention are needed. Based on the findings, the study proposes recommendations for policymakers, legal practitioners, and industry stakeholders. These recommendations aim to enhance the legal and regulatory frameworks for NFTs, ensuring better protection and compliance mechanisms. Given the global nature of NFTs, this paper emphasizes the necessity of a broader, supra-national approach to effectively address the multifaceted legal challenges they present. Ultimately, the paper aims to identify deficiencies in the current legal landscape and propose well-founded recommendations for harmonizing regulations and enhancing the protection of NFTs, contributing to a more coherent and robust legal framework.

3 Definition and Nature of NFTs

Non-fungible tokens (NFTs) are a class of digital assets distinguished by their unique characteristics, rendering them irreplaceable and indivisible (Li et al. 2023). In contrast to cryptocurrencies such as Bitcoin or Ether (ETH), which are fungible and can be exchanged on a one-to-one basis, NFTs possess unique identification codes and metadata that ensure their distinctiveness (Jordanoska 2021). NFTs are primarily derived from blockchain technology (Krishnan et al. 2023), particularly leveraging the smart contract capabilities of Ethereum. This technological framework allows for the creation, management, and transaction of these unique digital assets. (Wang et al. 2021). The uniqueness of each NFT is embedded within its metadata, which can encompass a wide range of digital content, from artwork and music to virtual real estate and gaming assets. This distinctiveness is not only essential for ensuring the integrity and authenticity of the asset but also for establishing ownership and provenance in the digital realm.

Numerous definitions have been proposed for NFTs, emphasizing various aspects of their nature and function. Collins Dictionary describes NFTs as "*a unique digital certificate, registered in a blockchain, that is used to record ownership of an asset, such as an artwork or a collectible*," highlighting their role in recording ownership (Collins Dictionary, 2024). The term was also announced as the word of the year in 2018 by Collins dictionary beating out words like 'Crypto' and 'Cheugy' (CNN 2021). This definition underscores the non-fungible nature of NFTs, distinguishing them from fungible assets like money or cryptocurrencies, which can be exchanged without any practical difference. As Sharma notes, "*non-fungible tokens are assets that have been tokenized* via*a blockchain, assigned unique identification codes and metadata that distinguish them from other tokens*" (Sharma 2022). This non-fungibility means that NFTs cannot be exchanged on a one-to-one basis without altering their value (Beckman 2021). An often-cited, unverified quote attributed to George Bernard Shaw illustrates this concept: '*If you have an apple and I have an apple and we exchange apples, then you and I will still each have one apple. But if you have an idea and I have an idea, and we swap them, we will each have two ideas.*' In this analogy, apples represent fungible items, as their exchange results in no practical difference. Conversely, exchanging NFTs results in each party holding a unique token with different encrypted data and an unchangeable record of ownership, thus highlighting their unique characteristics and the complexities they introduce to the legal landscape (Okonkwo 2021; Beckman 2021).

Thus, the distinct nature of NFTs, underpinned by their non-fungibility and unique digital characteristics, introduces a range of complexities and opportunities that extend beyond simple ownership records. As we delve deeper into the mechanics and implications of NFTs, it is essential to explore the legal frameworks governing digital assets, which are rapidly evolving to address the unique challenges and opportunities presented by NFTs. The subsequent discussion will focus on the specifics of ownership and transfer of NFTs, highlighting the legal nuances and transactional intricacies involved.

4 Legal Frameworks for Digital Assets

The legal landscape for non-fungible tokens (NFTs) is marked by significant complexities and gaps, particularly in the realm of intellectual property (IP) rights. Blockchain technology underpins the functionality and security of NFTs, providing a decentralized, immutable ledger for recording transactions and data, which is crucial for ensuring the integrity and chronological order of transactions (Twesige 2015). Its primary attributes, including speed, transparency, cost-efficiency, ease of use, and programmability, make it ideal for global information and financial asset transfers (Wang et al. 2021). The Ethereum blockchain, is pivotal in the creation and management of NFTs, ensuring each token's uniqueness and non-fungibility (Jordanoska 2021). This technological foundation, characterized by immutable data recording and programmable smart contracts, enables NFTs to securely and transparently represent unique digital assets.

One of the primary challenges is the absence of mechanisms to verify that the individual minting an NFT holds the necessary copyrights or licenses for the underlying content. Therefore the absence of established mechanisms for verifying the copyright ownership or licensing status of the underlying content is a major challenge (Çağlayan 2023). This deficiency in verification mechanisms engenders persistent ambiguity within NFT marketplaces regarding the legitimacy of tokenized assets. Additionally, the automated minting process, facilitated by blockchain technology, further complicates the matter by diminishing individual creative input, thus diminishing the possibility of generating new intellectual property (IP) rights solely through the act of minting (Clark and Burstall 2019). Consequently, any IP rights associated with NFTs pertain exclusively to the underlying work, rather than the NFT itself. Moreover, the current lack of specific legislative frameworks addressing NFT ownership and IP rights necessitates the resolution of conflicts within the framework of private international law (PIL) governing digital content, highlighting the urgent need for precise legal definitions and frameworks to regulate the creation, ownership, and transfer of NFTs.

Moreover, the anonymity afforded to participants in NFT transactions by blockchain technology presents another significant legal challenge, as it obstructs the identification of parties involved in legal proceedings, thereby complicating jurisdictional determinations. While blockchain transactions are traceable, the automatic disclosure of party identities is not guaranteed, emphasizing the importance of ensuring such disclosures for the proper application of legal procedures. Furthermore, recent strides in recognizing NFTs within the intellectual property framework, such as their inclusion in the Nice International Classification by the World Intellectual Property Organization (WIPO) as of January 1, 2023, represent a significant milestone. This classification allows for the registration of 'downloadable digital files authenticated by NFTs' under International Class 9, potentially paving the way for broader acknowledgment and regulation of NFTs within the realm of IP rights. However, these advancements underscore the ongoing need for academic and regulatory endeavors to adapt existing legal frameworks to address the unique challenges posed by NFTs, ensuring clarity and protection for stakeholders in this swiftly evolving technological landscape.

5 Ownership and Transfer of NFTs

The notion of property is inherently complex and multifaceted, with legal systems adopting varied approaches and definitions (Çağlayan 2023). The foundational principles of property law in any systems are deeply rooted in historical contexts and were not originally designed to seamlessly accommodate modern technological advancements such as digital assets (Çağlayan 2023). This incompatibility poses significant challenges, particularly because digital assets, lacking physical tangibility, do not meet traditional property criteria in many jurisdictions (Çağlayan 2023; Von Appen 2021). While some civil law countries, such as Germany, grapple with this non-compatibility (Von Appen 2021), common law jurisdictions have been more receptive to recognizing intangible crypto assets as a form of property (Beckman 2021). Despite this more inclusive approach, practical analytical issues remain unresolved.

It is insufficient to simply accept digital assets as property since a critical question remains unanswered: how should they be categorized? The Law Commission of England and Wales suggested adding a third category to better accommodate digital assets (Law Commission 2024). Traditionally, English law, as established in Colonial Bank v. Whinney (1886), classifies all personal property as either in possession or in action, with no intermediate category. Some scholars, like Low (2023), argue that digital assets can fit within the existing "chose in action" category without the need for a new classification. However, the UNIDROIT Principles on Digital Assets and Private Law, adopted in May 2023, took a different approach. These principles recognized digital assets as subject to proprietary rights without explicitly categorizing them as "property" under any state's law. This pragmatic stance highlights the need for practical solutions rather than prolonged theoretical debates. Ultimately, successful legislative frameworks must adapt existing rules to new technologies and create new definitions when necessary to achieve harmonization. The evolving judicial recognition of NFTs as property, as seen in recent injunction decisions across various jurisdictions, reflects a gradual but significant shift in legal interpretations, indicating a promising future for digital asset regulation.

Examining lay understandings and feelings regarding ownership reveals that the manner of acquisition often influences perceptions of ownership, subsequently informing legal constructions. Heller and Salzman (2023) propose that ownership instincts are governed by six hidden rules, which offer common bases for legal definitions of ownership and property. These rules include 'possession'—claiming ownership through current possession, 'first-in-time'—rewarding the first claim, 'labor'—earning ownership through work, and 'attachment'—claiming ownership over things attached to what is already owned. These principles underpin many legal and moral conceptions of ownership and property.

NFTs, as a form of ownership technology, have evolved since their conceptualization in 2012 with the notion of 'colored coins' on the Bitcoin blockchain, where unique and scarce tokens were used to represent ownership of various assets. The NFT ecosystem involves a range of stakeholders, from independent artists to large corporations and auction houses, with NFTs linked to various forms of digital media, including generative art, photography, music, and game items.

The table illustrates a comparison of popular NFT collections, highlighting their launch years, notable sale prices, and the platforms on which they were traded (Graves 2021).

Table 7.1 reveals the diversity and evolution of the NFT market over recent years. CryptoPunks, launched in 2017, set a significant precedent with a sale price of $69 million for CryptoPunk #7804 on the Ethereum platform, demonstrating the early high value attached to unique digital assets (Hayward 2021). Bored Ape Yacht Club and Beeple's "Everydays" in 2021 also achieved remarkable sale prices of $3.4 million and $69 million respectively, with Beeple's work auctioned at Christie's, signaling mainstream acceptance (Kyle 2021). Art Blocks and NBA Top Shot, both launched in 2020, show significant engagement in art and sports NFTs, with sales reaching $23 million and $208,000 respectively, the latter on the Flow Blockchain. Collections such as Cool Cats, World of Women, and Azuki, all launched in 2021–2022, further exemplify the growing variety and interest in the NFT space, with sales reaching up to $1.4 million on the Ethereum platform (Graves 2021). Notably, RTFKT x Nike's 'Cryptokicks' and The Sandbox LANDs showcase the involvement of major brands and virtual real estate in the NFT market, fetching prices of $134,000 and $4.3 million respectively, also on Ethereum (Graves 2021). This data underscores the rapid growth, high value, and diverse applications of NFTs across different sectors.

Several noteworthy examples highlight the diverse applications of NFTs across various sectors. Nike's SWOOSH platform utilizes NFTs to enhance customer engagement by enabling users to co-create digital assets, fostering a unique brand connection (MINTOLOGY BLOG, n.d.). McDonald's Singapore launched the Grimace NFT collection, leveraging NFTs for customer interaction and innovation in the food industry (blockapps.net 2024). Starbucks' Odyssey program combines NFTs with gamified experiences, significantly boosting customer loyalty through

Table 7.1 Comparison of Popular NFTs

NFT Collection	Launch Year	Sale Price	Platform
CryptoPunks	2017	$69 million	Ethereum
Bored Ape Yacht Club	2021	$3.4 million	Ethereum
Beeple's "Everydays"	2021	$69 million	Christie's Auction
Art Blocks	2020	$23 million	Ethereum
NBA Top Shot	2020	$208,000	Flow Blockchain
Cool Cats	2021	$1.1 million	Ethereum
World of Women	2021	$750,000	Ethereum
Azuki	2022	$1.4 million	Ethereum
RTFKT x Nike "Cryptokicks"	2022	$134,000	Ethereum
The Sandbox LANDs	2019	$4.3 million	Ethereum

collectible and tradable digital assets (blockapps.net 2024). Mediacorp's collaboration with Mintology for a Lunar New Year event achieved a 100% NFT claim rate, illustrating effective use of NFTs for event engagement (MINTOLOGY BLOG, n.d.). Finally, HelloVacay's Pilot Passport NFT offers exclusive rates and perks to holders, revolutionizing customer experience in the travel sector. These case studies exemplify the multifaceted applications of NFTs, underscoring their potential for innovation and enhanced customer engagement in various industries.

Despite the growing academic interest in NFTs since 2021, research has often been confined to specific disciplinary perspectives such as financial, technological, and legal, with limited studies on the social implications and experiences of NFT ownership. Financial scholars have explored the nature of NFT marketplaces, while critics highlight potential social inequities. Technologists investigate the operational aspects of NFT ecosystems and propose new platforms and standards. Legal scholars primarily address the copyright challenges posed by NFTs, noting widespread misconceptions about NFT ownership and usage rights. Research on NFT marketplaces for digital art examines issues of authenticity, commodification, and rarity, reflecting the potential for NFTs to explore broader concepts of ownership, possession, and identity. The scarcity of human-centered research on NFTs highlights the growing need to examine ownership and possession in the context of digital assets, particularly to understand how NFTs can offer meaningful experiences of ownership that extend beyond their speculative and legal dimensions.

6 Intellectual Property Considerations

It is evident from the discussions above that Non-Fungible Tokens (NFTs) have revolutionized the digital asset landscape by leveraging blockchain technology to address longstanding issues in intellectual property (IP) management (Tharun et al.

2023).Central to their appeal is the transparent and secure record of ownership and transaction history provided by blockchain, ensuring verification of ownership and provenance (Narayanan et al. 2016). The cryptographic security and immutability inherent in blockchain safeguard NFTs against tampering, duplication, or unauthorized modifications. Decentralization further enhances NFTs' appeal by promoting peer-to-peer ownership and trading without central authority control. Additionally, smart contracts on the blockchain automate processes such as royalty distributions and licensing agreements, streamlining digital asset management.

From an intellectual property perspective, examining the ownership rights associated with a specific NFT is crucial. Before completing a purchase, it is essential to verify whether the seller genuinely holds the NFT. Documented cases have shown individuals creating counterfeit NFTs by photographing non-circulating tax receipts or minting replicas (Okonkwo 2021). When an NFT is acquired, the purchaser receives the rights to use it but not the associated IP rights, which are specified in the metadata of the underlying smart contract (Okonkwo 2021). Despite the significant income opportunities NFTs offer, they face numerous challenges, including a market hype bubble driven by endorsements from public figures and the absence of specific regulations or guidelines governing their exchange. This regulatory gap necessitates careful research and verification to navigate the NFT market effectively. In creative industries, particularly music, proving ownership and combating copyright infringements have been persistent challenges. Traditional methods, such as registration with national copyright offices, offer limited protection and can be legally dubious. In contrast, NFTs simplify ownership verification and automate licensing through smart contracts, providing a more efficient and secure solution for managing IP.

However, a significant concern is whether the IP rights associated with NFT images are included in transactions, potentially affecting image licensing revenue. Typically, purchasing an NFT grants ownership of the token but not the underlying IP rights, which remain with the creator unless explicitly transferred. This practice mirrors that of physical art sales, where ownership of the physical item does not equate to ownership of its IP (United Kingdom Legislation 2022).Platforms like CryptoKitties and Foundation.app have implemented different approaches to IP rights, with some allowing limited commercial reuse and others restricting commercial rights altogether.

Copyright, an automatic right that arises upon the creation of a protectable work, grants the creator a bundle of exclusive rights, including reproduction and public communication (Stokes 2021).Jurisdictions vary in their requirements for originality, with the UK requiring works to result from labor, skill, and judgment (University of London Press v. University Tutorial Press 1916; Ladbroke v. William Hill 1964), and the US requiring a minimal degree of creativity (Feist Publications, Inc. v. Rural Telephone Service Co., Inc. 1991). Collaborative works and computer-generated art introduce additional complexities. Joint authorship requires consent from all authors for exploitation, and the copyright status of AI-generated works depends on jurisdictional nuances, with some regions recognizing rights for the arrangers of such works and others not recognizing non-human authorship.

Unauthorized use of copyrighted images for minting NFTs can lead to infringement and other legal issues such as passing off (Gould 2022). Buyers often misunderstand the extent of rights acquired with an NFT, typically receiving only a limited license for personal use rather than full IP rights (Frye 2022). Licensing structures for NFTs vary, ranging from personal display rights to limited commercial use and open-source models like Creative Commons (CC0) waivers. These evolving licensing practices reflect ongoing efforts to balance creator rights with market demands in the NFT space.

7 Regulatory and Legal Challenges

The NFT market is often seen as unregulated compared to the broader art market, yet both sectors are governed by numerous laws and regulations across various jurisdictions. Although NFTs currently lack a comprehensive regulatory framework, existing legal structures cover many aspects of their trade. The rapid growth of the NFT market has outpaced specific regulatory developments, necessitating adjustments in legal systems (Blemus 2018). Promising signs of increasing regulatory clarity are emerging, driven by commercial and legal pressures. NFTs face challenges across sectors such as law and politics, particularly in commodity markets, foreign exchange, and Know Your Customer (KYC) requirements. Understanding the regulatory landscape is crucial before engaging in NFT transactions, as countries like China and India have stringent regulations on NFTs and cryptocurrencies. Navigating these regulatory hurdles is essential for trading NFTs, with countries like Malta and France working on relevant regulations (Blemus 2018). In other regions, existing laws impose complex, sometimes contradictory requirements on buyers, necessitating thorough research before investing in NFTs to ensure compliance and mitigate risks.

Historically, contracts have been oral or written, establishing a legal relationship upon signature, with execution driven by one of the parties. In contrast, smart contracts are codified, anticipate legal relationships, and are self-executing (de Sousa 2022). They introduce elements of immutability and anonymity not found in traditional contracts. Generally, a contract requires mutual agreement, which must be objectively verified, with the parties' intentions manifested through actions or declarations. Additionally, the parties must have legal capacity, the contract's object must be lawful, and there must be motivation (de Sousa 2022). In computer terms, execution signifies the program's initiation on the computer, definitively establishing the link between the parties at that moment. Consequently, until execution, a contract lacks legal force. Thus, smart contracts may conflict with the traditional notion of will to perform, as they are automatically executed instruments (de Sousa 2022). These differences necessitate a reassessment of legal concepts within contract theory, as the use of blockchain and smart contracts becomes increasingly prevalent in various commercial sectors, including the international art trade (de Sousa 2022).

Another significant issue is the immutability of smart contracts. Once deployed and activated, they cannot be modified and will execute automatically. This rigidity can pose legal challenges, as unforeseen events that can be accommodated in traditional contracts might not be preemptively coded into smart contracts. Therefore, if a smart contract deviates from the parties' intentions, it might not be desirable, particularly when the code is not written by the parties themselves. This inflexibility is problematic in business relationships, where modification of obligations is often necessary. Additionally, the immutability of these contracts can result in an inability to terminate them. These contractual effects necessitate new interpretive methods within the legal framework, requiring a fresh understanding of digital phenomena and greater flexibility in interpreting current legal principles (de Sousa 2022).

Additionally, wash trading is a significant concern during NFT trading. Wash trading involves repeated trading of assets between a buyer and seller without market risk, creating misleading market information (Jordanoska 2021). This disrupts market order by artificially inflating transaction volumes and values, leading to doubts about transaction authenticity. This practice is also present in the NFT market. Research by Victor et al. indicated that 2.04% of their sample transactions were suspected wash trades, potentially inflating real transaction volumes by up to $149 million. To mitigate wash trading, increasing transaction fees might be effective. Higher costs could deter fraudulent activities if the costs outweigh the rewards. However, this approach could also discourage legitimate transactions.

The high level of anonymity and lack of verification for user identities and the authenticity of NFT works in the NFT marketplace create a fertile ground for fraudulent activities. A common issue is the sale of counterfeit NFTs without the original owner's consent (Jordanoska 2021). To mitigate scam risks, investors should adopt several precautionary measures: protect private keys by keeping cryptocurrency wallet keys and recovery codes confidential, as sharing this information can compromise security; verify NFT sellers by thoroughly investigating the seller's NFT marketplace account, looking for blue check verification marks, examining the seller's social media profiles, reviewing other listings, and seeking out online reviews; and check the transaction history of NFTs, being cautious of those frequently day-traded, as this can indicate suspicious activity (Hetler 2022). By following these strategies, investors can significantly reduce the risk of fraud and make more informed decisions in the NFT marketplace.

Recent litigation surrounding NFTs underscores the intricate intersection of established legal principles and emerging digital contexts. Key challenges include identifying defendants due to blockchain transaction anonymity, exemplified by a New York court's approval of serving court papers via a "service token" to an anonymous crypto wallet (Grossman LLP). Legal actions against NFT marketplaces like OpenSea highlight accountability concerns for preventing scams. Intellectual property disputes, such as Yuga Labs' trademark infringement suit against an artist and Hermès' litigation over "MetaBirkins" NFTs for trademark violations, underscore the evolving regulatory landscape's complexity (Grossman LLP). Beeple's $69 million sale of "Everydays: The First 5000 Days" at Christie's catalyzed discussions on digital art valuation and buyer rights, while Adidas' "Into the Metaverse" NFT collaboration navigated fashion and intellectual property integration (Wang et al. 2022). Concurrently, Rarible's copyright infringement allegations underscored the necessity for robust verification processes on NFT platforms to protect artists' rights and curtail unauthorized sales (Chinlund and Gordon 2021).

8 Taxation of NFT Transactions

Current regulations stipulate that goods linked to intellectual property (IP), such as books, artworks, and domain names, are subject to taxation. However, NFT-based transactions are not uniformly included under these regulations. While a few countries, such as the United States through the Internal Revenue Service (IRS), tax cryptocurrencies as property, the majority of global jurisdictions have yet to establish clear tax policies for NFTs (Nguyen 2022). This regulatory gap may lead to an increase in fraudulent financial activities masked as NFT trading. Many governments are interested in formulating consistent tax policies for NFT transactions. Tax liabilities arising from investment income on NFTs should be the responsibility of the individual investors. Additionally, tax policies need to address NFT exchanges, whether for other NFTs, IP assets, or cryptocurrencies like Ether. High-value assets such as rare books and art might benefit from more favorable tax rates. Given these complexities, it is recommended that individuals involved in NFT transactions seek advice from specialized tax departments to navigate these issues effectively.

9 NFTs and Dispute Resolution

We are currently experiencing an era of significant technological disruption that is reshaping the way transactions are conducted and redefining our perceptions of art and written works (Susskind 2013). The traditional methods of negotiating deals in person and finalizing them with a handshake are becoming obsolete (Schmitz and Rule 2017). The ability to distinguish human interaction from automated responses is diminishing. Nowadays, the majority of transactions are completed online through e-contracts, confirmed with a simple click of an 'accept' button rather than a handwritten signature. Corporate executives frequently utilize emails and text messages to negotiate agreements, which are eventually finalized electronically via platforms like DocuSign. Additionally, iterative technologies such as ChatGPT are increasingly being employed in the negotiation process.

The integration of computer code and algorithms has revolutionized many traditional contracting processes and asset concepts (Zaslowsky 2018). For instance, while e-contracts remain traditional contracts involving offer, acceptance, and consideration, they are digitally accepted through clicks on computer or mobile screens. Resolving disputes related to NFTs is challenging and often involves complex legal issues. Ownership disputes are particularly common, as determining the true owner of an NFT can be difficult, especially during transfers or sales. Although blockchain technology and smart contracts record these transactions, they still require legal interpretation to establish ownership rights (Handono et al. 2023). Additionally, the use of cryptocurrencies like Ethereum in NFT transactions complicates tracking and raises concerns about potential money laundering (Handono et al. 2023). These challenges underscore the need for clear legal frameworks to ensure the stability and growth of the NFT marketplace.

In the contemporary landscape characterized by significant technological disruption, our approaches to transactions and artistic evaluation are undergoing profound shifts (Susskind 2013; Schmitz and Rule 2017). Traditional methods of negotiating deals face obsolescence, with in-person interactions yielding to online modalities reliant on e-contracts and electronic signatures. Corporate negotiations increasingly occur via digital channels like email and text, culminating in agreements electronically endorsed through platforms like DocuSign. Concurrently, emerging technologies like ChatGPT are finding roles in negotiation processes. The fusion of computer code and algorithms has reshaped conventional contracting processes and asset conceptualizations (Zaslowsky 2018). While e-contracts retain the essence of traditional contracts, they are digitally accepted through clicks rather than ink signatures.

From the previous discussions it is evident that resolving disputes related to Non-Fungible Tokens (NFTs) presents notable challenges, especially in determining ownership amidst transactions. Despite blockchain records and smart contracts, legal interpretation remains crucial (Handono et al. 2023). The involvement of cryptocurrencies like Ethereum further complicates tracking and raises concerns about illicit activities (Handono et al. 2023). These challenges underscore the necessity for clear legal frameworks to ensure the stability of the NFT marketplace.

Moreover, NFTs typically use Smart contracts and therefore lack flexibility in offering creative or fair solutions. They are essentially digital agreements executed automatically by algorithms. When applied to Non-Fungible Tokens (NFTs), a smart contract either executes a transfer or doesn't, with little room for nuance (Dell'Erba 2018). Once a smart contract is deployed on the blockchain, its execution is typically automated and difficult to undo, if not impossible. Consequently, the concept of "do-overs" is largely absent. Smart contracts are inherently self-governing and self-executing, posing challenges for courts in addressing disputes (Dell'Erba 2018). This rigidity makes it problematic for courts to devise remedies in smart contract disputes.

Recognizing that blockchain-related disputes are often ill-suited for traditional court settings, many commentators and practitioners advocate for non-judicial remedies. Arbitration, both traditional and online (OArb), is seen as a viable solution for resolving disputes involving smart contracts and NFTs (Schmitz and Rule 2019). These arbitration systems, which can operate independently of courts, offer the potential for efficient, cost-effective, and private dispute resolution. Additionally, innovative models of arbitration are emerging, incorporating blockchain technology, crowd-sourced dispute resolution, AI-powered resolutions, and metaverse processes (Schmitz and Rule 2019). These developments aim to address the unique challenges posed by the decentralized and digital nature of blockchain and NFT transactions, providing a framework for effective dispute resolution outside conventional legal systems.

Interestingly, tokenized dispute resolution, a novel form of online arbitration (OArb), allows anonymous users to vote on the outcome of disputes using digital tokens (Yang 2022). Unlike traditional OArb, this innovative method leverages crowdsourcing, where selected voters resolve disputes based on the majority opinion (Yang 2022). Jurors cast their votes with tokens, typically cryptocurrencies, which they lose if they are on the losing side and gain rewards if they are on the winning side. This process relies on game theory, motivating jurors to vote accurately to "win." An example of this cutting-edge approach is Kleros, an online arbitration "court" built on Ethereum, designed to provide a fast, inexpensive, transparent, reliable, and decentralized system for resolving smart contract disputes (Lesaege, Ast, & George, 2018). By utilizing global jurors who deposit cryptocurrency to participate, Kleros showcases how blockchain technology can offer an efficient and equitable alternative for dispute resolution in the NFT space.

Artificial Intelligence (AI) and data analytics are increasingly utilized for decision-making across various fields, including the legal domain, where AI aids judges in setting bail and assists lawyers with legal research (O'Brien, 2019). In the context of smart contract and NFT disputes, AI holds potential for providing predictive analysis and swiftly suggesting resolutions that could be recorded on the blockchain. As machine learning algorithms become more sophisticated with additional data, the efficiency and fairness of AI-driven dispute resolution may improve. However, reliance on AI for legal determinations raises concerns about bias and the risk of "rubber stamping" AI decisions without human oversight (Law 2023). AI might eventually offer viable solutions for resolving low-stakes disputes. Nonetheless, the current state of AI and machine learning technology is not yet advanced enough to ensure consistently accurate and accepted "bot resolutions" for smart contract disputes, necessitating continued human involvement and scrutiny in the decision-making process.

Thus, the evolution of technology, particularly blockchain and AI, has significantly impacted dispute resolution in the context of smart contracts and Non-Fungible Tokens (Krishnan and Balas 2020). Traditional methods of negotiation and contract execution are being replaced by digital alternatives, while the rigidity of smart contracts presents challenges in achieving nuanced and equitable outcomes. To address these challenges, innovative approaches such as tokenized dispute resolution and AI-driven decision-making are emerging, offering potential solutions for efficient and equitable dispute resolution. However, while these technologies hold promise, they also raise concerns regarding bias and the need for human oversight. Therefore, as we navigate this technological landscape, it remains crucial to strike a balance between leveraging technological advancements and ensuring the integrity and fairness of the dispute resolution process.

10 Future Research Agenda for NFTs

Future research in the domain of non-fungible tokens (NFTs) should prioritize addressing several pivotal and underexplored areas. A primary objective is the development of rigorous, specialized research methodologies tailored to the unique attributes of NFTs (Chernikova 2024). Ensuring data availability and security is critical, requiring thorough investigations into the most effective consensus algorithms and mechanisms to enhance the usability and safety of NFTs (Gonserkewitz et al. 2022). In the art sector, research should focus on improving the detection and management of copyright infringements and examining the necessity for regulatory frameworks (Mehr and Shahim, 2023). Similarly, the metaverse offers a rich field for research, particularly regarding user data protection, content moderation, and addiction prevention (Gonserkewitz et al. 2022). Investigating these areas will not only advance the technological capabilities of NFTs but also ensure their responsible and beneficial integration across various sectors.

Moreover, examining the economic models that underpin NFT transactions, including pricing mechanisms and market dynamics, will provide crucial insights into their viability and sustainability (Borri et al. 2022). Interdisciplinary research involving law, economics, and technology is essential to gain a holistic understanding of NFTs. Legal scholars should explore the implications of smart contracts and the legal recognition of digital ownership. Economists can assess the broader economic impact of NFTs on traditional markets, while technologists should innovate to enhance the efficiency and functionality of NFT platforms. Collaborative efforts across these disciplines are vital for addressing the multifaceted challenges and opportunities presented by NFTs, ultimately guiding their evolution to maximize societal benefits.

11 Conclusion

In conclusion, the emergence of Non-Fungible Tokens (NFTs) has ushered in a new era of digital asset ownership and transactions, presenting both opportunities and challenges within the legal landscape. This paper has explored the multifaceted nature of NFTs, starting with an elucidation of their definition and unique characteristics. The analysis of the legal frameworks governing digital assets has underscored the complexities surrounding the ownership and transfer of NFTs, alongside pertinent intellectual property considerations. Additionally, the regulatory challenges inherent in the NFT ecosystem have been examined, particularly the intricacies of taxation and the evolving landscape of litigation and dispute resolution. These discussions illuminate the intricate interplay between technological innovation and legal regulation, highlighting the need for a nuanced understanding of the legal landscape to ensure the sustainable growth and integration of NFTs within broader economic and cultural spheres.

To effectively address these challenges, this paper presents several key scholarly suggestions. National and international regulatory bodies should collaborate to establish comprehensive frameworks that clearly define the legal status of NFTs, ensuring consistent treatment across jurisdictions. Intellectual property laws must be clarified to address the specific needs of digital assets, with new legal instruments potentially introduced to protect digital artists' rights. International harmonization of laws is essential to facilitate smoother transactions and reduce legal conflicts, while blockchain and smart contract technologies can enhance the enforcement of legal rights and obligations. Robust consumer protection mechanisms should be included in regulatory frameworks to safeguard against fraud and market manipulation.

Moreover, the development of effective dispute resolution mechanisms is crucial to address issues arising from NFT transactions. These mechanisms should be tailored to the unique characteristics of digital assets, providing efficient and fair resolution of conflicts related to ownership, authenticity, and contractual obligations. Continuous education and research are crucial to keep pace with the rapid evolution of the NFT ecosystem, ensuring that regulatory responses remain relevant and effective. By implementing these suggestions, stakeholders can foster a more transparent, equitable, and secure NFT market, promoting innovation while safeguarding the rights and interests of all participants.

References

Beckman M (2021) NFTs, digital artwork, and blockchain technology: the comprehensive guide. Skyhorse Publishing Company, New York, NY

Birgit Clark LKL for BMI and T.P and Ruth Burstall, SLC at J & JIE (2019) Crypto-pie in the sky? how blockchain technology is impacting intellectual property law. Stanf J Blockchain Law & Policy [online] Available at: https://stanford-jblp.pubpub.org/pub/blockchain-and-ip-law/release/1. Accessed on 16 July 2024

Blemus S (2018) Law and blockchain: a legal perspective on current regulatory trends worldwide. SSRN Electron J. https://doi.org/10.2139/ssrn.3080639. Accessed on 16 Jul 2024

Blockapps.net (2024) NFT projects with a purpose: case studies—blockapps Inc. [online] Available at: https://blockapps.net/blog/nft-projects-with-a-purpose-case-studies/. Accessed on 18 Jul 2024

Borri N, Liu Y Tsyvinski A (2022) The economics of non-fungible tokens. [online] papers.ssrn.com. Available at: https://papers.ssrn.com/sol3/papers.cfm?abstract_id=4052045

Çağlayan Aksoy P (2023) The applicability of property law rules for crypto assets: considerations from civil law and common law perspectives. Law Innov Technol 15(1):185–221. https://doi.org/10.1080/17579961.2023.2184140

Chernikova A (2024) The future of NFTs: will the market revive in 2024? Forbes. [online] 20 Feb. Available at: https://www.forbes.com/sites/digital-assets/2023/11/13/the-future-of-nfts-will-the-market-revive-in-2024. Accessed on 21 July 2024

Chinlund GJ, Gordon KS (2021) What are the copyright implications of NFTs? Reuters. [online] 29 Oct. Available at: https://www.reuters.com/legal/transactional/what-are-copyright-implications-nfts-2021-10-29/

Chuvaieva A (2023) Non-fungible tokens and failed promises of simple virtual ownership. NYU Journal of Intellectual Property & Entertainment Law 12:304

CNN, J.G. (2021). 'NFT' is Collins Dictionary's Word of the Year for 2021, beating out 'crypto' and 'cheugy'. [online] CNN. Available at: https://edition.cnn.com/style/article/nft-word-of-the-year-collins-scli-intl-gbr/index.html. Accessed on 16 July 2024

Colonial Bank v. Whinney [1886] 17 QBD 20

Collins Dictionary (2024) Collins dictionary | definition, thesaurus and translations. [online] Collinsdictionary.com. Available at: https://www.collinsdictionary.com/

Dell'Erba, M. (2018, May 17). Demystifying technology: Do smart contracts require a new legal framework? Regulatory fragmentation, self-regulation, public regulation (Working Paper No. 3228445). SSRN. https://doi.org/10.2139/ssrn.3228445

Feist Publications, Inc. v. Rural Telephone Service Co., Inc., 499 U.S. 340 (1991)

Frye BL (2022) Are cryptopunks copyrightable? SSRN Electron J. https://doi.org/10.2139/ssrn.4029323

Gonserkewitz P, Karger E, Jagals M (2022) Non-fungible tokens: use cases of NFTs and future research agenda. Risk Govance Control: Financ Mark Insts 12(3):8–18. https://doi.org/10.22495/rgcv12i3p1

Gould E (2022) NFTs: an overview of law and regulation in 2022 and beyond. Art Antiq Law XXVII(2)

Graves, D. / D.P., Stephen (2021) The 15 most expensive NFTs ever sold. [online] Decrypt. Available at: https://decrypt.co/62898/most-expensive-nfts-ever-sold

Handono, M., Widiyanti, I. D., & Andini, P. P. (2023). Dispute resolution for non-fungible token (NFT) businesses in Indonesia. International Journal of Social Science and Education Research Studies, 3(8), 1519–1526. https://doi.org/10.55677/ijssers/V03I8Y2023-02

Hayward, D. / A. (2021). Bored Ape Yacht Club Ethereum NFT breaks record with $3.4M sale. [online] Decrypt. Available at: https://decrypt.co/84474/bored-ape-yacht-club-ethereum-nft-breaks-sales-record. Accessed on 18 Jul 2024

Heller, M. A., & Salzman, J. (2022). Mine!: How the hidden rules of ownership control our lives. Knopf Doubleday Publishing Group.

Hetler A (2022) '8 ways to avoid NFT scams', WhatIs.com, TechTarget, 26 May. Available at: https://www.techtarget.com/whatis/feature/8-ways-to-avoid-NFT-scams. Accessed on 16 July 2024

Jordanoska A (2021) The exciting world of NFTs: a consideration of regulatory and financial crime risks. Butterworths J Int Bank Financ Law 10:716

Krishnan S, Balas VE, Kumar R (eds) (2023) Green blockchain technology for sustainable smart cities. Elsevier

Krishnan, S., Balas, V. E., Julie, E. G., Yesudhas, H. R., Balaji, S., & Kumar, R. (Eds.). (2020). Handbook of research on blockchain technology. Elsevier Science.

Kyle (2021) A gold fur ape sold for $3.4 million today At Sotheby's. [online] Bored Ape Gazette. Available at: https://www.theboredapegazette.com/post/a-gold-fur-ape-sold-for-3-4-million-today-at-sotheby-s. Accessed on 18 Jul 2024

Ladbroke v. William Hill [1964] 1 All E.R. 465

Law Commission (2024) Digital assets. [online] Available at: https://lawcom.gov.uk/project/digital-assets/

Law AJS (2023) Resolving NFT and blockchain disputes. Stanf J Blockchain Law & Policy Retrieved from https://stanford-jblp.pubpub.org/pub/resolving-nft-blockchain-disputes

Lee E (2023) Creators take control. HarperCollins Publishers, New York, NY

Lerner R, Rottermund A, Verano S (2022) Insights into blockchain tokens and crypto art's effect on the art market. PLI Curr: J PLI Press 6(1):7–8

Lesaege, C., Ast, F., & George, W. (2018). Kleros: Short paper v1.0.7. Kleros. https://kleros.io/assets/whitepaper.pdf

Li X, Yuan C Kent J (2023) Proceedings of the 6th international conference on economic management and green development. Springer Nature

Mehr AA, Shahim A (2023) NFTs and the art world: understanding the role of social media in the emergence of digital collections. UniversePG. [online] Available at: https://www.academia.edu/111343125/NFTs_and_the_Art_World_Understanding_the_Role_of_Social_Media_in_the_Emergence_of_Digital_Collections. Accessed on 24 Jul 2024

Mintology Blog (n.d.) NFT case studies. [online] Available at: https://blog.mintology.app/nft-case-studies/. Accessed on 18 Jul 2024

Narayanan A, Bonneau J, Felten E, Miller A, Goldfeder S (2016) Bitcoin and cryptocurrency technologies. Princeton University Press, Princeton, NJ

Nguyen AQ (2022) The mysteries of NFT taxation and the problem of crypto asset tax evasion. SMU Sci Technol Law Rev 25(2):323

Okonkwo IE (2021) NFT, copyright and intellectual property commercialization. Int J Law Inf Technol 29(4):296–304

Schmitz AJ, Rule C (2017) The new handshake: online dispute resolution and the future of consumer protection. American Bar Association, Washington, DC

Schmitz AJ, Rule C (2019) Online dispute resolution for smart contracts. [online] papers.ssrn.com. Available at: https://papers.ssrn.com/sol3/papers.cfm?abstract_id=3410450

Sharma TK (2022) Security tokens vs. utility tokens: a concise guide. Blockchain Council. Available online: https://www.blockchain-council.org/blockchain/security-tokens-vs-utility-tokens-a-concise-guide/. Accessed on 15 May 2024

Sousa FMde (2022) Token-Art system and the new international art market: the impacts of NFT technology and the legal aspects involved. J Law, Mark & Innov [online] 1(1):97–115. https://doi.org/10.13135/2785-7867/6674

Sterpi M (2022) The art law review: art disruption—art and technology in the twenty-first century. In: Kaye LM Spiegler HN (eds) The art law review, 3rd edn. London, UK: Clare Bolton

Stokes S (2021) Art and copyright, 3rd edn. Oxford, UK: Hart Publishing. https://www.bloomsbury.com/uk/art-and-copyright-9781509934270/

O'Brien, C. (2019). How artificial intelligence will affect the practice of law? (Honors Capstone Project, Sacred Heart University). Academic Festival. Sacred Heart University. https://digitalcommons.sacredheart.edu/cgi/viewcontent.cgi?article=1283&context=acadfest

Susskind R (2013) Tomorrow's lawyers: an introduction to your future. Oxford University Press, Oxford, UK

Tharun T, Vamshi A, Eswari R (2023) NFT application for music industry using blockchain smart contracts. In: 2023 4th international conference on innovative trends in information technology (ICITIIT), pp. 1–6. https://doi.org/10.1109/ICITIIT57246.2023.10068684

Twesige R (2015) Bitcoin: a simple explanation of bitcoin and block chain technology. Richard Lee Twesige. https://doi.org/10.13140/2.1.1385.2486

United Kingdom Legislation (2022) Copyright, designs and patents act 1988 (c. 48). Grangis LLC

University of London Press v. University Tutorial Press [1916] 2 Ch. 601

Von Appen A (2021) NFTs: how a technological trend redefined art ownership. Center for Art Law. Available online: https://itsartlaw.org/2021/04/27/nfts-how-a-technological-trend-redefined-art-ownership/. Accessed on 30 Aug 2023

Wang D, Ren Q, Li X, Qi Y, Zhou Q (2022) Defining consumers' interest and future of Nft fashion. Adv Soc Sci, Educ HumIties Res 653. https://doi.org/10.2991/assehr.k.220401.111

Wang Q, Li R, Wang Q, Chen S, Luo X (2021) Non-fungible token (NFT): overview, evaluation, opportunities and challenges. arXiv preprint arXiv:2105.07447. Retrieved from https://arxiv.org/abs/2105.07447

Yang Y-P (2022) The crowd's wisdom in smart contract dispute resolution: is crowdsourced dispute resolution arbitration? Contemp Asia Arbitr J 15(2):175–205

Zaslowsky D (2018) What to expect when litigating smart contract disputes. Law360. Retrieved from https://www.law360.com/articles/1028009/what-to-expect-when-litigating-smart-contract-disputes

Application of Blockchain in the Healthcare Industry Using Nanotechnology to Detect Cancer

Monica Apte, Prapti P. Shukla, Aishwarya R. Patil, Murthy Chavali, and S. M. Faheem

Abstract Blockchain has provided solutions to security issues and many other industries in which healthcare takes a major share, especially for storing cancer patient's records. Nanotechnology is one of the popular diagnostic methods to detect the early stages of cancer cells, possible to have vivo imaging as it gives a brilliant sensitivity-specific multiplexed measuring capacity. In comparison to conventional diagnostic techniques, cancer nanodiagnostics is a cutting-edge approach to the early detection and surveillance of cancer. Data security, privacy, and the smooth sharing of information among stakeholders, including patients, healthcare providers, and researchers, are some of the issues that this area continues to encounter as it develops. By overcoming these difficulties, blockchain technology, which is renowned for being decentralized and secure, has the potential to revolutionize cancer nanodiagnostics. Today, blockchain technology has developed to a point where it can be explored for use in a variety of other sectors, such as the management of cancer registries. It takes a lot of time and coordination between many levels of data-collecting jurisdictions to gather and handle the cancer diagnostic and treatment data as required

M. Apte (✉)
School of Computer Science, Dr. Vishwanath Karad MIT-World Peace University, Pune, India
e-mail: monicaapte@gmail.com

P. P. Shukla · A. R. Patil
SecureUp, Pune, India
e-mail: ppshukla2002@gmail.com

A. R. Patil
e-mail: aishrjpatil114@gmail.com

M. Chavali (✉)
Department of Research and Development, Dr Vishwanath Karad MIT-World Peace University, Pune, India
e-mail: ChavaliM@gmail.com

S. M. Faheem
School of Life Sciences, Manipal Academy of Higher Education, Dubai, UAE
e-mail: faheemsm@manipaldubai.com

M. Chavali
NTRC, MCETRC, Guntur, India

S. Krishnan et al. (eds.), *Unveiling Blockchain Enabled Business Applications*, Contributions to Management Science, https://doi.org/10.1007/978-981-95-2337-5_8

by law in many localities. Cancer therapies employ a variety of nanotools, including inorganic, organic, and polymeric nanomaterials. Nano methods have demonstrated many benefits: they are more effective, sensitive, and dependable; they avoid drug resistance; they only need small amounts of pharmaceuticals; they have low toxicity; and they are site-specific. A continuously updated and comprehensive picture of the patient's information made possible by blockchain can serve as a foundation for better decision-making, ultimately enhancing patient value and adding value to the healthcare industry. Decentralized applications with incorporated smart contract capabilities can be created on top of an interoperable blockchain-based IT system architecture for medical data and run well.

1 Introduction

One of the deadliest diseases for the human race is cancer. Accounting for around 10 million in 2022 (Jia et al. 2017). It has been estimated that there will be one death in 6 cancer patients. Prostate, lung, and colon, are the maximum common forms of disease. It is a large group of diseases where the rapid creation of abnormal cells that grow beyond the area in the body. Growth of this cell may lead to death from cancer (Jia et al. 2017).

A noteworthy advancement in cancer therapy over the past ten years has been the training of our immune system to target cancer cells while sparing healthy cells harm. The average life expectancy in India and throughout the world will continue to rise as a result of economic development (Zhang et al. 2019).

Early identification to fight against cancer is essential for effective therapy. However, by using the traditional methods it becomes more difficult to detect cancer at an early stage. By using the nanotechnology the extracellular cancer biomarkers and cancer cells can be detected as sensitivity, specificity and multiplexed measuring capability is provided by the nanotechnology. Nanoparticles are increasingly being used in cancer detection due to their unique properties and ability to interact with biological systems at the molecular level.

The creation of engineered objects at the atomic, molecular and macromolecular levels in the nanoscale range is known as nanotechnology. In the healthcare industry nanoparticles could be involved for healing and diagnostic purposes. For the diagnosis of deadly cancerous and infectious diseases, nanotechnology instruments are being developed, which will act as a catalyst in the identification of the disease's early stages.

On the other hand, the likelihood that the particles would injure the lungs grows as they become smaller generally. Therefore, it is necessary to research the safety of nanoparticles (Pavelić et al. 2023).

The development of incredibly small systems, gadgets, and particles is referred to as "nanotechnology"; this type of technology occurs at a very minute scale. One billionth of a meter is a nanometre the size of roughly five atoms. With a high degree of functional specificity, these materials and devices can be made to interact with cells and tissues at the molecular (e.g., subcellular) level. This makes it possible to integrate the technology with the biological system, which was previously not possible. Scientists from many different scientific disciplines, including physicists, chemists, engineers, and biologists, are involved in the developing topic of nanotechnology. This review has looked at the cutting-edge technology being created for use in medicine for diagnostic and therapeutic purposes.

Nanoparticles are being used to capture cancer biomarkers for cancer diagnosis, including exosomes, circulating tumour DNA, and proteins associated with cancer. The high surface area to volume ratio of nanoparticles compared to bulk materials is a key benefit of using them for cancer detection (Pavelić et al. 2023). This characteristic allows for the dense coating of nanoparticle surfaces with antibodies, small molecules, peptides, aptamers, and other compounds. Specific cancer compounds can be bound and recognized by these moieties. Multivalent effects can be produced by exposing cancer cells to a variety of binding ligands, which can increase an assay's specificity and sensitivity (World Health Organization 2020).

For storing medical reports like patient cancer records blockchain is an ideal. Patients can easily access its report by using blockchain. As the reports are stored on a blockchain, they can be accessed by the patients anywhere in the world. One of the most used blockchains in the Healthcare industry is MediLedger. The open network goal of MediLedger is to cover the whole pharmaceutical supply chain. Pharmaceutical companies Genentech, AbbVie and Pfizer are the founding members. In addition, two giant pharmaceutical wholesaler companies AmerisourceBergen and McKesson are the clients of MediLedger.

There is a need for reliable data sharing between many different entities or objects. Blockchain technology comes with a solution for information sharing. As data is stored on multiple different nodes or computers on blockchain. This in turn makes it difficult for a hacker to modify the data. They won't be able to alter data on all the nodes even if they manage to steal data in one of the nodes (Kasyapa and Vanmathi 2024).

By providing a tamper-proof, transparent and secure way to store and share cancer reports, blockchain can help to improve the privacy, security, efficiency and accuracy of cancer patients. Eric Garvin and Susanne Somerville collaborated while she served as VP of Supply Chain at Genentech/Roche. They teamed up to provide serialization consulting for businesses at the beginning of 2016 but were soon directed towards blockchain. The San Francisco-based Chronicled company served as their technological partner.

The convergence of nanotechnology and blockchain has the potential to revolutionize various industries by combining the precision and innovation of nanoscale technologies with the security and transparency of blockchain. This synergy can lead to advancements in secure data management, supply chain transparency, healthcare, IoT, environmental monitoring, anti-counterfeiting, and energy management, among other fields. As both technologies continue to evolve, their intersection will likely yield even more innovative and transformative applications.

2 Significance of the Study

Due to Nanotechnology researchers are able to study and modify macromolecules in real-time and examine the cancer cells at start state. Nanotechnology supports fast and perfect detection of chemicals connected to cancer cells, permitting the researchers to identify molecular changes even when they only affect a tiny section of cells. Additionally, nanotechnology has an ability to make completely new and powerful medicinal medicines (Jin et al. 2020).

In the end and specifically, the use of nanoscale materials for cancer relies on their ability to be quickly functionalized and easily tuned; to deliver and/or act as the therapeutic, diagnostic, or both; and, finally, their ability to passively accumulate at the tumour site, to be actively targeted to cancer cells, and to be delivered across conventional biological barriers in the body, such as the blood-brain barrier that highly refracts nanoparticles (Kemp and Kwon 2021).

Pain experienced by the cancer patients was illustrated in Fig. 8.1, in two categories amongst them 66% is from Metastatic or terminally ill patients coming from Deep bone and aching, and 55% contribute because of radiation therapy and other complementary therapies. There is an overlap of % pain experienced by more than two parameters, but primarily from the mentioned two (Fig. 8.2).

The following are some drawbacks of screening mammography. Early diagnosis of pre-cancerous lesions like ductal carcinoma in situ carries the risk of unneeded therapies or multiple biopsies for confirmation, even though it may not progress to invasive breast cancer for the remainder of the patient's life.

According to cumulative research by the USPSTF released in February 2016 and titled "False-positive biopsy recommendation after screening mammography," the

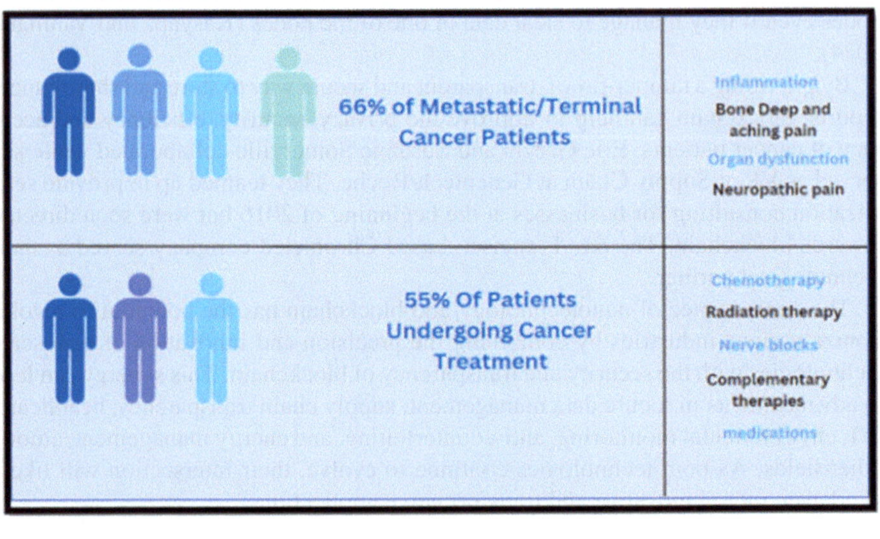

Fig. 8.1 Cancer patients treatment. *Source* Authors own compilation

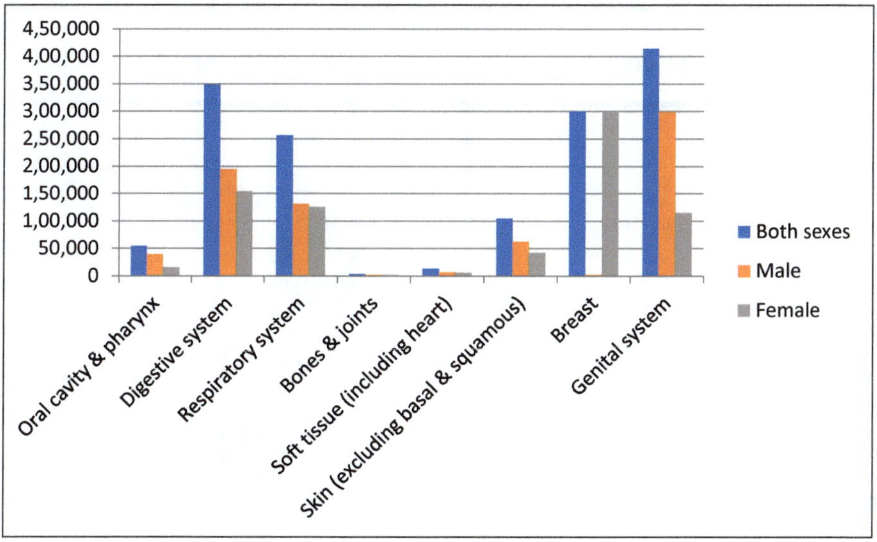

Fig. 8.2 Estimated number of new cases in 2023. *Source* World Health Organization

biopsy rate with annual and biannual screening is 7-9.4% among females aged 40–49, while it is 4.8–6.4% among women aged 50–59 (Eldridge 2023). The presentation and discussion of a study on the potential use of gold nanoparticles in mammography for improved image diagnosis and treatment. We quantify the increase in dose released to the tumour that has been enriched with Au-NPs relative to the nearby healthy tissues and discover that for X-rays, the increase can be as much as two orders of magnitude larger in intensity (Abu-Elezz et al. 2020). The treatment of brain tumours in the field of neuro-oncology continues to be a challenging issue despite significant efforts to create diagnostic tools and therapeutic options. The blood-brain barrier is one of the physiological barriers that prevents enough therapeutic agent accumulation at the location of a tumour to effectively destroy cancerous cells (Health 2023).

Before performing surgery with the help of nanotechnology a high resolution can be obtained. It is especially important for GBM also referred to as a grade IV astrocytoma. Astrocytoma is a brain tumour which is aggressive and rapidly growing, which is characteristically invasive. Such intrusiveness makes it relatively challenging to accurately define a clear tumour boundary by eye. Appropriate imaging of a tumour is critical for measuring the extent of tumour distribution preoperatively as well as for determining response to a treatment regimen postoperatively, both being necessary for the successful management of brain tumour patients (Health 2023).

The solution may be found in nanotechnology: equipping existing therapies with better targeting capability, increasing localized drug efficacy, limiting systemic toxicity, improving diagnostic sensitivity, enhancing imaging, and refining radiation therapy (Kemp and Kwon 2021) (Fig. 8.3).

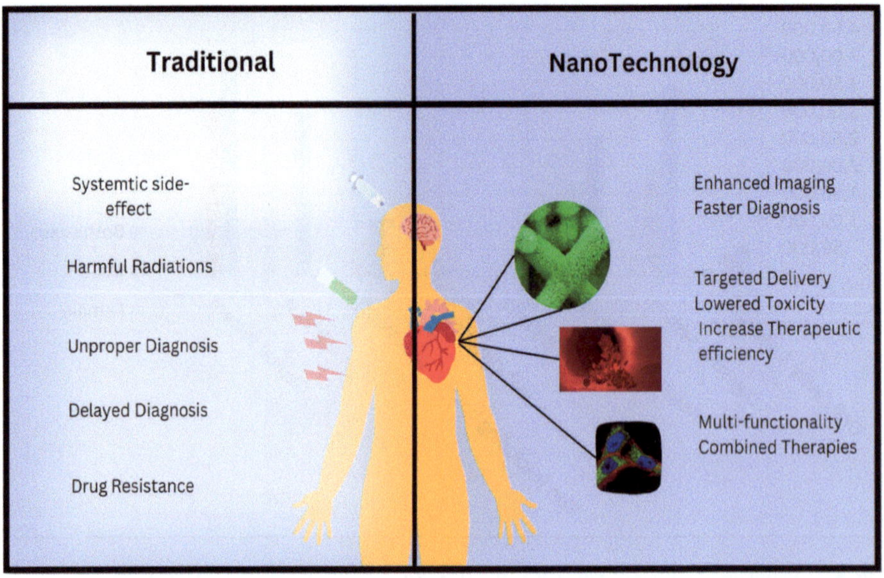

Fig. 8.3 Difference between infected cells using traditional way and nanotechnology. *Source* Authors own compilation

Nanotechnology provides many advantages over conventional anti-cancer drugs, radiation therapy, diagnostics, and imaging. Certain nanomaterials have intrinsic physio-chemical properties that enhance bioimaging, localize radiation therapy, facilitate early diagnoses, circumvent drug resistance, and enable multi-modal treatments (Kemp and Kwon 2021).

The nanotechnology fields are also contributing to the expansion of the worldwide market. In addition, the development of applications in the electronics sector and the expanding use of nanomedicine in the pharmaceutical sector are driving the growth of the global nanomaterials market in the years to come. Additionally, the sector is expanding because of the increased accessibility of various methods and instruments for synthesizing nanomaterials, such as covalent synthesis and self-assembly. Among its benefits are its high surface-to-volume ratio, high porosity, and better physicochemical properties (Padmaja et al. 2024) (Fig. 8.4).

From the above diagram, we can state that sample data of cancer cells undergo diagnosis by using nanotechnology. Nanotechnology bifurcates healthy cells and cancer cells with the help of their colour, shape and various other things which are diagnosed with nanotechnology using either quantum dots, nanoshells or gold nanoparticles. The affected cells should be treated as early as possible otherwise, they may hamper other healthy cells. For this also nanotechnology can be used which includes liposomes, carbon nanotubes, polymeric micelles, dendrimers or quantum dots.

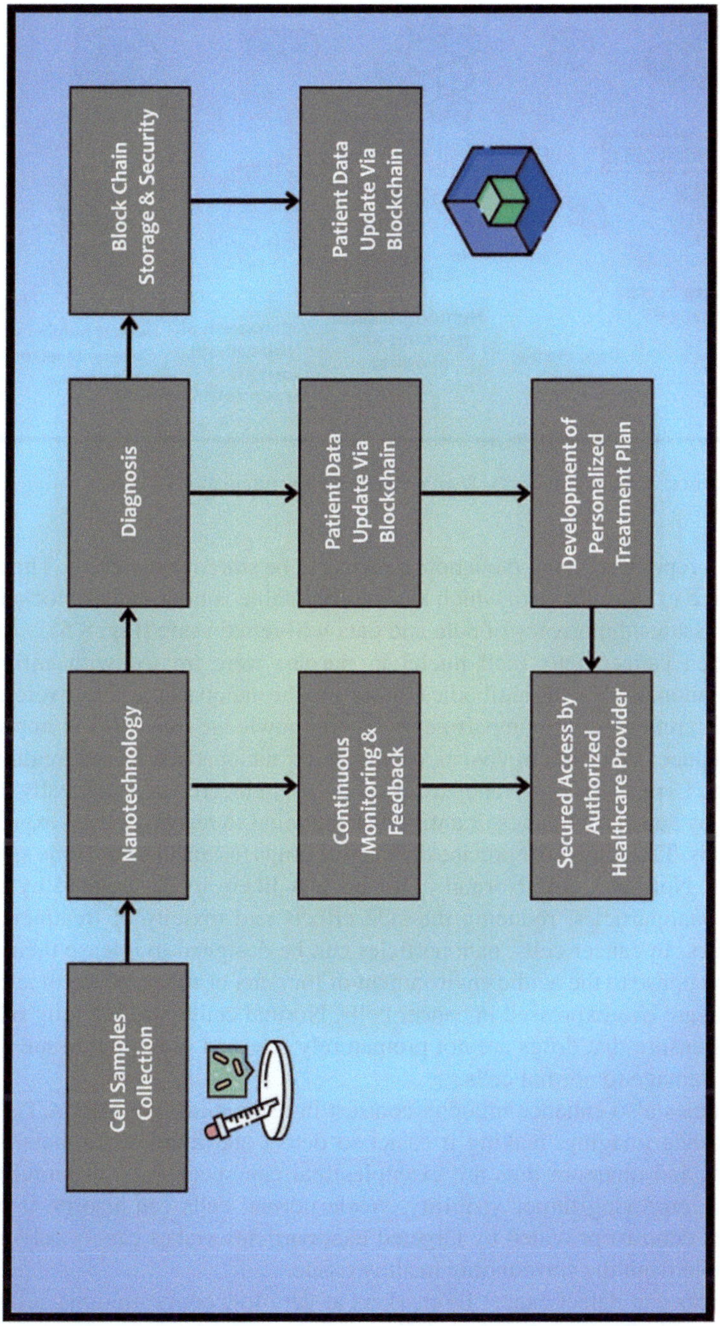

Fig. 8.4 Diagram for storing data of cancer patients in blockchain. *Source* Authors own compilation

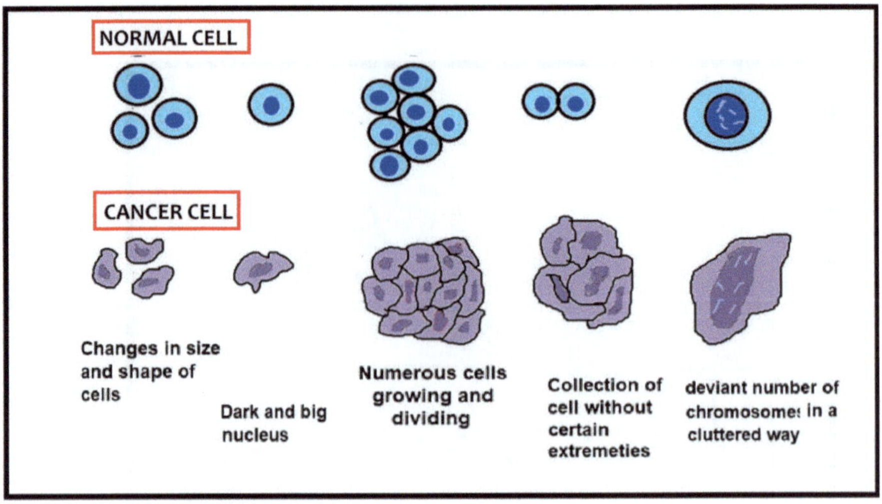

Fig. 8.5 Cancer cells and normal cells. *Source* Authors own compilation

After the proper treatment, the whole data has to be stored somewhere. The most reliable source to store the data which is also immutable is nothing but blockchain. Blockchain assures the privacy of data and data will remain safe (Fig. 8.5).

Here, when cancer cells (cell nuclei in purple) were treated with antibody-conjugated nanoparticles, the antibodies (blue) and the nanoparticle cores were separated into different cellular compartments. Such knowledge may lead to improved methods of cancer detection in vivo as well as better nanoparticle-based treatments. Targeting and Drug Delivery- Nanoparticles can be engineered to specifically target cancer cells by attaching ligands or antibodies that bind to receptors over-expressed on cancer cells. This allows for precise delivery of drugs to cancer cells while sparing normal cells. Normal Cells: Normal cells are less likely to be targeted by these specialized nanoparticles, reducing the side effects and toxicity of treatments on healthy tissues. In cancer cells, nanoparticles can be designed to release their drug payload in response to the acidic environment of tumours or the presence of specific enzymes that are overexpressed in cancer cells. Normal cellcontrolled drug release mechanisms ensure that drugs are not prematurely released in normal tissues, thus minimizing damage to normal cells.

Nanoparticles also enhance imaging contrast in techniques like MRI, CT scans, and fluorescence imaging, making it easier to detect and monitor tumours. Gold nanoparticles and quantum dots are examples that can specifically accumulate in cancer cells, improving tumor visibility. While normal cells can also be imaged, the enhanced contrast provided by targeted nanoparticles makes cancer cells more distinguishable from the surrounding healthy tissue.

Nanosensors can detect cancer biomarkers at very low concentrations, allowing for early diagnosis. These sensors can differentiate between the molecular signatures of cancer cells and normal cells. Normal cells generally do not express the same biomarkers at elevated levels, reducing the likelihood of false positives.

The treatment approach through nanotechnology can be nanoparticles, such as gold nanoshells or iron oxide nanoparticles, that can be used for hyperthermia treatment, where they absorb infrared light or generate heat in the presence of a magnetic field, selectively killing cancer cells. The targeted nature of these nanoparticles ensures minimal heating of normal cells, protecting them from thermal damage.

In one of the other treatment approaches of photodynamic therapy (PDT) nanoparticles can be loaded with photosensitizers that, upon exposure to light, produce reactive oxygen species (ROS) that kill cancer cells. These nanoparticles can be targeted to cancer cells to ensure that ROS generation is localized. In the case of normal cells minimizing light exposure to areas without cancer-targeted nanoparticles reduces the risk of damage to normal cells (Dinakaran and Wilson 2023).

The gene therapy for cancer cells nanoparticles can deliver genes or RNA molecules (e.g., siRNA, mRNA) that can silence oncogenes or activate tumor suppressor genes specifically in cancer cells. Normal cells' targeted delivery systems reduce the risk of off-target effects and genetic modifications in normal cells.

Immune System Modulation cancer cells nanoparticles can be used to deliver antigens or immune modulators to cancer cells, enhancing the immune system's ability to recognize and destroy them. In normal cells, this targeted approach minimizes the risk of systemic immune activation that could lead to autoimmune reactions against normal cells.

Hence we can say that nanotechnology enables the development of highly specific and effective tools for targeting, diagnosing, and treating cancer cells while minimizing the impact on normal cells. The precision of nanotechnology-based approaches improves the efficacy of cancer treatments and reduces side effects, offering a significant advantage over traditional methods.

3 Securing Patient Cancer Medical Data Through Digital Transformation Using Nanotechnology

To modernize operations, build more efficiency into the workflows, and processes and provide better patient experiences, healthcare practices across the nation have intensified their digital transformation initiatives in recent years. Although this digital growth is positive and vital. It also presents practices with some major difficulties. Healthcare providers have to be alert about security concerns as more of our healthcare processes move to digital formats.

Unfortunately, hackers frequently target healthcare organizations. A recent study indicated that more breaches are currently occurring in the healthcare sector than in any other area (Zhao 2019). Healthcare providers, businesses and organizations use hardware-secure networks and data security software to safeguard their resources. Health Insurance Portability and Accountability Act (HIPAA) standards are one of the most significant forces advancing healthcare data security. These standards offer broad guidelines for managing sensitive healthcare data (Baysal et al. 2023). Healthcare systems are concerned about three main categories of cybersecurity threats: internal, external and credentialing (Ahmad 2021).

Some well-established methods and procedures can be used if there is any uncertainty regarding the identity of the author, the date the document was created and whether it was altered. Specialists utilize to examine texts, both typed and handwritten.

3.1 Security Threats for Healthcare Electronic Documents and Their Solutions

Electronic records can be easily copied, transported over long distances, lost or destroyed because they are not tied to any one physical carrier. Such information cannot be retained securely on a single storage medium as being capable of effectively retrieving the data from the device cannot be completely trusted and becomes potentially destructive and unable to be used after only three to five years. Due to these reasons, data must be held concurrently. Backup, archived and transferred regularly into new storage media.

Threat simply refers to a potential source of harm or danger. A computer system and its contents (data) can be attacked in a variety of ways. The use of malware, or malicious software, in many of these attacks is possible. Research data and intellectual property are important to cybercriminals who want to steal it or hold it for ransom as well as to those who are pursuing knowledge for the benefit of society (Cremer et al. 2022).

Unauthorized access to critical data is when data theft or extortion starts. Through a range of purposeful and unintended methods, unauthorized access happens. Hackers access our systems and research by putting malware in a system, phishing for login information or stealing an unprotected device like a laptop or phone. Accidental unauthorized access and disclosure can also occur when a research assistant keeps their login information in a folder after they have finished working on the project, team members share the passwords, research data is sent in an unencrypted email or data is deleted inappropriately.

Data security is always a big concern for healthcare enterprises. Irrespective of the privacy of data. The data is stored on desktop or other computing devices. It's a policy of a system that permission to sensitive data will be granted to limited nodes or no access. We are very much aware that computer systems and the data they hold are liable to a variety of risks. Since computers began to be connected not just to LANs but also to the World Wide Web, these issues became more serious (Cremer et al. 2022).

The solution to the above problem is Blockchain technology which is a distributed ledger for data storage that uses encrypted blocks of fixed or variable size arranged in chains (Marbouh et al. 2022). This indicates that all parties (nodes) involved in, contributing to or accessing data in the blockchain collaborate to decide through the use of consensus algorithms, which blocks are accessed by various users. The data blocks contain information that has been independently validated as true and relevant. In the majority of systems, each node can have a copy of the entire blockchain, allowing them to track who accessed what data and when.

Decentralization, data transparency, privacy, and security are all the main components of blockchain technology. Decentralization architecture signifies the fact that all information contained in blocks is distributed over the blockchain network as opposed to being kept in a single location as we do in -client-server architecture. In addition, rather than being chosen by a central server, whether a block can be added to the chain is decided by all nodes collectively by a consensus mechanism.

Distributed ledger system that needs approval from its creators and organizers to use the data. This recent technology has the potential to greatly enhance data transmission in the healthcare industry, particularly in light of the COVID-19 pandemic's abrupt emergence and the resulting revelation of some systems' shortcomings in managing public health emergencies. Blockchain offers features like decentralization, trade, anonymity and its special immutable architecture. It is widely applied in the financial sector, government sector, supply chain management, agriculture record keeping and radiography (Stopfer et al. 2024).

Ethereum blockchain (EHR) is being widely adopted by medical professionals and healthcare institutions. Since EHRs contain confidential and identifiable patient data, a secure platform is required for data transfers between health systems and healthcare providers, making data sharing transparent. EHR systems are extremely challenging. Blockchain technology provides a means of enabling secure data interchange. The primary objective of this chapter was to ascertain whether blockchain technology could improve interoperability and secure repetitive tasks for efficient EHR exchange and viewing (Stopfer et al. 2024) (Fig. 8.6).

Four key characteristics of blockchain technology—decentralized storage, immutability, consensus mechanism and security can be used in the healthcare system. The blockchain contains characteristics that only allow system modifications when all parties are in agreement or consensus has been reached. It can also store information and distribute it to others with the originator's permission. Furthermore, once data is stored in the chain, it cannot be changed because it is immutable. Lastly, rather than being directly stored, data are normally accessed from the chain. Blockchain is therefore perfect for healthcare settings since information may be written once and seen numerous times, but it cannot be changed once published (Stopfer et al. 2024).

It is well acknowledged that medical mistakes pose a serious threat to patient safety. Patient safety is affected by several factors, including inefficient communication and the inability to communicate and exchange data. Blockchain technology can be used to enable a safe, open and decentralized data flow among stakeholders to address these issues (AbdelSalam 2023). There are several healthcare procedures which include third-party data storage, which is a major concern for healthcare stakeholders including patients. Blockchain technology will be able to address these issues

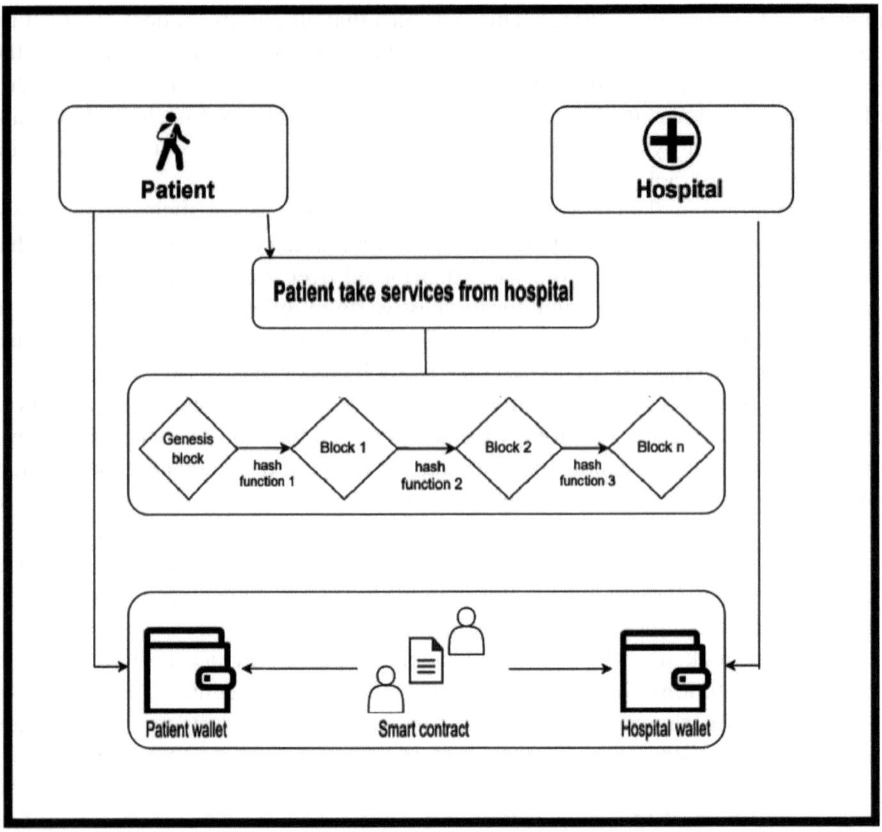

Fig. 8.6 Blockchain in healthcare domain. *Source* Authors own compilation

and provide transparency for health data management procedures while lowering the likelihood of data abuse or improper handling that might endanger patient safety (Haleem et al. 2021). Even though patient safety has received relatively little attention and study, blockchain is the solution for the healthcare industry (Andrew et al. 2023).

To be more precise, MeDShare closely monitors all operations on the data made possible by smart contracts and the keys associated with those contracts. As a result, if an uninvited user attempts to steal or tamper with data from reports produced by the smart contract, their rights to access the data will be revoked. Various studies have been conducted employing blockchain technology to store, track and manage medical records. The process encrypts and saves the user's health data using blockchain and the Internet of Medical Things (IoMT). Ethereum blockchain is a secure platform for dealing with different kinds of critical information in the healthcare industry.

Ethereum blockchain is a system that allows to generation of healthcare applications to give solutions to hospital and healthcare industry problems. It is a powerful tool to provide data security and non-tampering of data. In blockchain, a transaction is characterized by encryption to secure the records and each transaction is connected to previous transactions through a hash number. Blockchain transactions are validated on the nodes using smart contracts. A single person cannot make a transaction. Eventually, blockchains have transparency, allowing each user at any time to track the transactions. The smart contract is a code written in solidity or Python language which is executed when a node initiates a transaction. The third party such as bank involvement becomes negligible because of smart contracts. Ethereum platform gives a distributed environment running smart contracts. This helps developers to create markets to move funds according to instructions given in the smart contract. Once created it becomes nearly impossible to modify or update smart contracts. Hence integrity of patient records is maintained.

4 Blockchain-Based Smart Contracts for the Healthcare Industry

In this chapter, we have used Ethereum blockchain to create smart contracts. This entity is used to store the patients' records on the network within the individual nodes. There are many advantages of building a smart contract as it contains data integrity record ownership metadata and permissions. A healthcare blockchain system is characterized by encryption which uses cryptographically signed instructions. State-transition functions of the contract carry out policies, only by authentic transactions enforcing data alteration (Cointelegraph (n.d.) xxxx).

These procedures can be structured to enforce any set of rules regulating a specific medical cancer patient record as long as it can be computationally represented. For example, a policy may impose sending separate consent transactions from both patients and healthcare professionals before granting a third party viewing permission. We designed a system based on blockchain smart contracts for complex healthcare workflows. Smart contracts have been designed for different medical workflows and then managing data access permission between different entities in the healthcare ecosystem (Khatoon 2020).

A smart contract is an agreement which is automatically executed. It contains the terms of the deal between the parties which are encoded into individual lines of code. A smart contract is in electronic form and carries out the conditions of the agreement. American computer scientist and digital currency researcher Nick Szabo first introduced the idea of smart contracts in 1994. A smart contract, stored on blockchain technology, is coded using all the rules and conditions from obtaining permission between the nodes to granting access to the data. The smart contract also covers the rules required by the stakeholders which are required to access the data at different stages by different stakeholders and at different times. It could be designed to have all the conditions from managing different permissions to accessing data as shown in Fig. 8.2 which explains the stakeholders involved in healthcare have different entities and each entity has its role and responsibility. The conversation between doctors and patients becomes more transparent. The smart contracts contain rules for data authorization. With its unique id, it can track all the instructions given from source to destination.

All the functions and processes which are part of a smart contract are well-designed and explained through coding. Since smart contracts cannot be controlled by the central authority or servers. Hence it does not require approval for execution of its operations. Smart contracts reduce the overhead of managing the task and administrative processes.

The total medical record data is stored in the local database to maintain the performance and economic viability and the hash of the data is the data element of the block committed to the Blockchain.

In simple terms, smart contracts get executed when the conditions are fulfilled. The execution is in linear form following the protocols.

A smart contract is a key component of blockchain technology since it increases transaction security and efficiency while also serving as an initiation of transaction. It also makes different components, such as programs operating on these platforms, more accessible (Khatoon 2020). A certain class of Ethereum account is smart contracts. A transaction is executed only when there is a balance in the wallet which is carried by a metamask followed by the initialization of the smart contract. Smart contract works according to the code mentioned in it. They are not operated by the user. Then, user accounts can communicate with a smart contract by submitting transactions that carry out a smart contract function. Like a customary contract, smart contracts establish rules and have the system automatically enforce those rules. Smart contract interactions are irreversible by default and cannot be undone (Sookhak et al. 2021).

The automation processes that earlier required a mediator are one of the prime advantages of blockchain systems. A smart contract is the code written in solidity language which is initialized when we want to execute a code. It allows the transfer of money or documents. It's a financial transaction then no third party is involved like a bank or financial institute. There are no third-party charges to be paid by the customer. Hence transaction via aiblockchain takes less time and money.

The other benefit of smart contracts is the code that underlies its security. For instance, smart contracts on Ethereum are created using solidity or Python as a programming language. This ensures that no hacker will modify the constraints mentioned in the smart contract. The rules coded in smart contracts become part of the network code. Hence, blockchain networks provide a robust system which is tamper-proof (Sookhak et al. 2021).

A blockchain network is used to carry out the smart contract and several computers that make up the network duplicate the contract's code. This guarantees a more secure and transparent execution of the conditions of the contract.

A smart contract functions similarly to other blockchain transactions. These are the actions that must be taken:

1. A transaction is initiated by a user from their blockchain wallet.

2. Financial transfer involves authorization when a transaction is initiated.
3. The code defining the kind of transaction to be carried out is included in the transaction.
4. The block containing transactions is appended to the blockchain.
5. The same procedure is used to update any changes in the contract status.

4.1 Smart Contracts in the Healthcare Sector on the Ethereum Platform

Providing medication and healthcare device industrialists with a simpler and more price-operational alternative to the current recruitment of medical trials. Independent medical data suppliers often charge big amounts to get patient contact information and to execute ample pull-marketing drives.

An important objective is to permit users to execute medical trial-related smart contracts on an Ethereum blockchain network causing medicines without side effects and increasing general community awareness in clinical research. In this process, patient details, different protocol registrations, metadata logs, screening and enrollment data are stored via smart contracts (Ali et al. 2022).

The pharma sector retrieves metadata stored on the Ethereum blockchain network to identify potential patients for medical trials as shown in Fig. 8.7. The selected patients get the message from the company, including an application to read access to their medical records, and experimental study outcomes. If the patient gives permission, a pharma company bill will be handled via smart contracts, awarding part of the received fee to the patient, and another section to the labs that reported the appropriate test results for the patient (Ali et al. 2022) (Figs. 8.8, 8.9, 8.10).

A bank collaborates with a healthcare provider to offer financing for cancer treatment. It works as follows:

1. **Patient Enrollment**: A patient diagnosed with cancer at an early stage with the help of nanotechnology enrolls in a treatment program. A smart contract is created, detailing the treatment plan, milestones and cost associated with it.
2. **Funding Disbursement**: The bank provides a loan or credit line to the patient. The smart contract automatically disburses funds to the healthcare provider as treatment milestones are met (e.g., completion of chemotherapy sessions).
3. **Real-time Updates**: The patient's progress and treatment costs are updated in real-time on the blockchain. It ensures all parties have access to the latest information.
4. **Automated Payments**: The smart contract releases payments to the healthcare provider automatically upon verification of completed treatment milestones, reducing administrative overhead and ensuring timely payments.
5. **Insurance Claims**: If the patient has health insurance, the smart contract can also handle claims processing, automatically submitting claims, verifying coverage and processing payments.

Contract

Address	Balance
0*097531246f5abfc67cdb396c345b	0.00 ETH

CREATION TX
0*a9E763719362451290CADFF3854CFB322735AFFDC

STORAGE
{ 13 items
admin : {...} struct 1 item
doctor : {...} struct 1 item
patient : {...} struct 1 item
Doctors: {} mapping 0 items
Patients :{} mapping 0 items
Records : {} mapping 0 items
Dr_ids [] 0 items
Patient_id: {...} 3 items
AccountId : address "0*0000000000000000..."
Admin_id: address"0*56832abfd45987fc385..."
Get_patient_id : address "0*0000000000000000..."
Get_dr_id: address "0*00000000000000000..."

TRANSACTIONS
TX HASH
0*D4537962154763942FBC34DA986576FBCA432863DV4577

FROM ADDRESS	TO CONTRACT ADDRESS	Gas Used	Value
0*759325379621763942FBC34DA986576FBCA	Contract	184528	0

TX HASH
0*FABD5645379621763942FBC34DA986576FBCA432863D53

FROM ADDRESS	TO CONTRACT ADDRESS	Gas Used	Value
0*56BD5645379621763942FBC34DA986576FBCA432863D53	Contract	184528	0

Fig. 8.7 Transactions in Ethereum platform and Smart Contract between doctor and patient. *Source* Authors own compilation

1. **Automated Claims Submission**: Smart contracts can automate the process of submitting insurance claims. When a patient receives medical services, the healthcare provider can immediately trigger a smart contract that submits a claim to the insurance company. This reduces delays and administrative overhead.
2. **Verification and Approval**: The smart contract can automatically verify the details of the claim against the patient's policy and medical records. It checks for coverage, eligibility, and policy limits and if everything matches, the claim is approved instantly. This reduces the need for manual intervention and speeds up the process.
3. **Payment Processing**: Once a claim is approved, the smart contract can automate the payment process. It can trigger an automatic transfer of funds from the insurance company to the healthcare provider, ensuring timely payments and reducing the risk of errors.

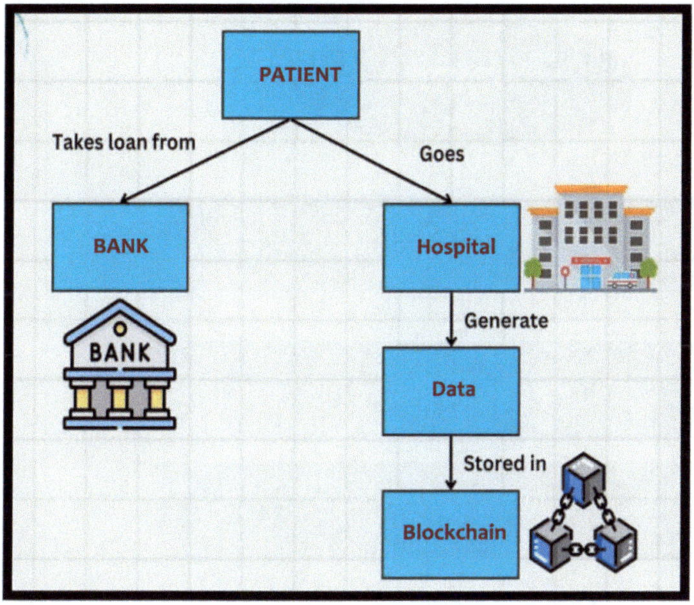

Fig. 8.8 A banking industry: Financing Cancer Treatment. *Source* Authors own compilation

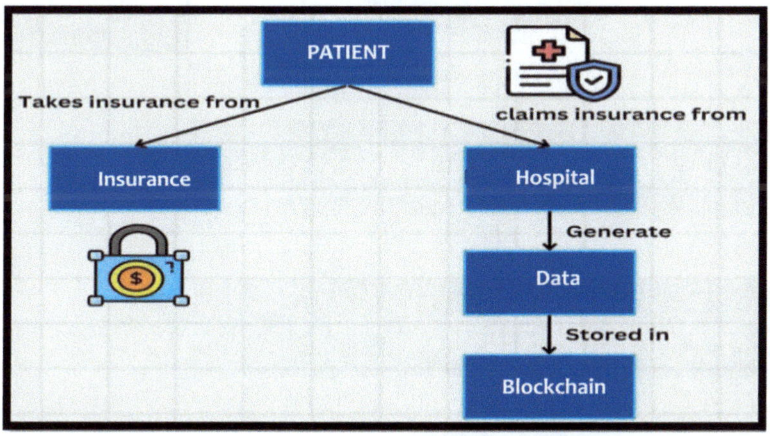

Fig. 8.9 Role of insurance companies in smart contracts. *Source* Authors own compilation

4. **Fraud Prevention**: Blockchain's transparency and immutability help prevent fraud. Smart contracts ensure that all conditions are met before processing claims, making it difficult to submit fraudulent claims. Any discrepancies or attempts at fraud are easily traceable.

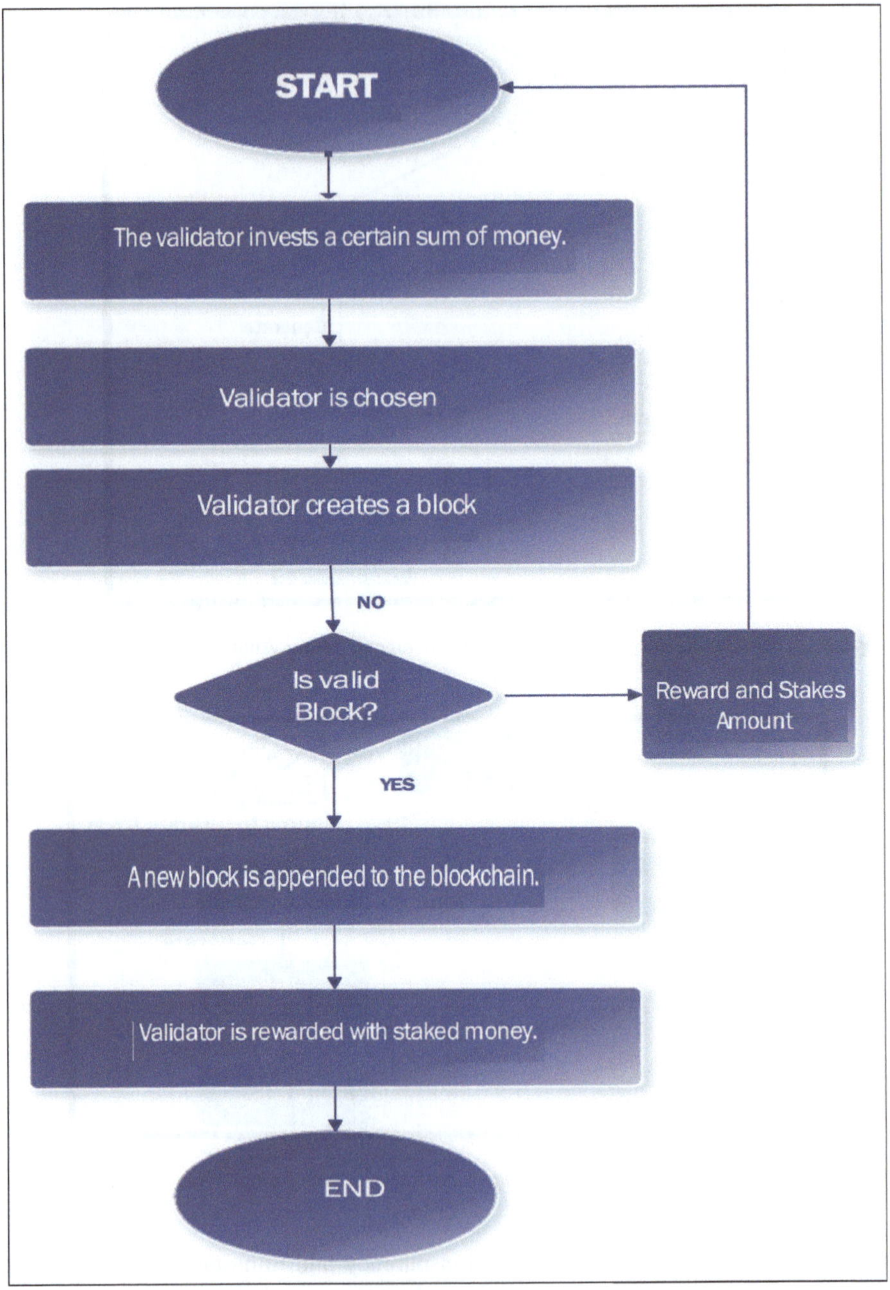

Fig. 8.10 Flowchart showing smart contract execution. *Source* Authors own compilation

Policy Management of Insurance Companies by Smart Contract

1. **Dynamic Policy Adjustments**: Insurance companies can use smart contracts to create dynamic policies that adjust based on predefined conditions. For example, a smart contract could automatically adjust premiums based on a patient's health data, such as reaching specific health milestones or reducing risk factors.
2. **Transparent Policy Terms**: Smart contracts can ensure that all policy terms are transparent and immutable. This builds trust with policyholders as they can independently verify the terms and conditions of their insurance policies on the blockchain.

Patient Data Management

1. **Secure Data Sharing**: Insurance companies can facilitate the secure sharing of patient data between healthcare providers using smart contracts. For instance, when a patient changes providers, a smart contract can grant the new provider access to ensure continuity of care, the patient's medical history.
2. **Data Integrity and Privacy**: By loading patient data on the blockchain, insurance companies can ensure data integrity and privacy. Control access to data smart contracts can to this data, ensuring that only authorized parties can view or update records.

Wellness and Preventive Programs

1. **Incentivizing Healthy Behaviour**: Insurance companies can create smart contracts to incentivize healthy behaviour. For example, a smart contract could reward patients with lower premiums or bonuses for achieving certain health goals, such as regular exercise, healthy eating, or quitting smoking and drinking.
2. **Real-time Monitoring**: Smart contracts can integrate with wearable devices and health apps to track a patient's health day–to–day basis. Based on this data, smart contracts can adjust coverage, premiums and benefits dynamically, encouraging preventive care and healthy living.

Benefits of Smart Contracts for Insurance Companies

3. **Efficiency and Cost Reduction**: Automation of claims processing, policy management and payments significantly reduces administrative costs and processing times.
1. **Enhanced Security and Trust**: Blockchain's encryption and immutability provide a secure environment for sensitive data and transactions, building trust with policyholders and healthcare providers.
2. **Improved Customer Experience**: Faster claims processing, transparent policy terms and personalized health incentives improve the overall customer experience, leading to higher satisfaction and loyalty.

3. **Regulatory Compliance**: Smart contracts can be programmed to ensure compliance with relevant regulations, reducing the risk of non-compliance and associated penalties.

Challenges and Considerations

1. **Technical Complexity**: Implementing blockchain and smart contracts requires significant technical expertise and investment in technology infrastructure.
2. **Regulatory Landscape**: Navigating the complex regulatory environment in healthcare and insurance is challenging. Ensuring compliance with data protection, privacy laws and industry standards is crucial.
3. **Interoperability**: Ensuring that smart contracts and blockchain systems can work seamlessly with existing healthcare and insurance systems is essential for widespread adoption.
4. **Data Privacy**: Managing sensitive patient data requires robust privacy measures. Ensuring that smart contracts comply with data privacy regulations, such as GDPR or HIPAA is critical.

4.2 Conclusion

Insurance companies have a significant role in generating and utilizing smart contracts within the healthcare industry. By adopting blockchain technology, they can streamline operations, reduce costs and enhance the security and transparency of their services. The integration of smart contracts can lead to more efficient claims processing, dynamic policy management, secure data sharing and incentivization of healthy behaviours. Despite the challenges, the benefits of smart contracts in improving the efficiency and reliability of healthcare insurance processes are substantial, making it a promising avenue for innovation in the industry.

Blockchain technology is in the lead of numerous current developments in the healthcare areas.

New opportunities are emerging for the administration of health data, as well as ease for patients to access and share their health data, this is only possible due to advancements in electronic health data, cloud data storage and patient data protection rules (Gordon and Catalini 2018).

Incompatible runtime environments are typically used to install smart contract programs. For instance, smart contract programs on the Ethereum blockchain are run by the Ethereum Virtual Machine (EVM). This enables the installation of decentralized applications such as smart contracts.

Medical records store all the patient data and treatment is done accordingly. On the blockchain, patient profiles can be created and stored using smart contracts, allowing doctors and other medical staff to examine previous medical records stored in the blockchain. Doctors will be able to develop the best strategy for treatment based on the patient's prior medical history and the results of those treatments.

A structure resembling this helps doctors to avoid problems with medical negligence and is also life-saving. Additionally, medical facilities can set up smart contracts to monitor health issues brought on by treatment side effects and convert it to supply the data to the pharma industry and other healthcare industries that are studying the negative effects of new treatments or effects of drug changes.

5 Consensus Algorithm in Blockchain

Consensus algorithms help blockchain technology to make choices to address certain issues.

A cell may develop into a cancer cell and will affect an organ. A consensus-aware blockchain platform is to reduce the possibility that the patient's organ may become infected with one of the cancer cells. The tumour traits can be decentralized and possessed via a consortium blockchain.

These consensus algorithms produce strict operational standards with the imposition of evidence of authority serving as a crucial instrument for block formation.

6 Implications

In several industries today, blockchain professionals are working to solve business complex problems. Few are as promising as the US pharmaceutical industry. The Drug Supply Chain Security Act (DSCSA), which has a deadline for returns in November 2019, gradually takes effect through 2023. The law mandates that medications be traceable and serialized. The advantage of blockchain is that it gives access to the information when required among the different parties and acts as a reliable source of data. Due to the cryptographic storage, it becomes difficult for a hacker to manipulate data. Hence it is quite preferable for healthcare blockchain MediLedger. Intruders won't be able to alter data on all the servers even if they manage to breach one of them.

With the use of blockchain, patient records can be maintained in a block of fixed size. As an advantage the record becomes secure. And the data can be fixed in one place (Shilpa and Apte 2022). 1 Duplicate copies of data are not necessary to be conserved. Hence the number of bytes required for data is saved. Another important advantage of this information tool is that less space is required, so it becomes cost-effective.

Numerous advantages are witnessed in the healthcare industry by saving medical data in a blockchain. The strong privacy and security it offers is one of the most important benefits. Blockchain guarantees that sensitive patient data is preserved securely and is resistant to unauthorized access and alteration through cutting-edge encryption techniques and decentralized consensus mechanisms. As soon as medical records are stored on the blockchain, they become immutable, which lowers the possibility of mistakes and fraud in patient records. Data integrity is another important advantage. Furthermore, the use of blockchain technology has the potential to significantly enhance data interoperability, facilitating the safe and secure exchange of medical data between various healthcare systems and providers.

Additionally, patients can increase their autonomy and privacy by controlling who has access to their data through smart contracts. Blockchain can support medical research by safely and openly exchanging anonymized data, possibly resulting in advancements in therapies and healthcare procedures. These favourable effects add up to a healthcare ecosystem that is safer, more effective, and patient-focused.

7 Role of Stakeholders

In the detection of cancer, insurance companies, pharmacy and banks impart crucial roles by ensuring financial accessibility, supporting advanced diagnostic machines and facilitating the development of robust healthcare infrastructure. Insurance companies make cancer screenings and diagnostic tests accessible through coverage and preventive programs. While banks provide the necessary funding and investment to advance medical innovations and support healthcare providers. Together they contribute to earlier detection. Pharmacies are indispensable stakeholders in the healthcare system, contributing to medication management, patient education, preventive care and chronic disease management. Their role extends beyond dispensing medications to actively participating in health screenings, providing immunizations and collaborating with other healthcare providers to deliver comprehensive care. By ensuring the availability of essential medications and offering valuable health services pharma companies enhance the overall healthcare delivery system and improve patient outcomes.

8 Conclusion

In conclusion, this research underscores the transformative potential of integrating blockchain technology with nanotechnology for early cancer detection in the healthcare industry. With the help of nanotechnology, it is possible and feasible to diagnose early cancerous cells which will help patients for further treatment. Also, it will escalate decision-making based on the cancer stage by doctors. Nanotechnology has promise for diagnosis by replacing current techniques with more sophisticated as well as less expensive with various alternatives. Nanomaterials have helped with better cancer detection and care. It is possible with nanotechnology to enable low toxicity by targeted medicine. The future of technology is quite promising because of blockchain and nanotechnology. Blockchain and nanotechnology working together can open up a world of intriguing future possibilities for the detection of cancer cells and securing patient's data. This connection may lead to advancements in medicine delivery and healthcare monitoring as well as a revolution in data storage and supply chain management. This will eventually help the healthcare industry to grow over time and will also save time in terms of data security. Which in turn will lead to smooth management and operations. Future developments and the healing advantages of nanotechnology will make them useful for the treatment of cancer.

By leveraging the strengths of blockchain, we can ensure secure, transparent, evidential and efficient handling of sensitive patient data, which is crucial to fight against cancer. The roles of stakeholders such as banks, insurance companies and the pharma industry are pivotal in this ecosystem. Banks provide essential funding, insurance companies facilitate access to advanced detection methods and pharmacies ensure the authenticity and availability of critical medications.

While the benefits are substantial, challenges such as technological integration, regulatory compliance and the need for widespread adoption must be addressed. Future research should focus on overcoming these barriers, advancing nanotechnological innovations and developing more sophisticated blockchain applications.

Collaborative efforts of all stakeholders promise a future where healthcare systems are more robust, cancer detection is highly precise and patient outcomes are significantly improved, demonstrating the profound impact of blockchain technological convergence.

Insurance companies and banks are integral to the healthcare industry each playing distinct but complementary roles. Insurance companies provide financial protection, promote access to care and encourage preventive health measures. Banks offer the necessary capital and financial services to support the infrastructure, innovation and efficient operation of healthcare organizations. Together, they help to ensure a robust and sustainable healthcare system capable of meeting the needs of the population and advancing medical science.

Appendix

Blockchain code in solidity language:

```
// SPDX-License-Identifier: MIT
//Software Package Data Exchange
pragma solidity ^0.8.7;
//Solidity version

contract DataStorage {
string public data;

event DataChanged(string newData);

function setData(string memory _data) public {
data = _data;
emit DataChanged(_data);
}

function getData() public view returns (string memory) {
return data;
}
}
```

References

AbdelSalam FM (2023) Blockchain revolutionizing healthcare industry: a systematic review of blockchain technology benefits and threats. Perspect Health Inf Manag 20(3):1b. PMCID: PMC10701638

Abu-Elezz I, Hassan A, Nazeemudeen A, Househ M, Abd-Alrazaq A (2020) The benefits and threats of blockchain technology in healthcare: a scoping review. Int J Med Inform 142:104246. https://doi.org/10.1016/j.ijmedinf.2020.104246

Ahmad S (2021) Cybersecurity threats and attacks in healthcare. https://doi.org/10.13140/RG.2.2.17256.80647/2

Ali A, Pasha MF, Ali J, Fang OH, Masud M, Jurcut AD, Alzain MA (2022) Deep learning based homomorphic secure search-able encryption for keyword search in blockchain healthcare system: a novel approach to cryptography. Sensors 22(2):528. https://doi.org/10.3390/s22020528

Andrew J, Isravel DP, Sagayam KM, Bhushan B, Sei Y, Eunice J (2023) Blockchain for healthcare systems: architecture, security challenges, trends and future directions. J Netw Comput Appl 215:103633. ISSN 1084–8045. https://doi.org/10.1016/j.jnca.2023.103633.

Baysal MV, Özcan-Top Ö, Betin-Can A (2023) Blockchain technology applications in the health domain: a multivocal literature review. J Supercomput 79(3):3112–3156. https://doi.org/10.1007/s11227-022-04772-1

Cointelegraph (n.d.) What is a smart contract, and how does it work?. Cointelegraph. https://cointelegraph.com/learn/what-are-smart-contracts-a-beginners-guide-to-automated-agreements

Cremer F, Sheehan B, Fortmann M, Kia AN, Mullins M, Murphy F, Materne S (2022) Cyber risk and cybersecurity: a systematic review of data availability. Geneva Pap Risk Insur Issues Pract. 47(3):698–736. https://doi.org/10.1057/s41288-022-00266-6. Epub 2022 Feb 17. PMID: 35194352; PMCID: PMC8853293.

Dinakaran D, Wilson BC (2023) The use of nanomaterials in advancing photodynamic therapy (PDT) for deep-seated tumors and synergy with radiotherapy. Front Bioeng Biotechnol 11:1250804. https://doi.org/10.1049/htl2.12008

Eldridge L (2023) Cancer cells vs. normal cells: How are they different?. https://www.verywellhealth.com/cancer-cells-vs-normal-cells-2248794

Gordon WJ, Catalini C (2018) Blockchain technology for healthcare: facilitating the transition to patient-driven interoperability. Comput Struct Biotechnol J 16:224–230. https://doi.org/10.1016/j.csbj.2018.06.003

Haleem A, Javaid M, Singh RP, Suman R, Rab S (2021) Blockchain technology applications in healthcare: an overview. Int J Intell Netw 2:130–139. ISSN 2666–6030. https://doi.org/10.1016/j.ijin.2021.09.005.

Jia S, Zhang R, Li Z, Li J (2017) Clinical and biological significance of circulating tumor cells, circulating tumor DNA and exosomes as biomarkers in colorectal cancer. Oncotarget 8(33):55632–55645. https://doi.org/10.18632/oncotarget.17184

Jin C, Wang K, Oppong-Gyebi A, Hu J (2020) Application of nanotechnology in cancer diagnosis and therapy—a mini-review. Int J Med Sci 17(18):2964–2973. https://doi.org/10.7150/ijms.49801

Kasyapa MSB, Vanmathi C (2024) Blockchain integration in healthcare: a comprehensive investigation of use cases, performance issues, and mitigation strategies. Front Digit Health 6:1359858. https://doi.org/10.3389/fdgth.2024.1359858

Kemp JA, Kwon YJ (2021) Cancer nanotechnology: current status and perspectives. Nano Converg 8(1):34. https://doi.org/10.1186/s40580-021-00282-7

Khatoon A (2020) A blockchain-based smart contract system for healthcare management. Electronics 9(1):94. https://doi.org/10.3390/electronics9010094

Marbouh D, Simsekler MCE, Salah K, Jayaraman R, Ellahham S (2022) Blockchain for patient safety: use cases. Oppor Open Chall Data 7(12):182. https://doi.org/10.3390/data7120182

Padmaja VM, Richelle MR, Pei LY, Dusan L, Mahaveer DK (2024) Unveiling cutting-edge advances in high surface area porous materials for the efficient removal of toxic metal ions from water. Prog Mater Sci 146:101314. ISSN 0079–6425. https://doi.org/10.1016/j.pmatsci.2024.101314

Pavelić K, Pavelić SK, Bulog A, Agaj A, Rojnić B, Čolić M, Trivanović D (2023) Nanoparticles in medicine: current status in cancer treatmen. Int J Mol Sci 24(16):12827. https://doi.org/10.3390/ijms241612827

Shilpa P, Apte M (2022) A study of security issues in blockchain and its preventive measures. Praman Res J 12(4):10089. ISSN 2249–2276

Sookhak M, Jabbarpour MR, Safa NS, Yu FR (2021) Blockchain and smart contract for access control in healthcare: a survey, issues and challenges, and open issues. J Netw Comput Appl 178:102950. ISSN 1084–8045. https://doi.org/10.1016/j.jnca.2020.102950.1

Stopfer L, Kaulen A, Purfürst T (2024) Potential of blockchain technology in wood supply chains. Comput Electron Agric 216:108496. ISSN 0168–1699. https://doi.org/10.1016/j.compag.2023.108496.

Verywell Health (2023) Cancer cells vs. normal cells: How are they different? Accessed on 5 December 2023. https://www.verywellhealth.com/cancer-cells-vs-normal-cells-2248794

World Health Organization (2020) Homepage—IARC. World Health Organization. https://www.iarc.who.int

Zhang Y, Li M, Gao X et al (2019) Nanotechnology in cancer diagnosis: progress, challenges and opportunities. J Hematol Oncol 12:137. https://doi.org/10.1186/s13045-019-0833-3

Zhao W (2019) Blockchain technology: development and prospects. Natl Sci Rev 6(2):369–373. https://doi.org/10.1093/nsr/nwy133

Revolutionizing Blockchain Healthcare System Engineering Using Artificial Intelligence

S. Muthukaruppasamy, G. Arun Sampaul Thomos, K. Saravanan, Seifedine Kadry, and Selvakumar Manickam

Abstract This article explores numerous dimensions of blockchain and artificial intelligence (AI) in healthcare engineering. The significance of machine learning in comprehending data patterns, doing analysis, and making informed judgments has been demonstrated across diverse industries. To make accurate conclusions, machine Learning necessitates enough data. The importance of data sharing and the dependability of data cannot be overstated in machine learning since both factors play a critical role in enhancing the accuracy of machine learning models. The decentralized database within the context of blockchain technology places significant emphasis on the data exchange. The prevailing agreement within the blockchain technology ensures the authenticity and security of data. The fusion of these two technologies can provide accurate results in machine learning, while guaranteeing the security and reliability provided by blockchain technology. Using blockchain with artificial intelligence, experts of healthcare management can get ingress for exhibiting the medical information about the patient. Mainly, AI implementation on blockchain can provide a remarkable algorithm to give a final optimistic conclusion in healthcare management. Through this current advancement of technologies, the healthcare system will get improvised care of patient efficiency at a diminished price. Blockchain with AI enhances capacity to store more secure information cryptographically and segregates data history for sharing. This study aims to provide a solid foundational understanding of these domains and thoroughly explore the potential benefits, challenges of data security, privacy issues, resource restraint, holding ability, data repository, and short

S. Muthukaruppasamy (✉)
Department of EEE, Velammal Institute of Technology, Panchetti, Tiruvallur Dt, India
e-mail: mksamy14@yahoo.com

G. Arun Sampaul Thomos
Department of AI, Anurag University, Hyderabad, India

K. Saravanan
Department of CSE, College of Engineering Guindy, Anna University, Chennai, India

S. Kadry
Department of Applied Data Science, Noroff University College, Kristian Sand, Norway

S. Manickam
National Advanced IPv6 Centre, Universiti Sains Malaysia, Penang, Malaysia

S. Krishnan et al. (eds.), *Unveiling Blockchain Enabled Business Applications*, Contributions to Management Science, https://doi.org/10.1007/978-981-95-2337-5_9

fall of standards, compatibility, regulations, and future research directions in the finest approach.

1 Introduction

Physicians, individuals, medical observatories, labs, clinics, computerized medical records, and patient's databases are part of contemporary health care systems. Additionally, they could involve blockchain to show patients' medical records. These amenities all need key-enabler technology to be controlled and managed. Implantable and wearable sensors are just some of the medical advancements made possible by digital innovation. With AI, we will be able to take things one step higher and have associated software that will offer knowledge and support for decision-making. Artificial Intelligence refers to the interdisciplinary field of study and engineering focused on developing intelligent computer systems. These systems are designed to effectively mitigate the workplace stress experienced by individuals without requiring external guidance or assistance. Artificial intelligence is capable of intricate computing tasks and effectively analyze extensive patient information, thereby exerting a beneficial impact on a patient's overall health and wellness. Artificial intelligence can identify early warning signs of disease by performing several cognitive processes. This observation demonstrates that AI has significantly boosted the medical community's ability to interpret unstructured and structured patient data for improved diagnosis. Artificial intelligence is used to analyze large amount of patient data, allowing for more precise sequencing and the development of highly individualized treatments (Raul et al. 2023; Shinde et al. 2022). Blockchain technology is a significant factor in the healthcare sector as it serves as a valuable tool for developing tailored, trustworthy, reliable, and safe medical services (Shine et al. 2023). Additionally, it enables the maintenance of accurate and confidential clinical records, thereby enhancing and monitoring of patients within a healthcare system. One of the several advantages of blockchain technology is the provision of a decentralized system, which empowers end users with enhanced transparency and control over the utilization of their data. The potential of blockchain technology to make substantial changes in the healthcare sector was observed during the global pandemic, as it brought novel approaches to securely conserving and handling healthcare records (Mohey Eldin et al. 2023). Novel developments like artificial intelligence (AI) and blockchain technology have the potential to impact the future of healthcare significantly. Healthcare providers can use blockchain to show patients' medical records. They make blockchain more accessible for extraction and follow-up. They can also use AI programs to help them make better decisions or analyze a lot of data to get the best results (Ahmad et al. 2022). The key components of blockchain and AI are outlined here.

(i) ***Explicit Data Source***

A lot of data is required to train an AI program. The blockchain is a trustworthy repository of improved data as it is the most transparent technology. The exact location of nodes makes it possible to verify the data's origins rapidly.

(ii) ***Self-governing Framework & Protection of Privacy***

The distributed ledger structure of AI applications assures no server handles all tasks. The decentralized management of AI operations and training is driven by the self- governing framework, which operates independently of supervision. The use of cryptography fortifies AI operations and learning. Because AI systems are more complicated and competitive, we can train and provide when we have strong privacy structure.

(iii) ***Distributed Processing Power & Securabality***

AI training and maintenance demand massive computing resources. Blockchain technology establishes the responsibility and facilitates its fulfillment. It also handles the space needs for storage, maintenance expenses, equipment, and software. The smart contracts on blockchain are insufficiently safe. The blockchain is based on the strictness of the contract; if there are any gaps, we may simply take advantage of them to damage the apps. AI is used to create secure contracts in order to reduce such weaknesses.

(iv) ***Accession***

AI examines data quickly and thoroughly. It also adds greater cognitive capacity to blockchain based enterprise systems, enabling them to comprehend and analyze data exponentially. Blockchain can help to expand AI for offering relevant insights by giving it access to bigger data sets, govern model sharing and data consumption, and establish a reliable and honest information market. By combining the latest developments in both fields, healthcare systems could enhance service efficiency, lower costs, and make healthcare more accessible to more people. The combo technologies such as blockchain and artificial intelligence (AI) can potentially modify healthcare systems and improve their efficiency and service quality. Incorporating several contemporary technologies, including artificial intelligence (AI) and blockchain, can provide an automated, intelligent, secure, and cost-effective computational environment. This environment forms the foundation of healthcare and has the capacity to enhance the functionality of the conventional healthcare system.

2 Role of Blockchain in Healthcare System

The blockchain has the potential to bestow a secure platform for enterprises to submit and share data by maintaining a consistent data chain and utilizing encoded links to store independent data such as radiography and other pictorial data. It enables the blockchain to function as a single transaction layer for secure data exchange. The utilization of smart contracts and established protocols could significantly facilitate the requirement of uninterrupted connectedness. From 2008 to 2023, an estimated 176 million incidents of breaches of data pertaining to healthcare records were recorded (Ben and Lahami 2020). The secure attributes of blockchain technology have the potential to enhance significantly the guarantee of health and fitness data. Every individual possesses a distinct public identifier, and an individual key that is exclusively accessible throughout the necessary timeframe. Moreover, the imposition of targeting particular users as a prerequisite for acquiring confidential data would restrict the practice of hacking. Consequently, blockchain can offer an unalterable record of medical data.

2.1 Medical Supply Tracking and Security

Blockchain technology always provides full disclosure and could secure and track the supply chain for medication. It can track the cost of labor and carbon emissions that are linked with the goods.

2.2 Storing Medical Data and Trailing Disease

The system has significant development in archiving medical records, but neither transactions nor guarantees are up to par. Due to the proliferation of system hackers around the globe, hacking of health records has replaced all other ways as the most common method of data breach. The software company, Meditab asserts that it is among the best providers of Electronic Medical Record (EMR) systems to healthcare facilities (see "Best EMR Software Company | Medical Software | EHR | IMS | Meditab"). Electronic faxes are still commonly used to share information about individual patients. At the Dubai-based cyber security firm Spider Silk, an Elastic search database comprising around 5.5 million patient repositories was run on a fax server (see "Cyber security firm Spider Silk- Home"). So, individuals could monitor live fax transmissions. Patients' names, addresses, phone numbers, fax numbers, and even social security numbers were all included in the transmitted faxes. The special properties of blockchain could facilitate present disease detail and disease model study, which in turn could help to identify the cause and mode of disease transmission.

3 Role of Artificial Intelligence in Healthcare System

Artificial Intelligence (AI) is intrinsically connected to developing novel computing systems. Most of current computer science can trace its roots back to Alan Turing and the work describes the World War II-German Enigma machine, which was the foundation for machine learning. It is designed to see if AI has reached the level of human brilliance (Toh and Brody 2021).

3.1 Machine Learning in Healthcare System

Machine learning, a subfield of AI, has found extensive use in the medical field. Due to the advent of the digital age, medical and healthcare industries have generated copious amounts of data (Klra 2016). This data includes patient data, records, and stages of medication. Research into machine learning has shifted primary attention to the medical industry, where it is being used to analyze complex medical data. The research analyzes how the healthcare industry utilizes neural network, deep learning for data examination and data mining. The ultimate aim is to develop smart medicine by developing methods for detecting human behavior, monitoring health, predicting sickness, and diagnosing illness. Machine learning is utilized to identify human actions. For this purpose, they devised a sensory deep learning system for identifying human actions (Gmaei et al. 2019). This system could function with multisensory data more brilliantly, assisting hospitals in providing care for the sick and old. A Machine learning system is implemented within the patient's health supervising system, collecting information through the Internet of Things. It analyzed the information by machine learning to track patient's health information. To forecast cardiac failure and produce circumstances for efficient medication, a disease prediction system is created using a combination of deep learning and the information gain technique (Ali et al. 2020).

3.2 Supervised Learning

Training data in supervised learning often takes the format of data with labels. The training data's labeled output has one or more inputs. Models employ labeled results to evaluate their performance during training to enhance their capacity to forecast fresh data. Classification and regression techniques are the heart of supervised learning systems. Classification problems arise frequently in the medical field. Clinicians typically make diagnosis by referring patient's reported symptoms. Forecasting quantitative outcomes from a group of data, such as vital indications, is a common theme in regression competitions (see "scikit-learn: machine learning in Python—scikit-learn 1.0.1 documentation"). A random forest consists of several independent decision trees. The estimates from multiple trees are typically averaged for more accurate results and predictions (see "sklearn.ensemble.RandomForestClassifier—scikit-learn 1.0.1 documentation"). A random collection of characteristics is selected. It enhances the model's ability to generalize. The model is constructed by integrating the outcomes of multiple trees. It will provide a forecasting that could be evaluated by test (see "Blockchain Applications in Healthcare").

3.3 Unsupervised Learning

Finding patterns in data without labels is the goal of unsupervised learning. Commonly, these algorithms are quite good at assembling data into meaningful groups, which facilitates finding concealed features. Nevertheless, they are more data and produce less accurate results. Deep learning and clustering is the common and widely recognized methods. But deep learning is utilized in a controlled way. Such algorithms also carry out an association function and they are analogous to clustering. These algorithms are unsupervised since they choose clustering features without human interaction.

4 Convergence of Blockchain and AI in Healthcare Industry

Blockchain and AI are potential technologies, which has changed the healthcare industry as a revolutionary field. The advent of blockchain and AI is creating exciting new possibilities in the healthcare sector, benefiting patients and therapists. When used together, both technologies can help healthcare providers to keep patients' personal information and medical records in protected way. Artificial intelligence can analyze patient records, gain knowledge from insights, and uncover hidden trends. Moreover, the blockchain can safely store sensitive patient information and health records while providing convenient access to them. Integrating the two systems is the most effective way to fix the problems with AI and blockchain technology (Campbell 2018). Artificial intelligence algorithms require data to train, understand, and create conclusion. Machine learning system functions well when information comes through a trusted source. A blockchain is a distributed ledger which securely keeps and trades information using cryptographic hash functions. It is verified by network nodes. The data kept on a blockchain has a high degree of probity and wholesomeness, also it is impossible to alter it. Smart contracts can be relied upon, and their findings accepted when they use machine learning algorithms to get conclusions and conduct examinations. When AI and blockchain are used together, the sensitive data is protected in a decentralized and irreversible system (Dubovitskaya et al. 2020).

Medical, private, trading, and legal data are the types of information, which is benefited from this method (Mackey et al. 2019). Intelligent decentralized autonomous agents can swiftly and easily validate data and asset transfers among various authorities when they are integrated with one another (Franklin and Muthukumar 2020). Artificial intelligence helps to spot trends in data. Strategic analytics using AI could strengthen operations and provide patient lifestyle and health information. Learning AI entails good understanding of algorithms. A computer determines the diagnosis, suggests treatments, and how long a patient should stay in the hospital when they present with a potentially fatal illness. The computer's precision comes from its ability to analyze a lot of data quickly. It will rely heavily on Google's formidable computing resources. Microsoft's AI technology, Hanover, uses AI to harvest data in the healthcare industry. Using this technique, machines could learn from medical research publications and then use that knowledge to identify new treatments for wide range of patients. It analyzes the individual's health records, compares them to data from published medical studies, and suggests treatments (Ismail et al. 2021). DeepMind, Google's AI division, is working on a healthcare data auditing system that utilizes blockchain and AI to facilitate real-time patient data tracking by hospitals and individuals. It will cooperate to keep things running efficiently. The blockchain will guarantee the system's safety, and the data can be transferred quickly and easily. Artificial intelligence will allow doctors to access prognosis and analytics based on individual patients' records. DeepMind is already experimenting on degenerative eye problems using block chain technology with artificial intelligence. This application is used to detect early warning signals of degenerative eye problems.

4.1 How Can Blockchain and AI Become a Great Trouble Shooter?

In industrialized nations, healthcare expenditures constitute a sizable percentage of GDP (Gross Domestic Product). However, inefficient practices and health data breaches within hospitals, contribute to ever-increasing costs. Blockchain technology could be a potential one for achieving efficiency gain. Reliable encryption of medical data and epidemic control are the best achievements of blockchain technology. For instance, Estonia was an early adopter in 2012. At the moment, 97% of health data and 98% of prescription information are all stored on the blockchain.

The paradigm of applying blockchain and AI in the healthcare sector is illustrated in Fig. 9.1. Each participant in the blockchain network will receive a valid certificate from a trusted certificate authority. It will serve as a form of identification for anyone conducting business on the network. The digital certificate serves as identification verification. After digitally signing the transaction, the user will submit it to the blockchain. The advantage of signing is that it enables users to confirm their identity on the blockchain before engaging in a transaction with them and confirming that the user is authorized to see the ledger in the light of the transaction.

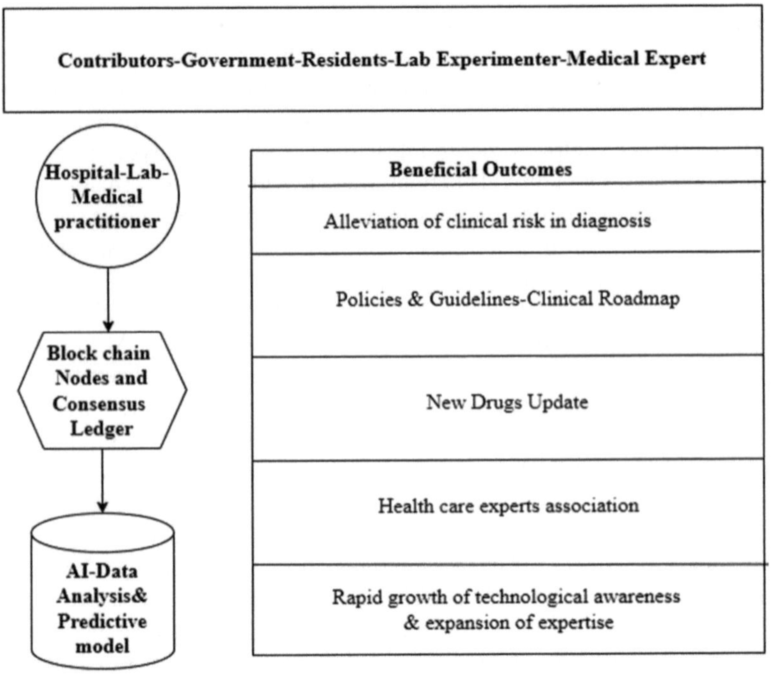

Fig. 9.1 Structure of Blockchain with Artificial Intelligence in Smart Healthcare Framework

The potential of blockchain and AI as a problem-solving tool performs the following functions:

- All authorized users will have access to a shared copy of the ledger. The problem with collecting data will now be solved. Directly feeding highly trustworthy data into machine learning models allows for easy extracting actionable insights.
- The model can be trained with actual data. It will lead to enhanced model efficiency and precision at a reduced cost.
- The patient might seek guidance on how to adopt healthier habits. Clinicians' recommendations can be used to train the model for treating patients with comparable conditions.
- A simple health inquiry from a patient triggers the posting of a competent model with an accompanying test to the blockchain model. This model is used to predict epidemics and provide advice to physicians. Diseases can be identified, and treatment is given using Natural Language Processing. According to patient's symptoms, the perfect trained model can provide medical guidance to specialists.
- It can train the model to predict epidemics. The algorithm can detect an epidemic and provide suggestions to the doctor; for instance, if a patient goes through testing, the data is sent to the blockchain network.
- The healthcare business depends on a wide range of machinery and equipment to carry out diagnostic procedures. There is an expected lifetime for any machine or

machine part. It can also foresee when maintenance on the machine or a specific component is due.

4.2 *Use Cases of Blockchain and AI in Healthcare*

In order to give a holistic and unified picture of a patient's record, blockchain technology facilitates the creation of electronic health records of patient. Patients can track the status of their data and offer explicit permission before each time they are sent. Insurance companies can bypass middlemen and acquire verified patient information quickly from these data. Other blockchain use cases are below.

- Supply chain transparency, wherein the immutable ledger of the system could aid in confirming the validity of medical items by tracing their origins.
- Improved remote patient monitoring is possible through blockchain cryptography, which ensures that only authorized parties have access to the personal information of the sick person.

Artificial intelligence transforms healthcare from a theoretical possibility to a practical reality across a wide range of medical disciplines. By analyzing data trends on time, AI's deep learning, machine learning, and natural language processing capabilities help medical experts and stakeholders to identify healthcare needs and solutions more precisely. Use cases of artificial intelligence in healthcare are below.

- AI simplifies the decision-making process for healthcare providers by analyzing massive volumes of data and various healthcare organization use it to determine which cases should be prioritized to reduce the likelihood of mistakes.
- Instead of starting from scratch, pharmaceutical companies can use artificial intelligence to construct streamlined drug development platforms that aid in repurposing and repositioning current medications.
- By analyzing both verbal and nonverbal signals, AI can help emergency medical experts to make a timely diagnosis that improves a patient's chances of survival.

This advancement can help to improve the healthcare industry to increase the patient satisfaction and modernize the system. The potential benefits of the blockchain and AI confluence on healthcare in the following areas are shown in Table 9.1.

(i) ***Data Management***

Specific blockchain implementations can be perfect for preserving private data, which may subsequently be used by powerful AI models to assess health data and make precise diagnoses using medical tests and records. Furthermore, innovative encryption methods, like homomorphic encryption, allow calculations to be performed on this data while maintaining data privacy. AI and blockchain technology can improve data management, privacy by allowing for the safe storage and exchange of patient records, medical research data, and other sensitive information. This can enable healthcare and rejuvenation researchers to work efficiently from various physical

Table 9.1 Distinct characteristics of AI and blockchain technologies

Qualities of AI	Qualities of Blockchain	Confluence of Blockchain and AI-Advantages
Unified and Centralized	Sggregated and Decentralized	Greater confidentiality of information
Changeable	Determinative	A greater trust in automated decisions
Random	Unalterable	Creating decisions that are supported by evidence
Unstable	Data reliability	Dispersed Intelligence
Information, expertise, and making decisions revolve around data	Springy attacks	Optimal efficiency

places while maintaining the greatest levels of data management and privacy. By using blockchain technology as a data storage basis, AI-driven diagnostic tools and specific treatment plans can be created with more data protection, resulting at effective and customized system for healthcare.

(ii) ***Encrypted and Compatible Records***

Clinical transcripts, medical imaging, and health records are among the types of medical data that AI can interpret and analyze. Patient data could be safely kept, shared across healthcare professionals, and safeguarded. While AI algorithms offer fast data analysis for better diagnosis, treatment planning, and research, blockchain ensures data integrity, traceability, and patient consent management. A patient's medical records accumulate more data with time, therefore it's critical to have the resources for protection.

(iii) ***Authentication and Traceability of Drugs***

It is necessary for the public to make, consume, and carry medication in a safe manner. Blockchain technology can aid in the fight against fake medications by verifying goods and tracking their journey from producer to consumer. It can be used in supply chain management to monitor costs, labor, and emissions of waste.

(iv) ***Clinical Investigations***

Researchers are always searching for novel drugs that they may provide to their patients. AI systems can expedite the process of finding new medications. Clinical studies, genetic data, and a plethora of scientific literature are all analyzed in the process. Clinical trial data may be made more valid and transparent by combining AI and blockchain. Blockchain-based smart contracts could streamline trial procedures including data exchange, consent management, and participant recruiting while protecting patient privacy.

(v) ***Therapy and Diagnostics***

AI can help healthcare workers to make diagnosis. Algorithms could analyze treatment images and predict patient results. The blockchain allows for the safe storage, sharing, and access of data used to train AI models. It also keeps an auditable record of AI model training and validation, which ensures openness and confidence in decision-making.

(vi) ***Patients' Data and Telemedicine***

Medical data gathered outside of medical situations has grown since the introduction of new technologies like telemedicine and health tracking devices. AI can analyze this acquired data to generate insights that are specific to each individual and their data. Blockchain allows for safe storage and exchange of this data, ensuring patient privacy.

(vii) ***Billing and Claims Administration***

AI-powered chatbots could aid with administrative tasks like billing, claims administration, and verification of insurance. Blockchain technology can assist to guarantee that these procedures are secured. It can assist with faster and more accurate refunds.

(viii) ***Acoustics Healthcare System***

Acoustics has seen the most current advances and revolutionary potential in machine learning, particularly deep learning. Unlike classical acoustics and signal processing, machine learning is driven by data. Thus, acoustic AI has significant promise for analyzing acoustic inputs in applications related to healthcare (Rufo et al. 2022). Acoustic medical data could be maintained on individual devices and use federated learning using Blockchain infrastructure to safeguard acoustic based on artificial intelligence applications in healthcare from hostile threats.

5 Real Mechanism of AI-Moderated Healthcare Data Exchange on the Blockchain

The integration of AI with blockchain technology has the ability to strengthen the fields of biomedical research and healthcare. Decentralized approach is provided for calculating the cost and benefit of sharing health records on the blockchain with the help of artificial intelligence. High-level overview of how artificial intelligence and blockchain technologies are applied to speed up biomedical review, and enhance predictive methods. Data collection can be made easier with the help of an AI-blockchain based system. It ensures equitable tracking of all data consumption activities and facilitates the beneficiary to send data to the computing system, sanctioning permission to use patient's data. The healthcare field can also benefit greatly from the integrated aspects of these technologies. Socially supportive robots can be useful in assisting with the care of elder people. One of the most promising technologies, socially assistive robots could operate as a juncture and assess the requirements of sick patients, which are given the growing demand for elder care. By providing assistance with routine exercise, lowering stress levels, and personal caregiving, robots want to foster a pleasant beneficiary experience to inspire patient, and increase their quality of life. There are serious challenges of infusion with respect to data collection and storage capacity, despite AI algorithms are effective at categorizing and analyzing massive datasets in the healthcare industry. Trusting a robotic agent's judgment or action during a crucial decision-making process poses a risk in the medical profession, where all medical data must be correct and safe. When it is applied to healthcare, blockchain and AI have the potential to improve trust in robotic decisions. It provides more individualized care based on a patient's trauma, location, climate, and previous medical conditions. Figure 9.2 provides high-level overview of the diagnostic, analytical, crucial decision-making and medical test report validation processes that could be aided by combining AI and blockchain technology in the medical profession.

In order to gather information, categorize patterns, and make predictions for future medication, graph database is used by the machine learning algorithm. The capacity of the Bayesian network to control probability and prognosis is crucial to its success. It can be used effectively to anticipate health data. An immutable transaction log can be used to record the interactions between the entities in a healthcare graph database, including doctors, drug manufacturers, patients, and so on, as well as the methods used to treat patients and their medications. Research in artificial intelligence and machine learning has shifted its focus to managing massive data sets, the exponential

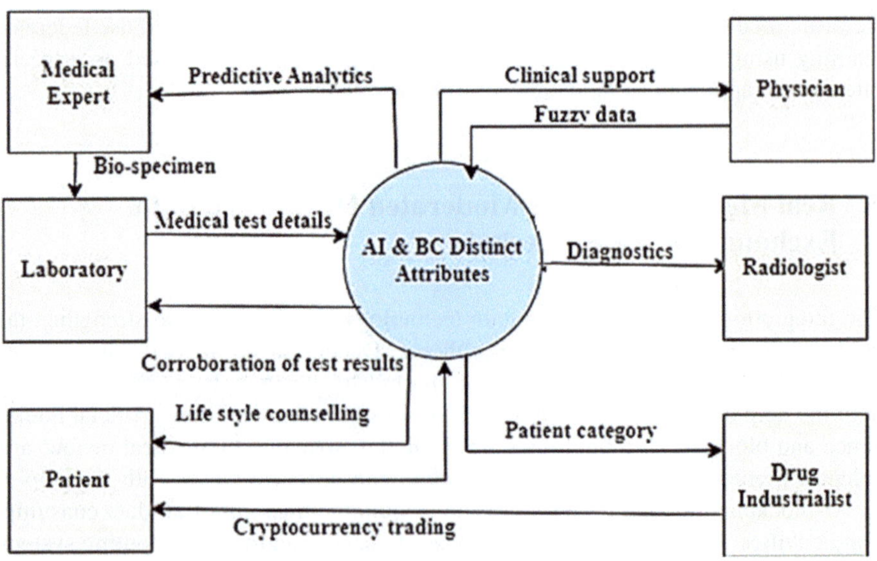

Fig. 9.2 Decentralized healthcare system

expansion of computer power (Ismail et al. 2022). Due to the massive amounts of data and processing power needed to train an artificial neural network, significant investments have been made in the development of high-performance data centers in order to amass such data (Tarafder et al. 2023). In order to combat the security dangers posed by the proliferation of bots and human-bot interactions on the Internet, Woods stresses the significance of merging AI approaches and blockchain infrastructure (Ismail et al. 2022). As the number of bots grows, it is anticipated that bot-bot communications may soon surpass those between humans and machines. Before any encounters, bots will need to be able to ask each other for identity and access their data and rating histories. Recording audit-related inquiries and data on a distributed ledger (blockchain) improves transparency and security. Combining blockchain and machine learning offers tremendous promise for rebalancing and bettering machine learning algorithms since it provides powerful, decentralized system, and self-sovereign identification.

6 Advancements and Applications: Tangible Health Sector Examples in the Confluence of AI and Blockchain Technologies

The healthcare industry has several technical advancements. The integration of AI and blockchain in healthcare has immense potential to drive positive transformations. We can realize the full potential of AI and blockchain through following advancement in healthcare industry which paves the way for a brighter and healthier future. The combination can provide the following advancement in healthcare sector.

6.1 Integration of AI Technology and Blockchain Technology in the Cancer Care Survivorship

Clinical oncology of artificial intelligence is one example of technological progress, providing new avenues for diagnosing disease and finding ways to lessen its impact. AI simulates intelligent behavior with little human interaction. Categorizing patients into suitable risk categories for improved therapy is a significant challenge in cancer care. The use of machine learning could result in solution to this significant problem in cancer management. Because, machine learning approaches use a wide range of analytical, prospective, and optimal ways to teach systems to acquire knowledge and find pattern from big and complicated oncology datasets. It is an appropriate tool for categorizing persons into more risk and fewer risk categories. Support vector machine and decision trees are just a few machine learning techniques. Those machine learning techniques are applied in cancer diagnosis and prognosis (Dlamini et al. 2020). When it comes to dealing with massive amounts of data that could be incomplete or corrupted, machine learning provides advantages over more conventional statistical approaches of prediction. However, utilizing machine learning approaches for the status of oncology and prognostication could enable personalized therapy and improve the treatment process of cancer care, even though entry to a comprehensive history of oncology information is limited owing to increased sick person movement among many infirmaries. Nevertheless, siloed data islands controlled by multiple medical organizations are the main obstacle to attaining personalized medicine or nursing. The most effective cancer care comes from early diagnosis, fast treatment, and close monitoring. Blockchain technology is a useful resource for storing healthcare data for the purposes of sharing, trading, and analyzing the data across various service providers. Data verifiability, transparency, and trust are only some advantages of using blockchain in cancer applications, along with decentralization, increased data security and privacy, patient ownership of clinical data, and patient empowerment. Since blockchain is a novel digital system, regarded as a decentralized ledger to guarantee the durability, trackability, and administration of healthcare information. There have been several efforts to apply it to the generation of full history of oncology patients (Dubovitskaya et al. 2020). The digital infrastructure provided by the blockchain could also serve as an interface to other forms of AI, such as machine learning (Mackey et al. 2019). Therefore, blockchain is comprehensive and adaptable, making it suitable for use in a wide range of cancer settings. When both technologies are combined, machine learning and blockchain improves data security and transparency, allowing oncology experts to uncover the hidden data islands. So, machine learning and blockchain technology nourishes the healthcare sector by integrating two data-driven innovations in cancer survivorship care.

6.2 Heart Diseases Prediction Using Blockchain and Machine Learning

Heart disease is the leading cause of death globally. The lack of a developed healthcare department infrastructure that can offer a secure manner of storing and transmitting data, which is the primary cause of the dramatic rise in the mortality rate related to heart disease (Franklin and Muthukumar 2020). Cardiac professionals have a hard time getting accurate early diagnoses due to data redundancy.

Monitoring the essential indications in the early phases, such as blood pressure, cholesterol level, body weight, and smoking addiction, can limit this quick rise in the mortality rate owing to heart disease. The sophisticated architecture allows cardiac professionals to get patient data in hospitals and clinics. In terms of trustworthiness, blockchain is unparalleled. Currently, the most accurate method for predicting a heart disease using machine learning algorithm is a sine–cosine weighted k-nearest neighbor. Blockchain technology has been employed in the study to ensure the safety of the data collection during the session and provide more reliable findings. This approach has been used to enhance the suggested dataset, improving the system's performance.

6.2.1 Mechanism for Patient Data Using Blockchain

Blockchain technology has been used in a number of different ways to manage medical records. Information about the patient will be maintained in linked blocks of the blockchain. The new patient's information will be appended to the existing data block. A key will be produced after the data has been given as input and spread over the blockchain network. After receiving consent from the patient, the cardiac professional and the patient can use the key to analyze the data and make illness predictions.

6.2.2 Managing Patient Record Using Blockchain

Once a block is full, any further patient information is placed in a new block, and the two blocks are linked together. The patient's information is spread out throughout the blockchain and secured using a key, stored in each block. Using inputs from patients and healthcare providers, patient's likelihood of developing heart disease has been predicted using machine learning algorithms to choose the most appropriate class classification.

6.3 Prognostication of Diabetes Mellitus Using Artificial Intelligence Blockchain

In Fig. 9.3, the entire architecture of diabetes prediction is shown. The following components are the primary building blocks.

(a) ***Diabetes Factors Monitoring***: The user engages in communication with the program in order to ascertain their susceptibility to the occurrence of diabetes. A user is linked to various health trackers and sensors, including a glucose monitor, blood pressure monitor, sleep tracker, and activity tracker. These sensors with gadgets collect and transmit the user's health data in relation to diabetes risk factors.

(b) ***Edge Server***: Edge servers convert risk factors data into a machine learning algorithm format. Then, data is transferred to blockchain network.

(c) ***Artificial Intelligence Diabetes Forecasting***: Glucose level, blood pressure, sleeping patterns, physical activity, smoking status, and alcohol intake are just a few diabetes risk indicators that can be uploaded by external users and used by artificial intelligence. The user can enter this information directly into the mobile application, or utilize information stored on the blockchain. In order to determine if

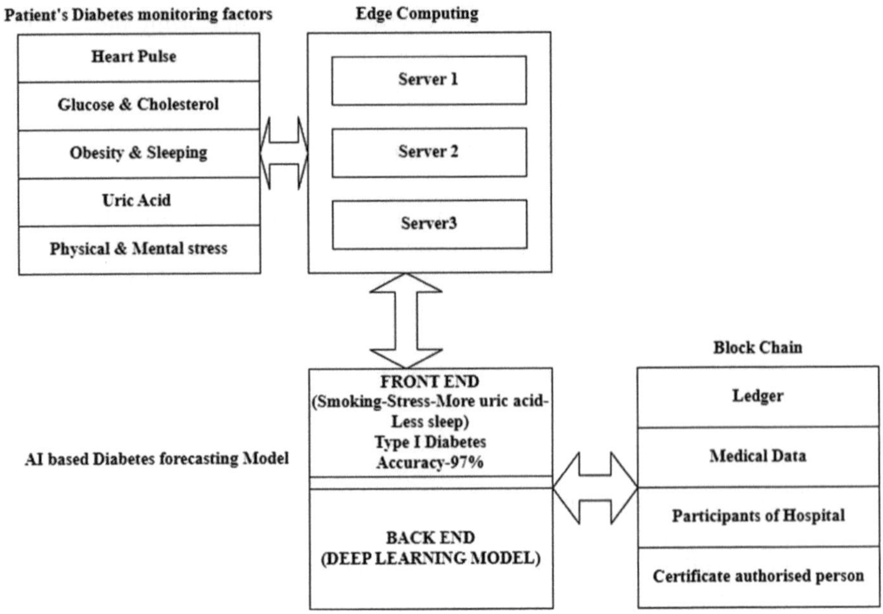

Fig. 9.3 Diabetes Mellitus Prediction: AI and Blockchain Monitoring System

a user is diabetic, the submitted data is processed by a machine learning model on the backend. Finally the predicted output is exhibited to the user.

(d) ***Blockchain***: It establishes decentralized peer-to-peer connections between all nodes in the network. Patients, pharmacologists, and physicians are all integral parts of this network. A certificate authority verifies the identities of all participants. Patients' health records and related information are entered into the distributed ledger system. Many medical facilities share a single copy of the ledger. Diabetes, characterized by an elevated glucose stage, is correlated with genetic, medicinal, and lifestyle factors (Ismail et al. 2021, 2022). The variables could be assessed by patients or external users, or through objective measurements, medical sensors, wearable gadgets and testing. The risk variables describe themselves, which include age, gender, ethnicity, smoking habits, alcohol intake, and a family history of diabetes.

6.4 Pulse's Role in the Blockchain-AI Revolution

PulseMed is at the forefront of blockchain with AI revolution by offering medical treatment. It is improved by AI and is powered by blockchain technology. Pulse solves a long-standing need in healthcare by providing easy access to the most recent diagnostic and pharmacological developments through centralized database that eliminates information barriers (Mosaddek et al. 2023). The AI-enabled diagnostics of pulse allow doctors to make more accurate diagnosis and provide better treatment in real time. Recorded data in a blockchain cannot be modified or faked since it is time stamped, cryptographically sealed, and immutable. It is the recommended next-gen solution. Prioritizing security in healthcare is essential to preventing unauthorized access to personal health information and ensuring the integrity of financial and pharmaceutical transactions. PulseMed is assisting other healthcare institutions in realizing the benefits of blockchain technology. Increased productivity for doctors is made possible by Pulse's use of smart contracts to digitize administrative tasks including patient invoicing, prescriptions, and insurance claims. Diagnostic AI aids the platform in maximizing the value of data aggregation. Pulse is likewise at the forefront of this technological revolution. The accuracy of diagnoses will be improved by merging patients' genetic information and medical history with the latest medical research and data. This improved knowledge will be included into Pulse. Artificial intelligence will improve clinical procedures and the company's efforts to create a worldwide medical information network. Better treatment with faster and more accurate diagnoses is made possible by its AI-assisted diagnostic system. It will help to enhance healthcare data. As the healthcare sector speeds up its digitization and electronic health records initiatives, established institutions must step up to provide access to intelligent health. At PulseMed, information technology has improvised the features of the healthcare industry. Innovative healthcare improves and alters existing medical systems in terms of individualization, efficiency, and service delivery, which can be achieved by integrating cutting-edge technology like blockchain and AI.

6.5 Current Trends on Blockchain with AI in Healthcare Industry

The healthcare sector undergoes a significant transformation if blockchain technology and artificial intelligence (AI) are combined. Here are important current trends on blockchain in AI with reference to healthcare industry.

(i) ***Supply Chain Control:*** By tracking the movement of medicines and medical equipment along the supply chain, blockchain ensures transparency and thwarts the production of fake goods. This supply chain data may be analyzed by AI systems to find possible problems and optimize processes.

(ii) ***Clinical Studies and Investigation:*** Transparency and traceability is ensured by using blockchain-based smart contracts to automate and make simpler several elements of clinical studies. Artificial intelligence (AI) can help with the analysis of large clinical trial information, speeding up the process of finding patterns and possible medication candidates.

(iii) ***Customized Medicine:*** Genetic data about patients is safely stored on a blockchain, and artificial intelligence (AI) algorithms can use this data to create individualized treatment protocols according to a patient's genetic characteristics. This increases the effectiveness of medical treatments.

7 Open Challenges

Many sectors are revolutionized using blockchain and artificial intelligence technology in healthcare. All this kind of evolution comes at a price. Novel innovation and execution necessitate exorbitant working costs, a short fall in adaptability, and the storage of enormous data on systems.

7.1 Privacy Issues

Patients' personal and private information is stored in healthcare systems, so protecting privacy is the most important. When people share information in public blockchains, lot of private problems will arise (Attaran 2022). The fact that blockchain is decentralized, open and makes it a big threat to data. Blockchain joins different people in an organization, so it can be hard to figure out who illegally used the data. Distributed consensus techniques have been suggested to address blockchain privacy problems.

7.2 Resource Restraint, Holding Ability, and Data Repository

The prerequisites for effectively operating a blockchain and implementing AI on a system exhibit notable distinctions, necessitating substantial data storage and system capacity. A constraint accompanied by the present state of blockchain technology on the maximum number of transactions that may be executed inside the network in a given second. Furthermore, it should be noted that processing transactions at a large scale requires significant computational resources. Furthermore, the issue of excessive energy use has been a persistent problem associated with blockchain technology. The energy consumption associated with the process of accord and protocols might impede the effectiveness of operation. As each task necessitates a substantial volume of data to be scrutinized to arrive at closure or attain a definite output with enhanced predictive capability, the latest progression in data categorization has augmented the necessity of data storage. The inclusion of advanced big data tools is necessary at a higher cost than conventional options, and these tools also come with some limitations. Artificial intelligence systems also operate on hardware that has been optimized for enhanced performance (Reegu et al. 2021). The majority of AI architecture encompasses intricate computations, whereby several algorithms need the processing of substantial volumes of data throughout each iteration, resulting in delay. While the enhancement in network speed may indeed contribute to reducing latency, it is important to note that several algorithms need to complete a full cycle before they can proceed to analyze the subsequent data set.

7.3 *Blockchain Security and Threats*

Recent studies have emphasized the inherent vulnerabilities in blockchain technology about security. A dearth shortage of recent developments in blockchain protocols has resulted in a heightened susceptibility for attackers to scrutinize and identify vulnerabilities inside the current system (Kumar et al. 2018; Thomas et al. 2023). Blockchain technology is susceptible to a range of potential risks, including but not limited to data manipulation and unauthorized data access. The malevolent assailants can provide fabricated data, which not only impacts the process of AI but also fundamentally alters the ultimate result of the model. This presents a significant risk to the well-being and health of the customers.

7.4 *Shortfall of Standards, Compatibility, and Regulations*

Before formulating an operational strategy for blockchain and AI applications within the medical industry, it is essential to consult and consider the relevant regulatory statutes. The current advancement of artificial intelligence in the medical sector is mostly confined to academic research with a tiny number of practical applications being realized. This restricted implementation is largely attributed to challenges related to interoperability. There is also a concern over the appropriate placement of AI diagnostic systems in the healthcare industry. AI can be used in two ways within the medical field as a supportive tool for physicians or as a complete replacement for doctors (Shinde et al. 2021; Thomas et al. 2024). Nevertheless, the use of both approaches raises concerns over the potential risks associated with "dangerous AI". The issue of ownership poses a significant obstacle in the widespread use of blockchain technology within the healthcare industry. The existence of private and public blockchain raises a pertinent inquiry about data ownership inside each blockchain framework. The presence of an accountable authority for the technology is often lacking, resulting in a multitude of legal and regulatory challenges. Additionally, a lack of trust among many stakeholders and restricted open standards can impede the seamless flow of information across diverse health organizations. Furthermore, the initial implementation of a blockchain-based system incurs significant expenses, rendering it potentially unviable for modest healthcare institutions.

8 Future Research Directions

Further investigation is required for the model incorporating deep learning and blockchain technology, since the domain is rapidly advancing. The use of deep learning techniques has significant potential in identifying myopia and other internal bodily ailments. By integrating deep learning with blockchain technology, the precision and reliability of patients' information can be enhanced (Sharma and Kshetri 2020; Muthukaruppasamy and Thomas 2024; Krishnan et al. 2020, 2021, 2023). There is much potential for enhancing the security and privacy aspects of AI-driven blockchain technology. Futuristic exploration can have focal point on advanced hash algorithms and the development of a more robust verification system. The implementation of multi-step authentication across all nodes on the blockchain has both benefits and drawbacks. However, a conclusive determination cannot be made within the scope of this study inquiry. Emerging consensus protocols have the potential to enhance the reliability and security of the AI-powered blockchain system significantly. It is suggested that cooperative computing speeds up the process of calculating a block's hash. The use of a heterogeneous array of innovations has the potential to enhance the operational efficiency of AI-driven blockchain systems. Furthermore, exploring the integration of such systems with pre-existing infrastructures is a promising avenue for future study. Enhancing security measures and safeguarding privacy are crucial considerations for the successful commercial implementation of AI-driven blockchain technology in healthcare.

9 Conclusion

This article concludes a comprehensive study of assimilation of artificial intelligence with blockchain technology. Integration of both technologies focuses on their potential applications within the healthcare industry. Numerous research endeavors have been undertaken to explore the utilization of artificial intelligence and blockchain technology within this particular domain. Additionally, it offers a more comprehensive understanding of the potential enhancements that artificial intelligence can bring to the identification and prompt response in diagnosing illnesses. Furthermore, it highlights the potential structural improvements that blockchain technology could bring to ensuring the safety of healthcare information and the preservation of record. The study has extensively examined the many attributes of blockchain technology that enable it to provide robust security measures for health records and other sensitive data. Artificial intelligence facilitates the automation of data processing and scanning processes. The artificial intelligence with blockchain technology would provide automated, safe, and decentralized system. It facilitates the storage and analysis of crucial and sensitive medical information. By incorporating the advancements in recent technology, the medical field enhances efficiency, decreases expenses, and promotes equal access to healthcare.

References

Ahmad I, Asghar Z, Kumar T, Li G, ManzoorA MK, Shah SA et al (2022) Emerging technologies for next generation remote health care and assisted living. IEEE Access 10:56094–56132

Ali F, El-Sappagh S, Riazul Islam SM, Kwak D, Ali A, Imran M, Kwak K-S (2020) A smart healthcare monitoring system for heart disease prediction based on ensemble deep learning and feature fusion. Inf Fusion 63:208–222

Attaran M (2022) Blockchain technology in healthcare: challenges and opportunities. Int J Healthc Manag 15(1):70–83

Ben Fekih R, Lahami M (2020) Application of blockchain technology in healthcare: a comprehensive study. In: The impact of digital technologies on public health in developed and developing countries: 18th International Conference, ICOST 2020, Hammamet, Tunisia, June 24–26, 2020, Proceedings 18. Springer International Publishing, p 268–276

Best EMR Software Company I Medical Software I EHR I IMS I Meditab. https://www.meditab.com/

Blockchain Applications in Healthcare https://www.news-medical.net/health/Blockchain-Applications-in-Healthcare.aspx.

Campbell D (2018) Combining ai and blockchain to push frontiers in healthcare. http://www.macadamian.com/2018/03/16/combining-ai-andblockchain-in-healthcare/

Cyber security firm Spider Silk- Home. https://spidersilk.com/

Dlamini Z, Francies FZ, Hull R, Marima R (2020) Artificial intelligence (AI) and big data in cancer and precision oncology. Comput Struct Biotechnol J 18:2300–2311

Dubovitskaya A, Novotny P, Xu Z, Wang F (2020) Applications of blockchain technology for data-sharing in oncology: results from a systematic literature review. Oncology 98(6):403–411

Franklin RG, Muthukumar B (2020) Survey of heart disease prediction and identification using machine learning approaches. In: 2020 3rd International Conference on Intelligent Sustainable Systems (ICISS). IEEE, p 553–557

Gmaei A, Mehedi Hassan M, Alelaiwi A, Alalman H (2019) A hybrid deep learning model for human activity recognition using multimodal body sensing data. IEEE Access 7:99152–99160

Ismail L, Materwala H, Al Kaabi J (2021) Association of risk factors with type 2 diabetes: a systematic review. Comput Struct Biotechnol J 19:1759–1785

Ismail L, Materwala H, Tayefi M, Ngo P, Karduck AP (2022) Type 2 diabetes with artificial intelligence machine learning: methods and evaluation. Arch Comput Methods Eng 1–21

Klra R (2016) A survey on machine learning techniques in health care industry. Int J Recent Res Asp 3(2)

Krishnan S, Balas VE, Julie EG, Yesudhas RH, Balaji S, Kumar R (eds) (2020) Handbook of research on blockchain technology. Academic Press

Krishnan S, Balas VE., Julie EG, Yesudhas HR, Kumar R (eds) (2021) Blockchain for smart cities. Elsevier

Krishnan S, Balas VE, Kumar R (eds) (2023) Green blockchain technology for sustainable smart cities. Elsevier.

Kumar T, Ramani V, Ahmad I, Braeken A, Harjula E, Ylianttila M (2018) Blockchain utilization in healthcare: key requirements and challenges. In: 2018 IEEE 20th International conference on e-health networking, applications and services (Healthcom). IEEE, p 1–7

Mackey TK., Kuo T-T, Gummadi B, Clauson KA, Church G, Grishin D, Obbad K, Barkovich R, Palombini M (2019) 'Fit-for- purpose?'—challenges and opportunities for applications of blockchain technology in the future of healthcare. BMC Med 17(1):1–17

Mohey Eldin A, Hossny E, Wassif K, Omara FA (2023) Federated blockchain system (FBS) for the healthcare industry. Sci Rep 13(1):2569

Muthukaruppasamy S, Thomas GAS (2024) Artificial intelligence and blockchain technologies for revolutionizing electric vehicles and charging infrastructure. In: A sustainable future with e-mobility: concepts, challenges, and implementations. IGI Global, p 130–149

Raul S, Das S, Murty Ch SVVSN, Devi K (2023) A review on intelligent health care system using learning methods. Recent developments in electronics and communication systems, p 154–159

Reegu F, Daud SM, Alam S (2021) Interoperability challenges in healthcare blockchain system-a systematic review. Ann Rom Soc Cell Biol:15487–15499

Rufo J, Zhang P, Zhong R, Lee LP, Huang TJ (2022) A sound approach to advancing healthcare systems: the future of biomedical acoustics. Nat Commun 13(1):3459

scikit-learn: machine learning in Python—scikit-learn 1.0.1 documentation. https://scikit-learn.org/stable/.

sklearn.ensemble.RandomForestClassifier—scikit-learn 1.0.1 documentation. https://scikit-learn.org/stable/modules/

Sharma R, Kshetri N (2020) Digital healthcare: historical development, applications, and future research directions. Int J Inf Manage 53:102105

Shinde R, Patil S, Kotecha K, Ruikar K (2021) Blockchain for securing ai applications and open innovations. J Open Innov Technol Mark Complex 7(3):189

Shinde R, Patil S, Kotecha K, Potdar V, Selvachandran G, Abraham A (2022) Securing AI-based healthcare systems using Blockchain Technology: a state-of-the-art systematic literature review and future research directions. arXiv preprint arXiv:2206.04793

Shine T, Thomason J, Khan I, Maher M, Kurihara K, El-Hassan O (2023) Blockchain in healthcare: 2023 Predictions from around the Globe. Blockchain Healthc Today 6

Tarafder Md M, Tarequzzaman Md, Anisur Rahman Md (2023) Development of wireless pulse rate monitoring system. In: 2023 7th International Conference on Computing Methodologies and Communication (ICCMC). IEEE, p 1606–1610

Thomas GAS, Muthukaruppasamy S, Sathish Kumar S, Saravanan K (2023) Green IoT use case approaches for blockchain technology taking industry 5.0 to the next level. In: Green blockchain technology for sustainable smart cities. Elsevier, p 119–141

Thomas GAS, Muthukaruppasamy S, Nandha Gopal J, Sudha G, Saravanan K (2024) Unleashing the power of XAI (explainable artificial intelligence): empowering decision-making and overcoming challenges in smart healthcare automation. In: Explainable AI (XAI) for sustainable development. Chapman and Hall/CRC, p 303–316

Toh C, Brody JP (2021) Applications of machine learning in healthcare. Smart manufacturing: when artificial intelligence meets the internet of things 65

Revolutionizing Healthcare: The Transformative Power of Blockchain Technology

Antoine Bechade

Abstract The integration of blockchain technology in the healthcare sector heralds a significant transformation in data security, interoperability, and efficiency. This chapter delves into how blockchain can revolutionize health data management, emphasizing applications such as electronic medical record management and drug traceability. It also discusses the profound benefits of blockchain, including fraud reduction and enhanced data management, while addressing challenges related to privacy and regulatory compliance. The chapter further explores the potential of blockchain to streamline patient care, enhance drug safety, and facilitate secure and efficient data sharing among healthcare providers. Case studies are presented to demonstrate successful blockchain applications in various healthcare scenarios, providing a glimpse into the future potential of this technology in the health sector.

1 Introduction

Blockchain technology is revolutionizing healthcare, addressing persistent challenges like data breaches and inefficiencies. This distributed ledger technology ensures unparalleled security, transparency, and immutability in managing healthcare data. It encrypts and decentralizes patient information across a broad network, greatly minimizing risks linked to centralized data storage and unauthorized access.

The introduction of blockchain in healthcare resolves longstanding issues of data integrity and accessibility. Traditional systems, often fragmented, obstruct the smooth exchange of essential health information. Blockchain promotes interoperability, offering a secure platform where data is shareable among authorized entities. This shift toward a more integrated approach enhances patient care by making medical histories and treatment records readily accessible.

Smart contracts, a key feature of blockchain, automate critical healthcare workflows, including patient consent and billing processes. These contracts activate based on set conditions, streamlining operations and reducing administrative burdens. This

A. Bechade (✉)
New York, NY, USA
e-mail: anto.bechade@icloud.com

S. Krishnan et al. (eds.), *Unveiling Blockchain Enabled Business Applications*,
Contributions to Management Science, https://doi.org/10.1007/978-981-95-2337-5_10

automation not only boosts efficiency but also supports a healthcare system that is both resilient and responsive.

Blockchain's adoption in healthcare started with improving electronic health records (EHRs) security, progressing to applications such as drug traceability, verifying healthcare professionals' credentials, and managing insurance claims.

From theoretical concepts to real-world applications, blockchain's journey illustrates its potential to address major healthcare challenges. Despite issues like technological complexity and regulatory demands, advancements suggest a future where blockchain underpins a secure, efficient, and patient-centric healthcare system.

This overview sets the stage for a deeper look into the technical mechanics that could make such a future achievable. The robust architecture of blockchain is crucial for the secure and reliable management of health data, ensuring each piece of information remains an immutable part of the digital ledger.

2 Technical Foundation of Blockchain

In the preceding overview, we recognized blockchain's potential to reshape healthcare, creating a secure, patient-focused ecosystem. As we pivot from the conceptual promise of blockchain to its practical applications, we anchor ourselves in the technical foundations that enable such a transformative vision.

2.1 Data Structures and Cryptography

Blockchain technology enhances healthcare systems with high-level security and interoperability. Its digital ledger securely links blocks of data into an unalterable sequence. Each block contains a unique hash and the hash of its predecessor, ensuring the integrity and historical accuracy of healthcare records. This feature is vital for maintaining informed medical decisions. Modifications to patient records and consent forms are permanently logged, creating a transparent audit trail that supports legal compliance and builds patient trust. Blockchain's implementation in healthcare data management offers robust security, facilitating the secure sharing and verification of essential health information, which improves patient care and outcomes.

A Merkle tree is a data structure in computer science that is used for efficiently and securely verifying the contents of large data structures. It organizes data into a binary tree, where each leaf node is a hash of data blocks, and each non-leaf node is a hash of its children.

Merkle trees enhance blockchain data verification by condensing transaction records into a single hash—the Merkle root—located in the block header. This efficient format allows for fast, streamlined authentication of records without accessing the entire blockchain. In healthcare, Merkle trees enable quick verification of medical

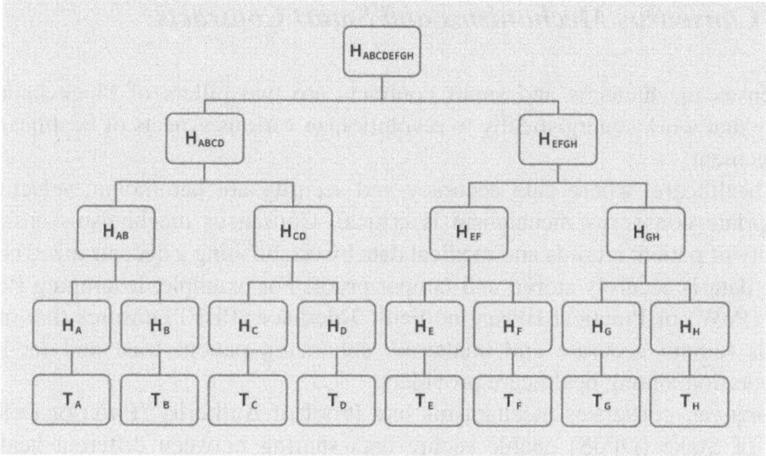

Fig. 10.1 Example of a Merkle tree. https://zebpay.medium.com/when-lambo-cryptos-supercar-craze-27535590ba4f

product histories, protecting the supply chain from counterfeit drugs. This method preserves data integrity and efficiency, essential in sectors where verifying item provenance is critical (Fig. 10.1).

Blockchain technology, with its advanced data structures and cryptography, significantly enhances healthcare beyond just securing records. At Mayo Clinic, each data block, including encrypted health records, is secured in a blockchain framework, cryptographically linked to maintain historical accuracy and integrity for medical decision-making. This immutable chain ensures a transparent audit trail for any modifications, boosting trust and aiding legal compliance.

Mayo Clinic also uses Merkle trees to efficiently validate health data. Patient consent forms are aggregated and hashed into Merkle roots for rapid validation without compromising privacy, ensuring data integrity in a field where reliability is paramount.

Further bolstering security, public-key cryptography protects medical data; information encrypted with a public key can only be decrypted with the corresponding private key, allowing access exclusively to authorized parties and ensuring patient confidentiality across healthcare networks.

Additionally, smart contracts at Mayo Clinic automate processes like insurance claims, activating upon treatment completion to streamline administration and reduce overhead.

These innovations exemplify a transformative approach to managing healthcare data, fostering a secure, transparent, and patient-centered environment. As blockchain technology evolves within healthcare, its potential to enhance patient outcomes, data security, and delivery processes is increasingly evident.

2.2 Consensus Mechanisms and Smart Contracts

Consensus mechanisms and smart contracts are two pillars of blockchain technology that work synergistically to revolutionize various aspects of healthcare data management.

In healthcare, where data accuracy and security are paramount, selecting the appropriate consensus mechanism is critical. Consensus mechanisms ensure the integrity of patient records and medical data by establishing a decentralized network where data is securely stored and tamper-proof. For example, leveraging Proof of Work (PoW) or Practical Byzantine Fault Tolerance (PBFT) ensures that medical records remain accurate and unaltered, enhancing patient trust and facilitating collaboration among healthcare providers.

Moreover, consensus mechanisms like Proof of Authority (PoA) or Delegated Proof of Stake (DPoS) enable secure data sharing between different healthcare entities, such as hospitals, clinics, and pharmacies. This fosters interoperability and improves patient care coordination across the healthcare continuum while maintaining data privacy and confidentiality.

Additionally, consensus mechanisms play a crucial role in ensuring the authenticity and traceability of pharmaceutical products within the supply chain. Implementing mechanisms like Proof of Authenticity (PoA) or Proof of Provenance (PoP) allows pharmaceutical companies to track the journey of drugs, mitigating the risk of counterfeit medications and ensuring patient safety.

Within healthcare, blockchain technology and smart contracts herald a new era, simplifying data management and securing patient records with unmatched precision. Smart contracts act as automated gatekeepers that execute predefined rules for data sharing and patient consent, thus enhancing privacy and reducing bureaucratic burdens. They transform insurance claim workflows into transparent and expedited processes, removing the need for intermediaries and enabling direct, error-free transactions.

These innovative contracts also extend their functionality to pharmaceutical supply chains, assuring stakeholders of drug authenticity and provenance, mitigating the risk of counterfeit medicines, and contributing to patient safety. This shift not only fosters patient autonomy but also propels the healthcare industry toward a more trust-oriented and patient-centric model.

Figure 10.2 in the provided material encapsulates this transformative approach. It depicts how a smart contract, grounded in blockchain, serves as a digital nexus among patients, healthcare providers, and pharmacists. This contract facilitates a seamless exchange of information, safeguarding the integrity and confidentiality of health data across various technological interfaces. Such a decentralized framework ensures that medical records are consistently accurate, transparently maintained, and readily available, marking a significant advancement in how healthcare data is preserved and accessed within a global context.

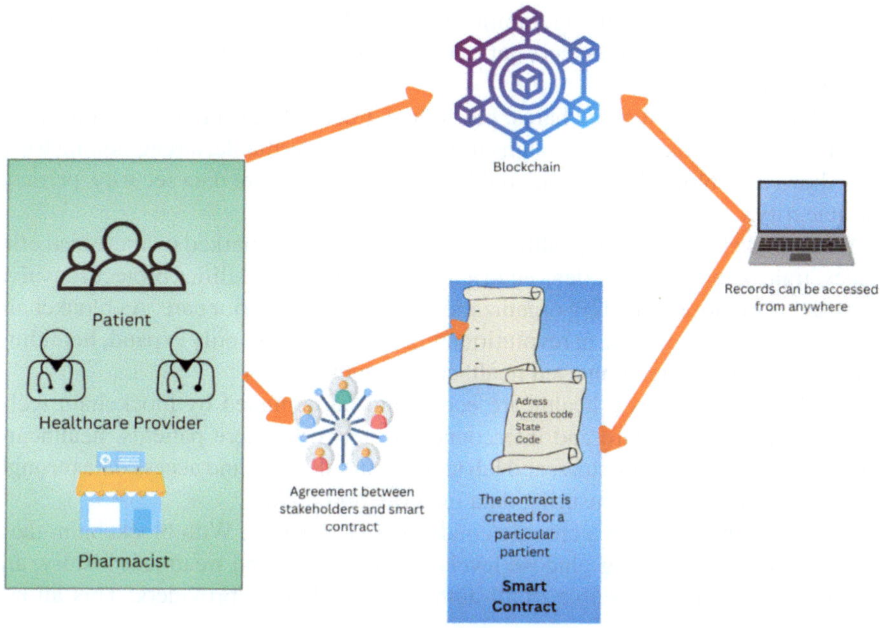

Fig. 10.2 Integrating Blockchain for security and accessibility of health data

2.3 Security Features Specific to Healthcare Data

Blockchain technology significantly enhances healthcare data security with its decentralized, immutable ledger, offering unique protections vital to managing medical data. Its ledger maintains data integrity by preventing unauthorized changes, while cryptographic methods encrypt sensitive information like patient records, bolstering privacy. Unlike traditional centralized systems susceptible to breaches, blockchain's decentralized architecture improves security.

Blockchain networks ensure that only authorized healthcare providers can access data by granting permission based on predefined criteria, preserving confidentiality. The transparency of transactions allows stakeholders to audit data modifications, increasing accountability. These security measures guarantee that patient information is secure and unalterable, thereby enhancing patient outcomes.

Exploring blockchain's technical pillars—robust data structures, cryptography, smart contracts, and consensus mechanisms—highlights a blend of security and functionality that transforms healthcare data management into secure, patient-focused, and interoperable ecosystems. For instance, secure medical data exchanges maintain patient record integrity and improve care coordination.

Imagine a system where a patient's medical records are securely stored on a blockchain network, accessible only to authorized providers. Each record is cryptographically secured, ensuring confidentiality. Smart contracts facilitate efficient consent management, reducing administrative tasks, while consensus models like Proof of Authority promote secure data sharing across healthcare entities, supporting interoperability.

In pharmaceutical supply chains, blockchain tracks drug origins, reducing counterfeit risks. Each transaction remains transparent and verifiable, strengthening trust. This showcases blockchain's potential in healthcare to ensure data security, privacy, and integrity.

With blockchain leading, healthcare is poised for an era marked by increased efficiency, transparency, and patient-centricity. Its security capabilities protect sensitive medical information, enabling organizations to provide superior care. As blockchain technology evolves, its role in revolutionizing healthcare will only expand, heralding a future of secure, interconnected healthcare systems.

The Fig. 10.3 shows a healthcare ecosystem underpinned by blockchain technology, depicting a secure and interconnected network where patients, healthcare providers, pharmacists, doctors, and lab technicians all share and access data through a centralized blockchain, ensuring data integrity and privacy.

Consider a patient with a chronic condition like diabetes. With blockchain, their medical records, including glucose levels, prescriptions, and treatment history, are securely stored and accessible to all authorized healthcare providers. This allows

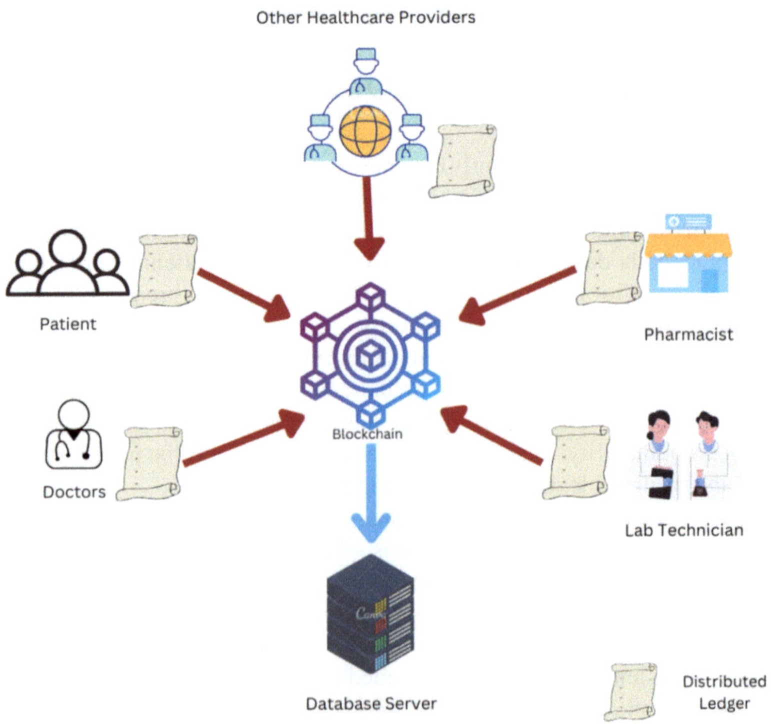

Fig. 10.3 Blockchain as the Nexus for medical data exchange and coordination

seamless coordination among specialists, pharmacists, and primary care physicians, ensuring timely interventions and personalized care plans tailored to the patient's needs. It creates another issue, the management of these data. The blockchain can also deal with this.

The technical foundation of blockchain forms the bedrock of a resilient healthcare infrastructure, ensuring the secure and seamless handling of sensitive medical data. This innovation heralds a new epoch in healthcare, characterized by heightened security, improved patient engagement, and a collaborative ecosystem for providers, ultimately enhancing the quality of care.

3 Use Cases of Blockchain for Healthcare

3.1 Healthcare Data Management

It's crucial to recognize the significant hurdles encountered in safeguarding healthcare data. From breaches to data integrity and confidentiality, these challenges underscore the importance of blockchain technology in reshaping healthcare data management. Through understanding these obstacles, we gain insight into how blockchain can drive transformative change in healthcare data security.

3.1.1 Current Challenges in Healthcare Data Security

In the current digital landscape, safeguarding healthcare data encounters numerous complexities, posing substantial risks to patient privacy and confidentiality. A primary concern is the persistent threat of data breaches, where malicious actors gain unauthorized access to sensitive patient information, potentially resulting in identity theft, fraud, and other detrimental outcomes. The widespread adoption of electronic health records (EHRs) and interconnected healthcare systems has broadened the attack surface, rendering healthcare organizations more susceptible to cyber threats.

Figure 10.4 presents a blockchain-based framework enhancing healthcare data security. It begins with patients and doctors generating data, recorded in Electronic Health Records (EHRs). Through customized user access, information is managed securely. The data moves to a centralized database and is also recorded on a blockchain, ensuring immutability and traceability, which helps detect unauthorized changes. Healthcare providers access this secured data, ensuring accuracy and currency. Cloud storage provides backup and data recovery support. This approach addresses security risks, ensures compliance, and maintains efficiency in healthcare data management.

Data governance in healthcare must meet stringent regulations like the U.S.'s HIPAA and Europe's GDPR, which protect patient information and impose penalties

Fig. 10.4 Streamlining healthcare: Interoperability and privacy through blockchain technology. https://www.mdpi.com/2304-8158/10/6/1289

for breaches. Balancing compliance with efficiency and accessibility is a complex task for providers.

Additionally, healthcare IT infrastructures face challenges with the growing volume and variety of data. Issues with legacy systems, interoperability, and human factors, such as insider threats and lack of staff security training, add to the complexity. Thus, enhancing data security in healthcare requires a multifaceted strategy combining technology, stringent security protocols, and continuous staff education.

3.1.2 The Role of Blockchain in Data Integrity and Confidentiality

Blockchain technology significantly boosts integrity and confidentiality in healthcare data management. Its decentralized design distributes data across many computers, deterring unauthorized alterations. Blocks linked via unique cryptographic hashes form an unbreakable chain, essential for safeguarding patient records.

Sophisticated encryption also protects data within blockchain. Public-key cryptography ensures that only intended recipients can access data, maintaining patient confidentiality. Smart contracts control data access by enforcing rules on who can view data and when.

Blockchain offers a solid answer to healthcare's data breach concerns, enabling secure data exchanges and storage. Its transparent, auditable system permanently logs transactions, providing a reliable basis for managing healthcare data. This ensures data remains confidential and secure, building trust with patients and providers.

The influence of blockchain goes beyond security—it's set to reshape data sharing and usage in healthcare, promising better patient outcomes and streamlined care. Blockchain empowers healthcare organizations to overcome traditional data management issues, paving the way for a more secure, interoperable, and patient-focused care model.

3.2 Interoperability in Healthcare Systems

Blockchain's advent into healthcare brings a pivotal shift in how patient data is managed across various platforms and systems. It heralds a new age of interoperability, enabling a seamless and secure exchange of health information, vital for comprehensive patient care and advanced medical research.

3.2.1 Issues with Current Interoperability Standards

Lack of Universal Standards

Healthcare data management is challenged by the lack of universal standards, obstructing smooth data exchange. Various EHR systems, each with proprietary formats and protocols, prevent reliable information transfer, risking patient safety with potentially incomplete or inconsistent records. While standards like HL7 and FHIR are steps toward interoperability, widespread adoption hasn't been achieved. The healthcare industry's failure to unite under common data standards remains a key obstacle to true interoperability.

For instance, when hospitals with different EHR systems transfer patients, critical information such as allergies might be lost due to incompatible standards. This issue underscores the necessity for universal standards like HL7 FHIR to ensure consistent data interpretation across diverse systems.

Figure 10.5 illustrates the protection of sensitive data in an Electronic Health Records (EHR) system, showing how security is reinforced by privacy protocols and adherence to regulations. It breaks down the dual strategies used to safeguard information: access management via control models and pseudonymity, and defense mechanisms through encryption, as well as thorough authorization and authentication measures.

Complexity of Healthcare Data

The healthcare industry manages diverse data types, from structured lab results and prescriptions to unstructured physician's notes and radiological images. Harmonizing this data, often in incompatible formats, requires complex translation to ensure usability across various systems. This complexity is magnified by the need to preserve the clinical context and data integrity, essential for continuous care. Advanced solutions like data mapping and natural language processing are developing to address these complexities, yet they require extensive industry integration. For example, a patient's health record includes straightforward structured data and complex unstructured notes, which can be misinterpreted without proper data harmonization techniques.

Privacy and Security Concerns

The push for interoperability must also tackle the essential need for privacy and security in healthcare data. Encryption and strict access controls are vital for preventing breaches and ensuring compliance with regulations such as HIPAA and GDPR. Breaches can lead to severe penalties and erode patient trust, pushing healthcare organizations to strengthen security infrastructures and training. A blockchain solution, for instance, could offer a secure alternative by decentralizing data storage and requiring consensus for data access, thus enhancing privacy.

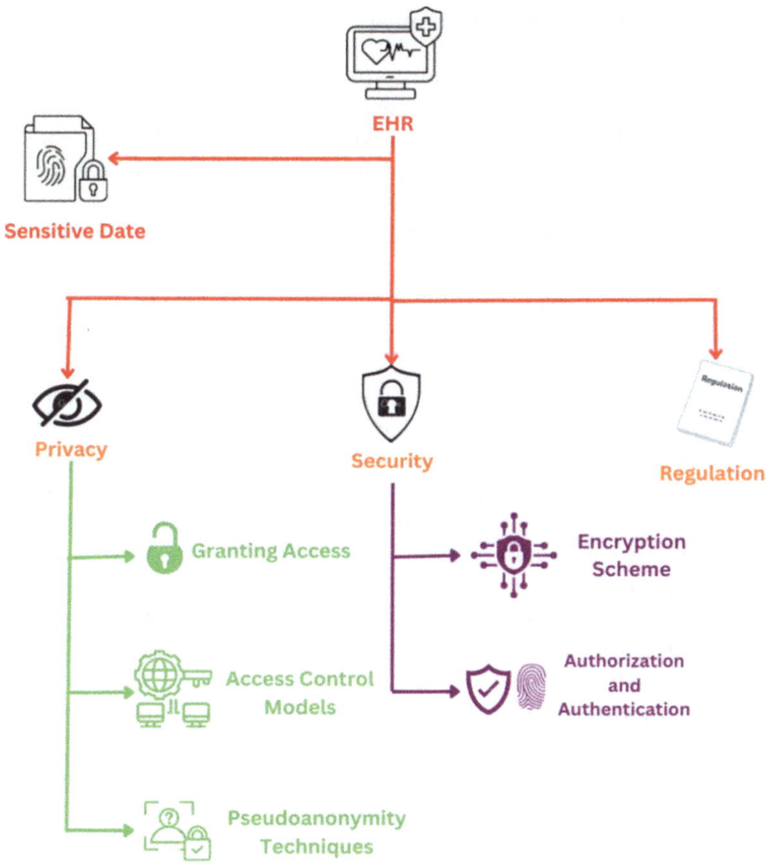

Fig. 10.5 Ensuring the integrity of EHRs: balancing privacy, security, and compliance

Technological Heterogeneity

A wide range of systems, from outdated legacy platforms to modern cloud-based solutions, challenges interoperability in healthcare. Bridging these technological divides often involves complex interfacing systems or middleware, potentially introducing failures and security risks. The financial and logistical burdens of updating or replacing legacy systems complicate this further.

Regulatory Barriers

The regulatory landscape in healthcare often doesn't keep pace with rapid technological advances, hindering the adoption of innovative interoperability solutions. Existing regulations can also create a complex maze of compliance requirements that vary by region or country, complicating the scalability of solutions like telemedicine across different jurisdictions.

Blockchain for Seamless Data Exchange

Blockchain technology is set to transform healthcare data exchange by providing a secure, immutable, and efficient method for sharing patient information. At its heart, blockchain uses a decentralized ledger that records transactions across multiple sites, avoiding the vulnerabilities of a single data repository. This is crucial in healthcare, where sensitive data flows between hospitals, clinics, insurance companies, and patients daily.

Blockchain resolves many interoperability issues by creating a secure platform for data exchange. Each transaction is encrypted and added to the ledger after all parties reach consensus, safeguarding against unauthorized access and ensuring data remains current and accurate.

Additionally, blockchain acts as a single source of truth for patient records, potentially reducing administrative burdens and avoiding redundant efforts. For example, a lab test recorded on the blockchain can be accessed by any authorized provider, preventing unnecessary repeat tests and minimizing errors. Smart contracts could automate the consent process, aligning data sharing with patient preferences and legal standards.

The transparency inherent in blockchain facilitates real-time data audit, building trust among stakeholders and granting patients more control over their information. Patients can easily manage access to their medical records, playing an active role in their healthcare.

However, transitioning to blockchain-based systems involves challenges, including the need for collaboration among healthcare stakeholders, investments in new technologies, and reevaluation of current data management practices. Despite these hurdles, the potential for improved data integrity, enhanced security, and increased operational efficiency positions blockchain as a valuable future solution for healthcare interoperability.

Pilot Projects and Emerging Standards

MedRec: Developed by researchers at MIT, MedRec is a pilot project that utilizes blockchain to create a decentralized content-management system for EHRs. It gives patients a comprehensive, tamper-proof record of their medical history, accessible to healthcare providers across different organizations.

Project Dovetail: This project by TIBCO Software aims to use blockchain for managing clinical and healthcare data across various stakeholders. It provides a framework for healthcare entities to rapidly adapt to new interoperability standards.

There is also Health Utility Network. It's a group of leading healthcare organizations, including IBM, Aetna, and Anthem, have collaborated on this project to improve data sharing and make healthcare more efficient and transparent using blockchain technology.

The integration of blockchain in healthcare is being guided by emerging standards such as FHIR, which enhances data interoperability, and efforts by HL7's Blockchain Workgroup to align blockchain with established health data standards. Additionally, the IEEE Standards Association's P2418.6 is crafting a comprehensive framework for blockchain's application in healthcare, addressing critical aspects of data format, security, and system interoperability, ensuring a unified approach to this transformative technology within the healthcare sector.

As the healthcare industry embraces innovative technologies, a suite of emerging standards is charting a new course for the field, heralding an era of enhanced interoperability, heightened security, and streamlined management of electronic health records through the integration of blockchain technology.

3.3 Patient-Centered Applications

Patient-centred applications in healthcare leverage blockchain technology to prioritize and empower patients. These innovative applications focus on enhancing patient autonomy, ensuring data security, and improving healthcare delivery through transparent and efficient access to medical information. As we delve into this segment, we explore how blockchain facilitates a more personalized and engaging healthcare experience.

3.3.1 Empowering Patients with Data Ownership

Patient empowerment is crucial for enhancing healthcare quality and improving patient experiences. Defined by the eHealth ERA project (Enhancing European eHealth research and cooperation), it posits that optimal health outcomes result from active patient participation, involving information, access, engagement, and agency. These elements empower patients to make informed decisions and manage their health effectively.

This empowerment shifts the traditional patient-provider dynamic toward a collaborative relationship characterized by mutual respect, benefiting both autonomy and clinical outcomes. For example, research in PLoS (Public Library of Science One shows that empowered patients tend to adhere better to medication regimes, and studies in the Journal of Diabetes Research have linked empowerment to improved outcomes in chronic conditions like diabetes.

Empowered patients also benefit healthcare providers. Active patient participation aligns patient-provider goals, enhancing satisfaction for both. This approach ensures high-quality care that respects and incorporates patient preferences.

Healthcare professionals can harness blockchain to enhance patient empowerment in the following ways:

- Customized Treatment Plans: Leveraging blockchain's secure data sharing for personalized healthcare, professionals can offer tailored treatment options.
- Encouraged Engagement: By fostering transparent communication channels on the blockchain, patients feel more inclined to ask questions and understand their care.
- Technology Integration: Blockchain-integrated patient portals ensure secure, immediate access to health records, enhancing informed decision-making.
- Health Literacy Advancement: Securely sharing educational materials over blockchain can help patients better understand their health and treatment options.
- Proactive Health Management: Blockchain can track and validate self-management actions like lifestyle changes or medication adherence, reinforcing the role of patients in managing their health.
- Seamless Collaboration: Blockchain facilitates a unified platform for healthcare providers and patients, streamlining coordination and care continuity.

By integrating blockchain, healthcare professionals can empower patients with secure data management and enhanced communication, fostering an environment where informed and engaged patients contribute to a more dynamic and efficient healthcare system.

3.3.2 Blockchain in Patient Consent Management

Blockchain technology is transforming healthcare consent management by providing a secure, transparent, and efficient way to handle consent data. Its key features—immutability and decentralized control—are particularly suited to address the challenges posed by evolving privacy laws and the increasing demand for personal data control.

Consent management, essential in managing personal identifiable information, allows users to give, revoke, or renew consent as needed. Traditional methods face issues like data aggregation from diverse sources, secure storage, transparency, and adapting to changing privacy regulations.

Blockchain offers a decentralized platform where consent actions are securely recorded on an immutable ledger, ensuring data integrity and building trust. This setup prevents unauthorized alterations and removes the risks associated with a single point of control, thereby enhancing security against breaches.

For example, in clinical research, patient consent is crucial. Blockchain can securely log each patient's consent, any changes to it, and its withdrawal, streamlining consent management across studies while ensuring regulatory compliance through transparent and auditable records.

Moreover, blockchain enables dynamic consent, allowing patients to adjust their preferences in real time—vital in personalized medicine. Patients can specify who accesses their data or for which studies, adjusting these settings as research evolves or personal circumstances change.

In conclusion, blockchain technology offers a robust solution to healthcare consent management challenges. By securing and transparently recording consent, it enhances patient control over personal information and eases compliance and management for healthcare providers and researchers.

3.3.3 Impact on Patient Privacy and Trust:

The advancement of interoperability, blockchain technologies, and patient-centered applications significantly influences patient privacy and trust in healthcare. These technologies enhance privacy protections but also introduce challenges that must be managed to maintain trust.

- Enhanced Control and Transparency: Patient-centered applications give patients greater control over their personal health information (PHI), increasing transparency in data usage and sharing across providers. This empowerment builds trust, as patients can consent to data sharing and understand who accesses their information and why. For instance, blockchain-based consent management systems allow patients to transparently and securely manage access to their PHI.
- Secure Data Exchange: Blockchain is crucial in secure data exchange, providing a decentralized, immutable ledger for health information. It ensures data is encrypted and traceable, minimizing unauthorized access and breaches. For example, blockchain enables secure sharing of patient records between hospitals and specialists, maintaining data integrity and enhancing trust.
- Regulatory Compliance and Data Protection: Integrating interoperability standards and blockchain helps healthcare organizations comply with strict regulations like HIPAA and GDPR. Using interoperable EHR systems that meet international security standards ensures PHI is managed respectfully and legally, reinforcing patient trust.
- Challenges and Concerns: While beneficial, these technologies pose privacy and trust challenges. Managing consent across platforms and potential PHI exposure from blockchain require vigilance and innovation in privacy technologies. Healthcare organizations must proactively address these issues to prevent trust erosion.
- Building a Trustworthy Ecosystem: The impact on patient privacy and trust hinges on collaborative efforts among healthcare providers, technology developers, and regulators to create a secure, transparent, and patient-focused ecosystem. Prioritizing patient privacy and involving patients in health information decisions fosters a trustful environment that supports ethical use of new technologies.

The application of blockchain technology in healthcare spans various areas, including data management, medical supply chains, and genomics, offering innovative solutions for improving efficiency, security, and patient care. Here are some noteworthy examples:

1. BurstIQ (Denver, Colorado) utilizes blockchain to manage large amounts of patient data securely, ensuring compliance with HIPAA and potentially aiding in the prevention of prescription drug abuse through comprehensive patient health information
2. Medicalchain (London, England) employs blockchain to maintain health records' integrity, providing a single point of truth accessible to doctors, hospitals, and laboratories while protecting patient identities
3. Blockpharma (Paris, France) provides a solution for drug traceability and counterfeiting, ensuring patients receive genuine medications through a blockchain-based supply chain management system

In conclusion, the intersection of interoperability, blockchain, and patient-centred applications presents a complex landscape for patient privacy and trust. While these technologies offer significant potential to improve healthcare delivery and patient empowerment, they also require careful management to ensure that patient privacy is protected and trust is maintained throughout the healthcare ecosystem.

3.4 Supply Chain Management

Supply Chain Management in Healthcare involves the strategic coordination and integration of all operations related to the flow of medical goods and services. This encompasses everything from the procurement of raw materials and medical equipment to the delivery of care services to patients. Effective supply chain management aims to enhance efficiency, reduce costs, and ensure the timely availability of necessary healthcare products and services. It plays a crucial role in maintaining the quality of patient care and operational effectiveness within healthcare institutions.

Based on the overview of pharmaceutical supply chain vulnerabilities, the five worst vulnerabilities that present significant risks to the supply chain's resilience and efficiency are in Table 10.1.

These vulnerabilities underscore the importance of strategic measures to enhance resilience, improve transparency, and adopt advanced technologies to mitigate risks and ensure the reliability of pharmaceutical supply chains.

Table 10.1 Different vulnerabilities of the pharmaceutical supply chain

Global Dependency and Concentration of Production	This vulnerability makes the pharmaceutical supply chain highly susceptible to geopolitical tensions, trade disputes, and localized disruptions, posing a significant risk to the global availability of medicines
Lack of Transparency and Traceability	The absence of transparency and traceability not only facilitates counterfeiting and fraudulent activities but also significantly compromises patient safety and trust in the pharmaceutical supply
Cybersecurity Threats	As the supply chain becomes increasingly digitalized, it faces greater risks of cyber-attacks, which can disrupt operations and lead to substantial financial losses and erosion of trust
Cold Chain Logistics and Storage	The specialized logistics and storage requirements for temperature-sensitive pharmaceuticals pose a complex challenge, with any breach in cold chain protocols potentially leading to compromised drug efficacy and safety
Reliance on Just-In-Time (JIT) Inventory Models	JIT models, while efficient under normal conditions, can lead to severe shortages and disruptions when unexpected events occur, demonstrating a critical vulnerability in ensuring drug availability

Antoine Bechade on Canva

3.4.1 Blockchain for Drug Traceability and Authenticity

The integration of blockchain technology into the pharmaceutical supply chain addresses critical vulnerabilities by enhancing drug traceability and authenticity. This approach mitigates risks associated with counterfeit drugs, supply chain inefficiencies, and regulatory compliance challenges. By leveraging blockchain, each transaction from manufacturing to distribution to the consumer is recorded in a transparent, immutable ledger, ensuring drug distribution integrity and improving pharmaceutical traceability. As a result, stakeholders can verify drug authenticity at any supply chain point, reducing the risk of counterfeit medications entering the market and improving patient safety.

An example of blockchain's application in this area is the pilot project initiated *(FDA DSCSA Blockchain Pilot)* by the U.S. Food and Drug Administration (FDA) under the Drug Supply Chain Security Act (DSCSA). This act aims to develop a system for tracing certain prescription drugs distributed in the United States. The pilot project explores how blockchain can enhance the security and integrity of the drug supply chain, ensuring safe and efficient pharmaceutical distribution and compliance with regulatory standards. Through blockchain, each drug's journey is securely recorded, creating a verifiable trail from production to consumption and enhancing the ability to detect and prevent the distribution of counterfeit drugs, thus safeguarding public health.

This example highlights the transformative potential of blockchain in enhancing the resilience of the pharmaceutical supply chain, offering a robust solution to the industry's long-standing vulnerabilities.

3.4.2 Real-World Applications and Demonstrated Outcomes

From the detailed exploration of real-life blockchain use cases in supply chains, we can observe a diverse range of applications that showcase the technology's potential to enhance transparency, safety, and efficiency across various industries. Let's summarize the key insights from each case:

1. Pharmaceutical Supply Chain Safety with Merck, IBM, KPMG, and Walmart: This collaboration highlighted blockchain's capacity to meet FDA requirements for tracking and tracing prescription medicines and vaccines in the U.S., demonstrating an increase in patient safety and paving the way toward industry standards for interoperable blockchain systems.
2. Saleable Drug Returns with FFF Enterprises: Using the MediLedger Network, FFF Enterprises showcased how blockchain can verify the return of saleable drugs, ensuring compliance with the Drug Supply Chain Security Act (DSCSA) regulations for verifying product identifiers before resale, thus preventing counterfeit drugs from entering the supply chain.
3. Ensuring Fresh Produce with Naturipe Farms: Naturipe Farms leveraged SAP's blockchain service to track blueberries, improving traceability from harvesting to the consumer, and addressing challenges related to the timeliness of shipments, which is critical for perishable goods.
4. Cold Chain Management for Meat: Though not detailed in the summaries, the implication is clear—blockchain technology is crucial in ensuring the integrity of temperature-sensitive products like meat throughout the supply chain, from farm to table.

These examples collectively demonstrate the transformative potential of blockchain across various sectors. By ensuring the authenticity of products, enhancing transparency, and improving the efficiency of supply chains, blockchain technology is not only optimizing operational processes but also contributing to more ethical and sustainable business practices. The diversity of applications—from pharmaceuticals and food to ethical sourcing in the chocolate industry—underscores blockchain's versatility and its growing importance in the digital transformation of supply chains worldwide.

3.5 Regulatory and Compliance Aspects

Navigating the complex regulatory landscape in healthcare blockchain applications is essential for ensuring that innovations comply with legal standards and protect patient privacy. As blockchain intersects with healthcare, understanding these regulatory frameworks is crucial. The interaction between evolving technology and legal requirements presents unique challenges and opportunities for providers, developers, and policymakers. This section explores how blockchain solutions are being adapted to meet strict regulatory demands, ensuring that technological advancements in healthcare improve efficiency and security while adhering to essential legal and ethical standards.

3.5.1 Blockchain in the Context of HIPAA, GDPR, and Other Regulations

Blockchain technology in healthcare must navigate complex legal frameworks that protect patient privacy and ensure data security, notably the Health Insurance Portability and Accountability Act (HIPAA) in the U.S. and the General Data Protection Regulation (GDPR) in the EU.

HIPAA: Established to protect sensitive patient health information, HIPAA mandates that Protected Health Information (PHI) be securely encrypted and only shared among authorized parties. Blockchain's encryption and immutable data recording align with HIPAA's requirements for PHI confidentiality, integrity, and availability. However, blockchain's decentralized nature poses compliance challenges, especially concerning the amendment or deletion of data under HIPAA's Privacy Rule.

GDPR: GDPR enhances privacy and data protection in the EU and EEA, emphasizing individuals' rights over their data, including access, rectification, and the right to be informed of data breaches. While blockchain supports GDPR's data integrity and security demands, its immutability conflicts with the "right to be forgotten," necessitating innovative solutions to enable data deletion on blockchain networks.

For instance, Mount Sinai Health System uses blockchain to ensure PHI confidentiality across institutions, promoting interoperability in line with HIPAA. In the EU, Estonia's e-Health system employs blockchain to allow citizens to manage their data, aligning with GDPR's protection rights. Furthermore, initiatives like the Synaptic Health Alliance project illustrate how blockchain can foster EHR interoperability and compliance with diverse legal standards, such as the 21st Century Cures Act in the U.S.

Other Regulations: Blockchain applications must also adhere to additional regulations like those set by the ONC in the U.S., which govern health data privacy, security, and interoperability. These standards ensure that blockchain applications in healthcare are secure, efficient, legally compliant, and ethically responsible.

In conclusion, integrating blockchain into healthcare demands a thorough understanding of and compliance with various regulatory standards. Developers and providers must innovate to harness blockchain's benefits while meeting legal and ethical requirements, such as the ePrivacy Directive in the EU that specifies regulations on electronic communications.

3.5.2 Legal Considerations and Ethical Implications

The integration of blockchain in healthcare confronts a dynamic legal and ethical terrain. The technology's immutable nature offers groundbreaking possibilities for enhancing data integrity and patient privacy but also presents challenges in aligning with established privacy laws such as HIPAA in the United States and GDPR in Europe. These regulations mandate the protection of patient data and grant individuals' rights over their personal information, including the right to have it deleted—a requirement that seems at odds with blockchain's permanent record-keeping.

One of the primary legal considerations involves ensuring that blockchain applications in healthcare comply with these privacy regulations, which aim to secure sensitive patient information from unauthorized disclosure. For instance, HIPAA requires that Protected Health Information (PHI) be encrypted and shared solely among authorized entities, emphasizing the confidentiality, integrity, and availability of PHI. Meanwhile, GDPR emphasizes the rights of individuals within the EU to control their data, underscoring the need for healthcare blockchain applications to navigate the "right to be forgotten" within an immutable ledger system.

Beyond compliance with privacy laws, the ethical design of healthcare blockchain systems is paramount. Ethical considerations must be integrated into the technology's development phase, focusing on patient respect and individual preferences beyond mere regulatory adherence. This includes creating ethical design frameworks that reflect organizational values, privacy and security policies, and sustainability practices, ensuring that blockchain technology respects individual interests and adheres to healthcare ethical principles.

Additionally, the use of blockchain in conjunction with AI in healthcare decision-making introduces autocracy concerns, where decision-making shifts from humans to algorithms. This transition raises ethical issues regarding consent, transparency, and the potential for bias, necessitating careful consideration of how blockchain and AI technologies impact patient care and privacy.

In summary, as blockchain technology continues to evolve within the healthcare sector, navigating its legal and ethical implications becomes increasingly complex. Balancing innovation with adherence to privacy regulations and ethical standards is crucial for ensuring that blockchain enhances rather than compromises patient rights and trust in healthcare systems.

For detailed insights, please refer to the discussions on the ethical and legal implications of blockchain in healthcare as presented by the EU Blockchain Observatory and Forum reported on CIFS Health and explored by Deloitte.

3.6 *Challenges and Limitations*

As blockchain technology forges pathways into healthcare, it faces pivotal hurdles. Scalability concerns and the integration with existing IT infrastructures emerge as significant barriers, challenging the seamless adoption and operational efficiency of this transformative technology within the healthcare sector.

3.6.1 Scalability Concerns

The integration of blockchain in healthcare navigates a complex legal and ethical landscape. Its immutable nature offers significant potential for improving data integrity and patient privacy but challenges compliance with privacy laws like HIPAA in the United States and GDPR in Europe. These laws protect patient data and give individuals control over their personal information, including the right to have it deleted—a concept that conflicts with blockchain's permanent records.

A key legal challenge is ensuring that blockchain applications in healthcare meet these privacy standards, which safeguard sensitive patient information. HIPAA mandates encryption and limited sharing of Protected Health Information (PHI), focusing on confidentiality, integrity, and availability. GDPR, on the other hand, ensures that individuals in the EU can control their data, complicating blockchain's use due to the "right to be forgotten."

Ethically, designing healthcare blockchain systems must prioritize respect for patients and individual preferences, incorporating these values early in the development process. This involves crafting frameworks that align with organizational values and healthcare ethical principles, ensuring that the technology respects individual rights.

Furthermore, combining blockchain with AI in healthcare introduces concerns about automated decision-making, which shifts control from humans to algorithms. This raises ethical questions about consent, transparency, and potential bias, highlighting the need to consider the effects of blockchain and AI on patient care and privacy.

In conclusion, as blockchain technology evolves in healthcare, it must navigate intricate legal and ethical issues. Balancing innovation with strict adherence to privacy laws and ethical standards is essential for ensuring that blockchain technology supports rather than undermines patient rights and trust in healthcare systems.

3.6.2 Integration with Existing Healthcare IT Infrastructure

The integration of blockchain technology with existing healthcare IT infrastructure poses unique challenges that are essential to address for its successful adoption in the healthcare sector. This integration is crucial for creating a seamless and efficient ecosystem that enhances data management, patient privacy, and overall care quality.

However, this process requires overcoming substantial technical, organizational, and regulatory hurdles.

A key technical challenge is the compatibility between blockchain technology and existing healthcare IT systems, such as electronic health records (EHRs) and laboratory information systems (LIS), which are often complex and heterogeneous. Effective integration necessitates developing interfaces or middleware that facilitate communication between these diverse systems and the blockchain, ensuring secure and efficient data sharing.

Organizational challenges are also significant, as healthcare institutions must adjust their processes and workflows to include blockchain technology. This adjustment may require staff retraining, redefining data management protocols, and gaining support from all stakeholders, including healthcare providers, patients, and regulators. Achieving consensus on standards and protocols for blockchain usage among various healthcare entities adds to the complexity.

Regulatory compliance is crucial, as integration must comply with strict privacy and security regulations like HIPAA in the United States and GDPR in Europe. Blockchain solutions must meet these standards while protecting the integrity and confidentiality of patient data.

Addressing these challenges involves collaboration among technology developers, healthcare professionals, regulatory authorities, and other stakeholders. Focusing on interoperability, standardization, and regulatory compliance can pave the way for blockchain to enhance healthcare IT infrastructure, leading to better data management, enhanced security, and improved patient outcomes.

4 Conclusion and Future Directions

4.1 Conclusion and Future Directions

The integration of blockchain is poised to revolutionize healthcare data management by fortifying security, ensuring privacy, and broadening interoperability. Its decentralized framework provides a safeguard against unauthorized access and data breaches, with immutable records establishing a foundation of trust between patients and healthcare professionals.

Blockchain's innovation, particularly through smart contracts, automates healthcare procedures, thereby reducing administrative burdens and enhancing accuracy. These contracts underpin a range of applications from supply chain oversight to the facilitation of patient-focused services, signaling a shift toward a cohesive healthcare infrastructure.

Addressing blockchain's scalability and integration with existing healthcare IT infrastructures is the focus of ongoing research. Potential solutions like "editable" blockchains may reconcile GDPR's 'right to be forgotten' without compromising data immutability, while advancements in consensus algorithms could streamline data exchange processes.

The healthcare sector is increasingly adopting blockchain through pilot projects, demonstrating its viability from local clinics to extensive health systems. This technology is instrumental in managing health crises, enabling efficient data sharing and swift, collaborative responses on a global scale.

Collaborative efforts between regulators and technologists are essential for crafting regulatory frameworks that respect blockchain's characteristics and protect patient rights. Looking forward, the convergence of blockchain with AI, IoT, and genomics is set to propel healthcare into an era of personalized medicine and anticipatory care strategies, signifying an epoch marked by innovation and centred on patient welfare.

Glossary

DPoS (Delegated Proof of Stake): A consensus algorithm where stakeholders delegate voting powers to a limited number of nodes to validate transactions and create blocks.

DSCSA (Drug Supply Chain Security Act): Legislation that outlines steps to build an electronic system to identify and trace certain prescription drugs as they are distributed in the United States.

EHR HIPAA: Regulations about the use of Electronic Health Records within the standards and privacy rules set by the Health Insurance Portability and Accountability Act.

EHRs: Digital systems for recording, storing, and managing patients' personal and healthcare information electronically.

FDA (Food and Drug Administration): A federal agency of the United States Department of Health and Human Services.

FHIR (Fast Healthcare Interoperability Resources): A standard for electronic exchange of healthcare information, focusing on interoperability and ease of integration.

GDPR: The General Data Protection Regulation, a legal framework that sets guidelines for the collection and processing of personal information within the EU.

Hash in Crypto: A cryptographic function that converts an input (or 'message') into a fixed-size string of bytes, typically used for secure data handling.

HL7 (Health Level Seven International): An organization that provides standards for the exchange, integration, sharing, and retrieval of electronic health information.

IEEE (Institute of Electrical and Electronics Engineers): A professional association dedicated to advancing technology for the benefit of humanity, often involved in setting industry standards.

Merkle Tree: A data structure used in cryptography to securely verify content in large sets of data.

ONC (Office of the National Coordinator for Health Information Technology): A division of the Department of Health and Human Services that oversees national efforts to implement and use the most advanced health information technology and the electronic exchange of health information.

PACS (Picture Archiving and Communication System): A medical imaging technology that provides economical storage and convenient access to images from multiple modalities.

PBFT (Practical Byzantine Fault Tolerance): A consensus mechanism used in distributed computing to reach an agreement on a single data value among distributed processes or systems.

PHI (Protected Health Information): Any information in a medical record that can be used to identify an individual and that was created, used, or disclosed in the course of providing a health care service.

PoA (Proof of Authority): A consensus mechanism in blockchain networks where transactions and blocks are validated by approved accounts, known as validators.

PoP (Proof of Participation): An algorithm that incentivizes and rewards participants' involvement and contribution to a network.

Proof of Work: A consensus algorithm in a blockchain that requires a computationally intensive process to validate transactions and produce new blocks.

References

Agbo CC, Mahmoud QH, Eklund JM (2019) Blockchain technology in healthcare: a systematic review. Healthcare 7(2):56. https://www.mdpi.com/2227-9032/7/2/56

Ahmad RW, Bin Ibrahim R (2020) Blockchain technology in healthcare: a systemic review. J Med Syst 44(8):1–11. https://link.springer.com/article/10.1007/s10916-020-01597-8

Ahmed K, Khalil A, Soursou G (2019) Applications of blockchain technology in medicine and healthcare: challenges and future perspectives. Cryptography 3(1):3. https://www.mdpi.com/2410-387X/3/1/3

Gordon R, et al (2020) Blockchain applications in health care: a systematic review. Healthcare 8(3):218. https://www.mdpi.com/2227-9032/8/3/218

Halamka JD (2021) Blockchain technology applications in healthcare: an overview. J Am Med Inform Assoc 28(4):806–810. https://academic.oup.com/jamia/article/28/4/806/6192682

IBM (2022) Blockchain healthcare and life sciences solutions. IBM J Res Dev 66(2/3):1–12. https://ieeexplore.ieee.org/document/9811275

Khara P, Kumar S (2024) Healthcare smart contracts market latest report. Int J Blockchain Cryptocurrency Res (IJBLCR) 6(1):55–68. https://www.igi-global.com/article/healthcare-smart-contracts-market-latest-report/336621

Khezr S et al (2019). Blockchain technology in healthcare: a comprehensive review and directions for future research. Appl Sci 9(9):1736. https://www.mdpi.com/2076-3417/9/9/1736

Kim H, Lee K (2023) Blockchain-inspired secure and reliable data exchange architecture for cyber-physical healthcare system 4.0. J Supercomput 79(5):4408–4425. https://link.springer.com/article/10.1007/s11227-022-04203-9

Krawiec N et al (2017) Blockchain: opportunities for health care. Deloitte University Press. https://www2.deloitte.com/content/dam/Deloitte/us/Documents/public-sector/us-blockchain-opportunities-for-health-care.pdf

Krawiec RJ (2020) Blockchain for healthcare. Blockchain in Healthcare Today 3:1–8. https://blockchainhealthcaretoday.com/index.php/journal/article/view/134

Liu Y, Liu Q, Liu X, Chen H (2023) Blockchain-secure patient digital twin in healthcare using smart contracts. PLoS ONE 18(10):e0286120. https://doi.org/10.1371/journal.pone.0286120

Martin F, Le Moing M (2020) Utilisation de la blockchain pour les données médicales et le contrôle d'accès en e-Santé. Revue Francophone d'Informatique Médicale 10(2):53–64. https://www.jle.com/fr/revues/rfim/e-docs/utilisation_de_la_blockchain_pour_les_donnees_medicales_et_le_controle_dacces_en_e_sante_317391/article.phtml

MediLedger (2020) FDA DSCSA pilot project complete. MediLedger. https://www.mediledger.com/dscsa-fda-pilot-project

Mettler D (2016) Blockchain technology in healthcare: the revolution starts here. IEEE 18th International Conference on e-Health Networking, Applications and Services (Healthcom), 1–3. https://ieeexplore.ieee.org/document/7749510

Pwc (2021) Blockchain in healthcare: 17 examples to know. Healthc 9(3):277 (Basel, Switzerland). https://www.mdpi.com/2227-9032/9/3/277

Reddy S, Gupta A (2021) Healthcare organizations form new blockchain company. Int J Healthc Manag 15(2):113–120. https://www.tandfonline.com/doi/abs/10.1080/20479700.2021.1912342

Saeed H, et al (2020) Blockchain technology in healthcare: a systematic review. PLoS One 15(10):e0239408. https://journals.plos.org/plosone/article?id=10.1371/journal.pone.0239408

Taherdoost H (2023a) Blockchain and healthcare: a critical analysis of progress and challenges in the last five years. Blockchains, 1(2):73–89. https://www.mdpi.com/2673-2301/1/2/6

Taherdoost H (2023b) Privacy and security of blockchain in healthcare: applications, challenges, and future perspectives. Sci 5(4):41. https://doi.org/10.3390/sci504004

Tippett A, Ratnarajah T (2023) 5 blockchain healthcare use cases in digital health. Int J Healthc Inf Syst Inform (IJHISI) 15(3):17–31. https://www.igi-global.com/article/5-blockchain-healthcare-use-cases-in-digital-health/312561

Various Authors (2023) Blockchain in Healthcare Today. https://blockchainhealthcaretoday.com/index.php/journal

Wolfson R (2019) Cigna and Sentara healthcare join IBM's blockchain health utility network to improve data sharing. Forbes 23(4):76–80. https://www.forbes.com/sites/rachelwolfson/2019/02/16/cigna-and-sentara-healthcare-join-ibms-blockchain-health-utility-network-to-improve-data-sharing/?sh=7c311ea054bb

Zhang Y, Xue W, Zhou Y (2021) Blockchain-based privacy and security preserving in electronic health: a systematic review. J Med Syst 45(7):1–12. https://link.springer.com/article/10.1007/s10916-021-01715-x

Transformative Potential of Blockchain in Supply Chain Management: Enhancing Transparency, Traceability, and Trust

Malla Jogarao and J. Siva Durga Prasad

Abstract Interconnected supply chains and enhanced efficiency, transparency, and security have become paramount in the modern business world. In response to this demand, blockchain technology has emerged as a disruptive influence positioned to fundamentally reshape supply chain management. This chapter delves into the transformative potential of blockchain in supply chain management, exploring its multifaceted applications and profound impact on various supply chain stages. It also highlights modern supply chains' complexities and challenges, emphasizing the need for innovative solutions. Subsequently, blockchain enhances transparency and traceability, enabling seamless tracking of products throughout the supply chain. Blockchain optimizes inventory management, streamlining procurement and reducing costs. The chapter also outlines how blockchain streamlines logistics and transportation, simplifies customs clearance, and reduces delays. Additionally, it explores how blockchain empowers sustainable practices within supply chains, enabling transparent tracking of environmental impact data. It also delves into how blockchain fosters collaborative innovation, providing a secure platform for sharing data and insights among supply chain partners. Real-world examples from various industries, including gems, food, and pharmaceuticals, demonstrate the practical applications of blockchain in supply chain management. The chapter discusses blockchain's transformative potential in revolutionizing supply chain management. It emphasizes the transformative impact of blockchain in improving transparency, streamlining logistics, empowering sustainable practices, fostering collaborative innovation, and increasing adoption of blockchain technology, as well as its significant potential to revolutionize supply chains, rendering them more interconnected, streamlined, and reliable networks.

M. Jogarao (✉) · J. Siva Durga Prasad
School of Maritime Management, Indian Maritime University, Visakhapatnam, India
e-mail: mjogarao@gmail.com

J. Siva Durga Prasad
e-mail: jsdurgaprasad@imu.ac.in

S. Krishnan et al. (eds.), *Unveiling Blockchain Enabled Business Applications*,
Contributions to Management Science, https://doi.org/10.1007/978-981-95-2337-5_11

1 Introduction

In an era where globalization has linked organizations on a worldwide scale, the efficiency and dependability of supply chains have become critical. However, traditional supply chain management approaches frequently need help with opacity, inefficiency, counterfeit items, and trust deficits. The emergence of blockchain technology is a revolutionary development that provides transparency, traceability, and trust in supply chain ecosystems. This chapter explores blockchain technology's revolutionary capacity in supply chain management, focusing on its ability to improve transparency, traceability, and Trust. Firstly, the existing difficulties encountered by conventional supply chain management systems emphasize the urgent need for revolutionary advancements. Next is a comprehensive analysis of blockchain technology, examining its core concepts and potential use in supply chain management.

The focal point of this chapter revolves around analyzing how blockchain technology tackles the inherent deficiencies of conventional supply networks. To improve transparency and traceability through a decentralized ledger, which enables the continuous tracking of goods from inception to ultimate distribution. In addition, blockchain technology promotes confidence by maintaining unchangeable records and verifiable data, thereby addressing problems related to counterfeit items and lack of Trust. Blockchain is crucial in enhancing the honesty and protection of supply chain management. This chapter will focus on how blockchain technology mitigates fraud risks, strengthens data integrity, and improves cybersecurity measures in supply chain networks. The capacity for automation in smart contracts demonstrates how they simplify contractual responsibilities and ensure compliance, ultimately enhancing the efficiency of supply chain operations. Blockchain technology presents a ledger system that records transactions across multiple nodes in a network. This technology's potential to improve transparency and Trust has garnered significant attention across industries, including supply chain management (Salha et al. 2019).

Blockchain technology can improve supply chain management by enhancing transparency, traceability, Trust, and efficiency. It allows for decentralized, transparent, immutable record-keeping, ensuring secure and reliable transactions (Yontar 2023). The technology allows for efficient sharing of information, process integration, organized data management, and prompt reactions, resulting in increased efficiency and sustainability in supply chains (Wang et al. 2023). Real-world case studies demonstrate blockchain's successful implementations and practical implications in supply chain management (Franke and Fischer 2023). Nevertheless, obstacles include unclear legal standing, compatibility with other technologies, and power disparities in supply chains (Lau and Yip 2023). Despite the current limitations and the need for further development, blockchain technology has notably impacted supply chain capabilities and flexibility, providing firms with a competitive advantage. Blockchain technology can completely transform supply chain management by enhancing transparency, efficiency, and reliability.

2 Understanding Traditional Supply Chain Challenges

Traditional supply chain challenges encompass both strategic and operational aspects. Strategic challenges include buyer dominance, diverging SKU (Stock et al.), sustainability issues, and lack of formal education dependency (Jana 2022). Operational challenges vary across geography and markets and include long lead time, multiple iterations during product development, low technology penetration, and limited digitization in manufacturing (Sali et al. 2023). Additionally, traditional supply chains face intermediaries, Trust, and subpar performance issues (Taschner and Charifzadeh 2020). Management accounting faces collaboration, resource and task allocation, information asymmetry, and performance measurement difficulties in a supply chain environment (Pal et al. 2017). Furthermore, traditional supply chains are vulnerable to cyber security risks, which can be mitigated through implementing frameworks for cyber security and its practices in supply chain management guidelines (Hayes 2023).

2.1 Discussion on the Prevailing Challenges in Traditional Supply Chain Management

There are numerous existing challenges that traditional supply chain management encounters, including demand forecasting difficulties, inventory management issues, supply chain visibility, supplier risk management, and the need for effective coordination and collaboration among all stakeholders. Operational challenges encompass long lead time, multiple iterations in product development, low technology penetration, and limited digitization in manufacturing (Taschner and Charifzadeh 2020). Additionally, there must be more overdependence on management philosophies, restricted use of scientific methods, and confusion between artisanal and industrial production (Taschner and Charifzadeh 2022). In a supply chain environment, management accounting encounters governance and coordination differences, information asymmetry, and performance measurement complexities (Schiffer and Dörr 2040). In the realm of distribution logistics, the visibility of the supply chain is hindered by unsuitable technologies and processes, a lack of communication and trust, and insufficient monitoring and decision-making metrics (Melnyk et al. 2021). Demand for more intricate supply chain networks, shorter delivery times, rising costs, and specific client demands are all factors that will pose difficulties for supply chain managers in the years to come. Recent cybersecurity attacks have highlighted the importance of cybersecurity in impacting the supply chain.

2.2 *Issues of Transparency, Traceability, and Trust (TTT) in Supply Chain*

These are prevalent in various industries such as gems, pharma, food, fashion, and textile industries. Insufficient visibility and accountability in the supply chain management systems lead to compromises in product quality and increase the risk of fraud and counterfeit products (Chen et al. 2021; Bhatnagar et al. 2020). This interest undermines the sustainability efforts of brands and discourages investment in social and environmental sustainability (Aravendan 2023). To address these issues, blockchain technology is proposed to enhance security, prevent data tampering, and improve transparency in supply chains (Ospital et al. 2022; Changhak 2021). By implementing traceability systems and providing comprehensive product information, such as the source, manufacturing processes, and sustainability conditions, Trust can be built among all stakeholders. This allows customers to confirm the genuineness of items, report forgeries, and make informed purchases, thereby reducing the proliferation of counterfeit goods and improving total transparency and Trust in the industry.

3 Blockchain Technology

Blockchain is a tech that allows safe and reliable data storage and sharing. (Zhou et al. 2023). It is based on a decentralized ledger, meaning that data is not stored in a central location but distributed across multiple network nodes. This decentralization ensures that the data is immutable and cannot be tampered with (Sarao et al. 2023). Multiple consensus mechanisms, such as proof-based and enterprise-based, regulate blockchain transactions' validation and inclusion (Barj et al. 2023). The validation and addition of transactions to the blockchain are dictated by several consensus processes, including proof-based and enterprise-based. The needs and use cases of the blockchain system dictate the selected consensus mechanism. The distinctive capabilities of blockchain technology to solve the problems encountered by conventional supply chain management systems are based on several core principles. These ideas include decentralized ledgers, immutability, and consensus methods (Xia et al. 2024).

Decentralized Ledgers: Blockchain technology relies on a distributed database known as a decentralized ledger, which is accessible through a network of computers referred to as nodes. Unlike traditional centralized databases, which centralize all information in one place and are managed by an external organization, distributed ledger technology, commonly called blockchain, allows for shared access to data by all network participants. The decentralized nature of the system requires more central authority and eliminates potential weak points. All participants keep their copy of the ledger, ensuring that transactions are documented openly and unchangeably.

Immutability: The immutability of data once recorded on the blockchain—the impossibility of editing or erasing it- is another crucial aspect of blockchain technology. A blockchain is a distributed record-keeping system that logs transactions chronologically, with each block cryptographically connected to the one before. Adding a transaction to the blockchain makes it permanent and makes it impossible to go back and modify it. The data maintained on the blockchain is impervious to fraud and tampering because of its immutability.

Consensus: All nodes in a blockchain network must agree on the legitimacy of each transaction for the network to function, and consensus mechanisms are the protocols that make this possible. Even in decentralized systems, these steps ensure that every node can reach a consensus on the current state of the ledger. Blockchain networks employ several consensus procedures, which consist of practical Byzantine fault tolerance (PBFT), proof of stake (PoS), and proof of work (PoW). Each mechanism requires its own set of rules to verify transactions and add them to the blockchain.

It can ensure secure ownership records, automate licensing and royalty payments, and identify infringing uses in the context of intellectual property (Nguyen 2023). In digital health, blockchain can address data security, automation, interoperability, and patient data ownership (Kjäer 2023). It also aids in ensuring the accuracy, audit trail, and reliability of data objects and message flows in inter-organizational process management (Mazouzi et al. 2023). In the context of the Internet of Vehicles (IoV), blockchain technology can help address privacy and security concerns by leveraging its decentralized and distributed architecture (Naje et al. 2022). Furthermore, blockchain technology has the potential to improve security and privacy in cloud computing by ensuring the integrity, authenticity, and confidentiality of stored data.

4 Role of Blockchain in Enhancing Transparency and Traceability

Through the establishment of an unchangeable digital documentation of every transaction and movement throughout the supply chain, blockchain technology enables immediate monitoring and openness (Cozzio et al. 2023). It monitors a product's source, whereabouts, and past, allowing for increased transparency and traceability. Through blockchain, consumers can gain Trust, leading to positive attitudes and behaviors toward the sharing experience. However, suppliers may still be hesitant to adopt blockchain due to concerns about data sharing and the need for intra-organizational support (Subramanian et al. 2023). Adopting blockchain technology in the distribution chain can enhance performance and transparency. Blockchain can significantly increase the food supply chain's efficiency, transparency, and traceability by reinventing how transactions are carried out and providing a complete record of each product's journey.

4.1 Implementation of Blockchain for Enhanced Traceability

Blockchain technology has been utilized in different scenarios to improve visibility. An example of a traceability system based on blockchain technology was suggested to enhance the protection of original achievements such as patents and copyrights. The system used distributed data storage, peer-to-peer transmission, consensus mechanisms, and encryption algorithms to store, secure, authenticate, and monitor registrations and transactions tied to initial accomplishments. In the agri-food sector, blockchain technology has been applied to traceability systems to address challenges in acquiring complete traceability information and mitigate risks of data tampering. These systems enable decentralized, transparent, and reliable tracking of food products, contributing to improved management and reductions in food loss and recall incidents (Ambika 2023). Integrating blockchain and IoT (Internet of Things) has also been explored to enhance information provenance and integrity.

5 Establishing Trust in the Supply Chain

Trust is critical for efficient and secure transactions in an increasingly digital world. Blockchain technology emerges as a potential game-changer, fostering Trust through its unique architecture of verifiable data. It explores how blockchain empowers trust-building mechanisms in supply chain systems, examining the advantages and possible challenges. The technology of blockchain can increase levels of trust within supply chains. It has the ability to promote Trust among supply chain partners, consumers/public, and supply chain units (Yavaprabhas et al. 2023). Lack of trust in supply chains can impede data sharing and hinder implementation (Al-Rakhami and Al-Mashari 1759). Trust and commitment play a vital role in supply chain management, and understanding the antecedents and consequences of trust and commitment is vital for building and implementing trust (Paluri and Mishal 2020). The existing manufacturing systems need comprehensive identity forensics and trust maintenance in crosswise collaborations with suppliers and manufacturers (Weng and Becker 2022). Trust can be established over time through learning from previous transactions in a pharmaceutical supply chain, leading to the elimination of sellers with low-quality products (Ackland et al. 2019). Blockchain technology and learning from experience play a significant role in establishing Trust in supply chains.

6 Strengthening Supply Chain Integrity and Security

Blockchain technology is crucial in fortifying supply chain integrity, reducing fraud risks, and enhancing cybersecurity (Wang 2023). By combining artificial intelligence (AI) and blockchain technology (BT), supply chain management can achieve information sharing, risk sharing, and data interaction, leading to a safer and more efficient supply chain (Gomes 2023; Strohmier 2023). Additionally, blockchain-based authenticity identification systems can help detect and prevent the sale of counterfeit products, ensuring product identification and consumer trust (Karumanchi et al. 2023).

7 Automation and Smart Contracts in Supply Chain Management

Smart contracts are self-executing programs that automate the enforcement of contract terms using blockchain technology and present a contemplative pathway toward mechanizing and optimizing diverse facets of supply chain management. Smart contracts improve compliance enforcement in supply chain networks, decrease dependence on intermediaries, and expedite contractual obligations through the automatic execution of predefined actions to fulfil particular conditions (Prajapati et al. 2023). They are instrumental in supply chain processes for automating contractual obligations and compliance enforcement. Smart contracts can streamline supply chain processes by automating payment triggers, order verification, and inventory updates and ensuring requirements and compliance with regulations are met (Reza et al. 2023). They can also be used to automate the delivery phase of projects by linking project management tools like Building Information Modeling (BIM) with contract conditions, thus facilitating the issuance of payments when specific parameters are met (Kamel et al. 2023).

Using smart contracts in supply chains can lead to more transparent, reliable systems without third-party interference and can enforce traceability constraints. They can be programmed to trigger actions automatically when predefined conditions are met, which is beneficial for managing complex transactions and ensuring compliance with legal and regulatory requirements (Bucur and Miclea 2023). Compensations and non-monotonic requirements can be specified with the help of smart contracts, which opens the door to easing some compliance standards to deal with unusual cases (Elgammal et al. 2016).

7.1 Application in Automating Supply Chain Processes

Smart contracts, a precursor to blockchain technology, have brought about a significant paradigm shift in supply chain operations by automating diverse undertakings and guaranteeing adherence to contractual commitments. Without intermediaries, these self-executing contracts are programmed to autonomously execute and enforce the terms of an agreement when predefined conditions are met.

Procurement: Purchasing operations are streamlined by utilizing intelligent agreements, which automate various duties, including order placement, invoice verification, and payment processing. An instance of this would be when a purchaser initiates a transaction with a provider by inputting an order into which a smart contract can autonomously validate the receipt of commodities or services and initiate payment upon confirmation of delivery; this would prevent the necessity for human intervention and diminish the duration of the processing.

Logistics: Smart contracts facilitate the administration and tracking of goods throughout the supply chain in logistics. By integrating Internet of Things (IoT) devices and sensors, smart contracts can autonomously initiate processes informed by real-time data. These processes may include, but are not limited to, updating inventory levels, scheduling shipments, and optimizing transportation routes. This automation increases logistics operations' efficacy, decreases duration, and mitigates their proneness to error.

Payments: In supply chains, smart contracts are also instrumental in expediting payments and settlements. By incorporating payment terms into the contract code, smart contracts can electronically transfer funds between parties upon the fulfillment of predetermined milestones or the delivery of products. By eliminating the complications and conflicts that arise from conventional payment methods, this system guarantees transactions that are conducted promptly and transparently.

Compliance: Smart contracts facilitate adherence to contractual agreements and regulatory requirements through the automated enforcement of conditions and rules inscribed within the contract code. They can also guarantee adherence to customs regulations, trade agreements, and financial regulations in international trade by autonomously validating documentation, computing tariffs, and initiating necessary approvals.

Smart contracts provide substantial benefits in automating supply chain procedures, lowering expenses, enhancing productivity, and fostering openness and confidence among participants in the supply chain. We anticipate an even greater integration of smart contracts across various supply chain functions as the technology develops and acquires traction, resulting in additional innovation and optimization in international trade and commerce. In business process compliance, smart contracts can be used to ensure that all blockchain requirements are met synchronised, ensuring that transactions are only triggered if all conditions are met (Bellavista et al. 2021). They can be formalized using deontic logic to express permissions, obligations, and prohibitions. They have the potential to align with corporate objectives and plans to enhance comprehension of adherence across various levels, including operational procedures and business goals (Hashmi et al. 2018). Overall, smart contracts offer a promising approach to enhancing efficiency, and the performance of supply chain processes is improved, leading to a decrease in the requirement for manual involvement and potentially lowering costs associated with traditional contract management (Kouhizadeh and Sarkis 2018).

8 Case Studies and Use Cases

8.1 Case Study: Diamond Certification

De Beers, a diamond company, uses blockchain technology to track diamonds from mine to market, which is part of a broader trend in the industry toward greater transparency and ethical sourcing. Blockchain's ability to provide a secure and immutable ledger makes it an effective tool for ensuring that diamonds are conflict-free and ethically sourced. The technology helps to establish a chain of custody that can be verified by all stakeholders in the supply chain, from miners to retailers, which enhances consumer trust in the products they are purchasing (Kongmanas et al. 2023). This approach aligns with the increasing consumer demand for responsible business practices and sustainable production methods. Fighting against the trade of conflict diamonds involves various strategies, including the Kimberley Process Certification Scheme (KPCS). The government, businesses, and NGOs have united to establish the KPCS to end the trade of conflict diamonds. However, the scheme has faced criticism for its limitations and the difficulty in fully guaranteeing the origin of materials (Carrigan et al. 2017). Despite these challenges, the diamond industry, including companies like De Beers, has been adapting to changes in the market environment and customer demands, moving away from a cartel-like structure to one that is more open and competitive (Berger et al. 2016). The Responsible Jewellery Council (RJC) is another industry-led initiative that requires members to adhere to auditable industry best practices, covering various aspects such as conflict minerals, human rights, and labor standards (Bloomfield 2017). More and more, ethical product demand and consumer awareness are propelling the industry toward more responsible sourcing and production practices, of which these initiatives are a part.

8.2 Case Study: Pharmaceutical Industry

Blockchain technology is increasingly being adopted in the pharmaceutical industry to enhance the security and traceability of drug supply chains. The technology provides a decentralized ledger that records every transaction or movement of drugs, ensuring that the data cannot be altered or deleted, which helps combat counterfeit drugs (Ghosh et al. 2023). Blockchain solutions like zero-knowledge proof protocols and reputation scoring systems based on Markov models are being proposed to ensure the integrity and reliability of the nodes participating in the pharmaceutical distribution system. Moreover, blockchain can be used to document the testing of new drugs, which could speed up their delivery to the market and allow patients to control their data, leveraging the transparency of the technology (Zoughalian et al. 2022). Blockchain's potential in healthcare extends beyond pharmaceuticals, with applications in electronic health records and medical research data management, offering secure and trustable data storage and sharing (Haddad et al. 2022). However, the success of blockchain in healthcare also depends on the accurate and complete entry of data into the system, the development of interoperability standards, and the resolution of intellectual property issues.

8.3 Case Study: IBM Food Trust

In the food industry, a notable case study is the IBM Food Trust platform, which Walmart has effectively utilized to enhance traceability for leafy green vegetables. Walmart utilized this platform to achieve a notable 20% improvement in traceability time, resulting in a substantial decrease in foodborne illness outbreaks. The advantages of using blockchain in the food industry are significant. Improved traceability guarantees that the path of food items from the farm to the table can be precisely monitored, thus enhancing food safety protocols. Consumers also develop greater confidence in buying things when they have straightforward and confirmable information about their origins. Blockchain simplifies regulatory compliance by offering a detailed record of every transaction and movement in the supply chain.

Implementing blockchain in the food business poses unique problems. Scalability is a significant difficulty, especially for intricate global supply chains that include many stakeholders and geographical regions. Integrating blockchain with current systems and technology in the food supply chain can be challenging and demands meticulous design and implementation. Valuable insights can be derived from this case study. Stakeholder collaboration is essential for effectively integrating blockchain technology into the food business. Through collaboration, stakeholders may address difficulties and tailor the technology to match the industry's requirements.

Moreover, concentrating on specific applications, like traceability for leafy green veggies, enables precise and efficient use of blockchain technology. Standardizing data formats and protocols is crucial for ensuring interoperability and smooth integration of blockchain technology throughout the agricultural supply chain. The IBM Food Trust platform showcases how blockchain technology can transform traceability and safety in the food business. Stakeholders can maximize the potential of blockchain technology in improving the food supply chain by overcoming obstacles, utilizing advantages, and drawing insights from successful examples.

8.4 Case Study: Maersk and IBM TradeLens

TradeLens, an initiative developed in partnership with Maersk and IBM, signifies an innovative undertaking within the shipping sector that seeks to transform worldwide commerce via blockchain technology. The shipping industry has historically encountered challenges such as inefficiencies and delays stemming from fragmented supply chain operations and manual documentation processes. To tackle these obstacles, TradeLens was created to digitize trade documentation and enable supply chain participants to view cargo in real-time, in real-time, and securely, securely, and transparently exchange data, many essential attributes and advantages. Its principal aim is to give stakeholders comprehensive visibility into supply chain operations, enabling them to monitor the progress of dispatches in real time, from the point of origin to the final destination. Enhanced visibility reduces the risk of delays and disruptions, improves operational efficiency, and facilitates more informed decision-making. In addition to bills of lading, certificates of origin, and customs declarations, TradeLens digitizes trade documentation to eliminate the need for paper-based procedures and manual data entry. Digitization serves to optimize workflows associated with trade documentation, minimize errors, and expedite the clearance of products at ports and border crossings.

TradeLens employs blockchain technology to enable secure data exchange among participants in the supply chain, thereby safeguarding confidential trade information against unauthorized intrusion or modification. Real-time data sharing between participants and dependable partners facilitates communication and collaboration throughout the supply chain. The adoption and utilization of TradeLens have yielded numerous advantages for all stakeholders within the international commerce ecosystem. Enhanced visibility, efficiency, and transparency are among the most significant benefits of TradeLens. The platform facilitates the movement of products through ports and customs checkpoints, reduces paperwork, and streamlines trade processes, resulting in reduced costs and shorter lead times for logistics providers and traders. The provision of real-time shipment visibility empowers stakeholders to monitor the whereabouts and condition of merchandise across the entire supply chain, thereby facilitating the proactive detection and resolution of potential disruptions or delays. TradeLens' provision of a transparent and auditable ledger of trade transactions fosters collaboration and confidence among participants in the supply chain, thereby facilitating more streamlined and effective trade operations.

8.5 Blockchain Revolutionizing Supply Chains: Lessons Learned from Case Studies

Blockchain technology has been recognized as a game-changing factor in supply chain management, offering enhanced traceability, transparency, and efficiency. Case studies from the Gartner Supply Chain Top 25 companies reveal various blockchain applications across supply chain processes. For instance, in sourcing and manufacturing, blockchain facilitates traceability, allowing companies to track parts or products from their origin, ensuring data integrity, and reducing the time and cost associated with conventional tracking methods (DiFrancesco et al. 2023).

The technology's core characteristics, such as smart contracts, mutual trust, scalability, and transparency, assist supply chain partners in achieving more efficient agreements and payments. However, challenges like security concerns in unauthorized blockchain networks, scalability, and latency issues have been noted (Shoaib et al. 2023). Blockchain's potential to enhance the sustainability of supply chains has also been explored, with applications developed to guarantee food safety and bring transparency to new levels, exposing negative practices such as child labor and unethical resource use (Venkatesh et al. 2020). Regarding strategy and potential, blockchain is seen as a tool to improve transparency, Trust, data security, and information sharing in decentralized supply chain networks. Integrating blockchain with other technologies like RFID and IoT is suggested to enhance production in manufacturing organizations.

The case studies indicate that while a blockchain offers significant advantages for supply chain management, its adoption involves technical and organizational challenges that must be addressed before its full potential can be realized. The IBM Food Trust platform has facilitated quicker traceability, improved food safety, and simplified regulatory compliance for Walmart in the food business, demonstrating the benefits of cooperation, particular applications, and standardization of data. De Beers' platform in the diamond business has enhanced its brand reputation and consumer confidence through its focus on ethical sourcing, transparency, trust-building, and privacy concerns. MediLedger has enhanced product authenticity, regulatory adherence, and patient well-being within the pharmaceutical sector, highlighting the necessity for clear regulations, compatibility, and precise data management frameworks.

Key points involve concentrating on particular use cases, working together throughout the supply chain, tackling scalability and interoperability issues, pushing for regulatory transparency, and emphasizing data protection and security. Blockchain technology will further enhance supply chain management, increasing global supply networks' security, transparency, and efficiency. Furthermore, delving into different sectors and examining blockchain's economic and ethical consequences will advance comprehension and acceptance of this revolutionary technology.

9 Future Challenges

Blockchain technology in supply chain management faces scalability, regulatory hurdles, and interoperability issues. Scalability refers to the capacity of blockchain systems to manage increasing workloads or expand to support such growth. Regulatory hurdles involve compliance with various laws and regulations that differ by region and can affect the deployment of blockchain solutions. Interoperability issues refer to the challenges in enabling different blockchain systems and applications to work together seamlessly (Cao et al. 2022).

Scalability: Blockchain networks frequently face challenges in efficiently managing high transaction volumes. Supply chains face a massive issue in scalability due to the multiple transactions over complicated networks. Creating blockchain solutions that can handle substantial transaction volumes while maintaining decentralization and security is essential.

Regulatory Hurdles: Rules regarding using blockchain technology in supply chain management are constantly evolving. Multinational supply networks face compliance problems due to differing legislation and norms between jurisdictions. Establishing unambiguous and uniform regulatory structures ensures legal clarity and encourages blockchain-based supply chain management progress.

Interoperability Issues: Ensuring compatibility between various blockchain platforms and traditional systems poses a significant obstacle to smoothly incorporating them into current supply chain networks. Supply chains frequently include several stakeholders utilizing different technologies and protocols, highlighting the importance of interoperability for efficient data transmission and cooperation. Establishing standards and protocols for interoperability will ease the incorporation of blockchain technology into current supply chain infrastructures.

Data Privacy and Security: Blockchain technology provides built-in security measures like cryptographic encryption and immutability, but maintaining data privacy is still a worry. Supply chain data frequently includes confidential information that must be safeguarded from unauthorized access and manipulation. Establishing strong data privacy controls and compliance systems to tackle privacy issues and foster stakeholder confidence is crucial.

Cost and Complexity: Integrating blockchain solutions into supply chain management necessitates substantial investment in technical infrastructure, skilled personnel, and resources. Blockchain technology's intricacy, specialized knowledge, and training requirements might present difficulties for organizations, especially small and medium-sized firms (SMEs). To achieve general acceptance, it is crucial to find economical and easy-to-use solutions that cater to the distinct requirements of various stakeholders.

Environmental Impact: Critics have raised concerns about blockchain networks' significant energy consumption and potential environmental consequences, particularly those that rely on proof-of-work consensus mechanisms such as Bitcoin. With the increasing use of blockchain technology, it is crucial to focus on reducing the environmental effects and developing more energy-efficient consensus methods to support sustainability objectives.

9.1 Potential Future Developments and Innovations

Future advancements in supply chain management using blockchain will involve combining blockchain with other technologies like IoT and cloud computing to improve traceability and efficiency. There is also a focus on developing new consensus mechanisms that are less expensive and more suitable for large industries like aviation and shipping. Research is encouraged to explore blockchain applications beyond tracking and tracing, such as environmental sustainability, economics, and finance. Additionally, there is a need for more empirical research to verify the propositions related to blockchain's impact on supply chain management (Mezquita et al. 2023).

Scalability Solutions: Enhancing blockchain networks' scalability through innovations such as sharding, sidechains, and layer-two scaling solutions is a primary focus of research and development. The innovations enhance transaction throughput and decrease latency while maintaining decentralization and security.

Regulatory Clarity: Collaboration among industry players, legislators, and regulatory bodies is crucial for establishing precise and uniform regulatory frameworks for blockchain-enabled supply chains. Regulatory sandboxes, pilot projects, and industry standards initiatives allow discussion and experimentation to tackle regulatory obstacles and promote innovation.

Interoperability Standards: Efforts are being made to create standard protocols and norms for blockchain networks and applications to ensure they can function smoothly. Interoperability standards will allow smooth data sharing and compatibility among various blockchain platforms, promoting cooperation and integration within supply chain ecosystems.

Integration with Emerging Technologies: Blockchain technology is used with other new technologies, such as the Internet of Things (IoT), Artificial intelligence (AI), and big data analytics, to improve supply chain visibility, automation, and decision-making. Collaborating to create compatible solutions that utilize various technologies would enhance creativity and efficiency in supply chain management.

10 Conclusion

Blockchain technology revolutionizes supply chain management by providing exceptional transparency, traceability, and trust in intricate global supply networks. Conventional supply chain management encounters obstacles, such as lack of transparency, inefficiencies, counterfeit products, and trust issues, all of which blockchain technology seeks to resolve. Blockchain improves transparency and traceability in supply chain management by utilizing decentralized ledgers and unchangeable records. This technology boosts confidence by providing verifiable data, ultimately enhancing the integrity and security of the supply chain.

This chapter examined how blockchain technology improves traditional supply chains by increasing transparency, traceability, and Trust. Case studies from industries like food, diamond certification, and pharma have demonstrated blockchain technology's successful use and tangible benefits in enhancing supply chain processes. Blockchain technology has shown considerable promise to transform supply chain capabilities and flexibility, providing enterprises with a competitive edge in the global economy, although facing problems like scalability, regulatory obstacles, and interoperability issues.

In the future, advancements and innovations in blockchain-based supply chain management show potential for improving transparency, efficiency, and reliability. Efforts to tackle scalability solutions, regulatory clarity, interoperability standards, and integration with new technologies such as IoT and AI are essential for fully realizing the potential of blockchain in supply chain management. Blockchain technology can revolutionize supply chain management by creating a more secure, transparent, and efficient global supply network. Ongoing collaboration and innovation are crucial for addressing issues and optimizing the benefits of using blockchain technology to manage supply chains.

References

Ackland GJ, Chattoe-Brown E, Hamill H, Hampshire K, Mariwah S, Mshana G (2019) Role of trust in a self-organizing pharmaceutical supply chain model with variable good quality and imperfect information. J Artif Soc Soc Simul. https://doi.org/10.18564/JASSS.3984

Al-Rakhami M, Al-Mashari M (2021) A blockchain-based trust model for the internet of things supply chain management. Sensors 21(5):Article 1759. https://doi.org/10.3390/S21051759

Ambika N (2023) Enhancing traceability in food supply chains using blockchain and IoT technologies. Advances in computational intelligence and robotics book series. https://doi.org/10.4018/978-1-6684-8851-5.ch015

Aravendan M (2023) Impacts of transparency and traceability on fashion supply chain system. Intelligent Information Management. https://doi.org/10.4236/Tim.2023.153006

Barj S, Ouaddah A, Mezrioui A (2023) A survey and a state-of-the-art related to consensus mechanisms in blockchain technology. https://doi.org/10.1007/978-3-031-29857-8_21

Bellavista P, Esposito C, Foschini L, Giannelli C, Mazzocca N, Montanari R (2021) Interoperable blockchains for highly integrated supply chains in collaborative manufacturing. Sensors 21(15):Article 4955. https://doi.org/10.3390/s21154955

Berger R, Lamond D, Gavish Y, Herstein R (2016) The evolution of management from a trust to arm's length model in family-run businesses: the case of the diamond industry. J Manag Hist 22(3):341–362. https://shura.shu.ac.uk/13373/7/Berger%20evolution%20of%20management.pdf

Bhatnagar RS, Jha SM, Singh SS, Shende R (2020) Product traceability using blockchain. https://doi.org/10.1109/ICACCCN51052.2020.9362807

Bloomfield MJ (2017) Global production networks and activism: can activists change mining practices by targeting brands? New Polit Econ, 1–16. https://doi.org/10.1080/13563467.2017.1321624

Bucur V, Miclea LC (2023) Entering the metaverse from the JVM: the state of the art, challenges, and research areas of JVM-based Web 3.0 tools and libraries. Future Internet 15(9):Article 305. https://doi.org/10.3390/fi15090305

Cao S, Foth M, Powell W, Miller T, Li M (2022) A blockchain-based multisignature approach for supply chain governance: a use case from the Australian beef industry. Blockchain: Research and Applications 3(4):Article 100091. https://doi.org/10.1016/j.bcra.2022.100091

Carrigan M, McEachern M, Moraes C, Bosangit C (2017) The fine jewelry industry: Corporate responsibility challenges and institutional forces facing SMEs. J Bus Ethics 143:681–699. https://doi.org/10.1007/s10551-016-3071-4

Changhak HL (2021) Blockchain-based traceability for anti-counterfeit in cross-border e-commerce transactions. Sustainability. https://doi.org/10.3390/SU131911057

Chen C, Shang X, Tsaur W, Weng W, Deng Y, Wu C, Cui J (2021) An anti-counterfeit and traceable management system for brand clothing with hyperledger fabric framework. Symmetry. https://doi.org/10.3390/SYM13112048

Cozzio C, Viglia G, Lemarié L, Cerutti S (2023) Toward an integration of blockchain technology in the food supply chain. J Bus Res. https://doi.org/10.1016/j.jbusres.2023.113909

DiFrancesco RM, Meena P, Kumar G (2023) How blockchain technology improves sustainable supply chain processes: a practical guide. Oper Manag Res 16(2):620–641. https://doi.org/10.1007/s12063-022-00343-y

Elgammal A, Turetken O, van den Heuvel W, Papazoglou M (2016) Formalizing and applying compliance patterns for business process compliance. Softw Syst Model 15(1):119–146. https://doi.org/10.1007/s10270-014-0395-3

Franke P, Fischer FF (2023) Will blockchain technology revolutionize supply chain management? Sri Lanka J Mark. https://doi.org/10.4038/sljmuok.v9i1.118

Freichel SLK, Rütten P, Wörtge JK (2022) Challenges of supply chain visibility in distribution logistics. Ekonomski Vjesnik. https://doi.org/10.51680/ev.35.2.16

Ghosh PK, Chakraborty A, Hasan M, Rashid K, Siddique AH (2023) Blockchain application in healthcare systems: a review. Systems 11(1): Article 38. https://doi.org/10.3390/systems11010038

Gomes JR (2023) The importance of cybersecurity in supply chain. https://doi.org/10.1109/isdfs58141.2023.10131834

Haddad A, Habaebi MH, Islam MR, Hasbullah NF, Zabidi SA (2022) A systematic review on ai-blockchain-based e-healthcare records management systems. IEEE Access 10:94583–94615. https://doi.org/10.1109/ACCESS.2022.3201878

Hashmi M, Governatori G, Lam HP, Wynn MT (2018) Are we done with business process compliance? Are we state-of-the-art, and what are the challenges ahead? Knowl Inf Syst 57(1):79–133. https://doi.org/10.1007/s10115-017-1142-1

Hayes M (2023) Challenges regarding protein provision for the growing global population: improving the environmental impact of traditional protein supply chains and maximising use of coproducts and alternative, new resources, Global challenges. https://doi.org/10.1002/gch2.202300059

Jana P (2022) Challenges in the traditional fashion and textile supply chain. Textile science and clothing technology. https://doi.org/10.1007/978-981-19-2108-7_2

Kamel MA, Bakhoum ES, Marzouk MM (2023) A framework for smart construction contracts using BIM and blockchain. Scientific Reports 13(1): Article 10217. https://doi.org/10.1038/s41598-023-37353-0

Karumanchi MD, Sheeba JI, Devaneyan SP (2023) An efficient integrity-based multi-user blockchain framework for heterogeneous supply chain management applications. Int J Comput Appl. https://doi.org/10.1080/1206212x.2023.2199966

Kjäer M (2023) Analyzing the use of blockchains for challenges in inter-organizational business processes. https://doi.org/10.1109/ICSA-C57050.2023.00039

Kongmanas Y, Pournader M, Seuring S (2023) Blockchain is the 'trust-building machine' for supply chain management. Ann Oper Res 327(1):49–88. https://doi.org/10.1007/s10479-022-04868-0

Kouhizadeh M, Sarkis J (2018) Blockchain practices, potentials, and perspectives in greening supply chains. Sustainability 10(10): Article 3652. https://doi.org/10.3390/su10103652

Lau Y, Yip TL (2023) Roles of blockchain technology in supply chain capability and flexibility. Sustainability. https://doi.org/10.3390/su15097460

Mazouzi H, Khannous A, Amechnoue K, Rghioui A (2023) Security challenges facing blockchain based-IoV network: a systematic review. Int J Adv Comput Sci Appl. https://doi.org/10.14569/ijacsa.2023.0140526

Melnyk SA, Schoenherr T, Speier-Pero C, Peters C, Chang JF, Friday D (2021) New challenges in supply chain management: cybersecurity across the supply chain. Int J Prod Res. https://doi.org/10.1080/00207543.2021.1984606

Mezquita Y, Podgorelec B, Gil-González AB, Corchado JM (2023) Blockchain-based supply chain systems, interoperability model in a pharmaceutical case study. Sensors 23:1962. https://doi.org/10.3390/s23041962

Naje A, Shukla VK, Gupta D, Ojha DB (2022) Addressing cloud security challenges through blockchain technology. https://doi.org/10.1109/ICRITO56286.2022.9964758

Nguyen AM (2023) Challenges of blockchain applications in digital health: a systematic review. arXiv.org. https://doi.org/10.48550/arXiv.2304.04101

Ospital P, Masson D, Béler C, Legardeur J (2022) Toward product transparency: communicating traceability information to consumers. Int J Fash des, Technol Educ. https://doi.org/10.1080/17543266.2022.2142677

Pal O, Srivastava V, Alam B (2017) Cyber security risks and challenges in supply chain. Int J Adv Res Comput Sci. https://doi.org/10.26483/IJARCS.V8I5.3385

Paluri AR, Mishal A (2020). Trust and commitment in supply chain management: a systematic literature review. Benchmarking: Int J. https://doi.org/10.1108/BIJ-11-2019-0517

Prajapati S, Sawant T, Solanki S (2023) Strengthening supply chain integrity with blockchain-based anti-counterfeiting measures. https://doi.org/10.1109/ICIDCA56705.2023.10100264

Reza S, Salameh HK, Kashani Zadeh H, Vasefi F (2023) Toward an intelligent blockchain iot-enabled fish supply chain: a review and conceptual framework. Sensors 23(11):Article 5136. https://doi.org/10.3390/s23115136

Salha R, El-Hallaq M, Alastal A (2019) Blockchain in smart cities: Exploring possibilities in terms of opportunities and challenges. J Data Anal Inf Process 7:118–139. https://doi.org/10.4236/jdaip.2019.73008

Sali N, Biwalkar S, Swamy A, Khadilkar N, Sahu S (2023) Blockchain in food supply chain: benefits, challenges and review. Int J Sci Technol Eng. https://doi.org/10.22214/ijraset.2023.50661

Sarao PK, Kaur N, Sharma MK (2023) Blockchain: a review study of protocols and enterprise frameworks. https://doi.org/10.1109/cictn57981.2023.10141434

Schiffer M, Dörr DM (2020) Development of the supply chain management 2040—opportunities and challenges. https://doi.org/10.15488/9644

Shoaib M, Zhang S, Ali H (2023) A bibliometric study on blockchain-based supply chain: a theme analysis, adopted methodologies, and future research agenda. Environ Sci Pollut Res 30(6):14029–14049. https://doi.org/10.1007/s11356-022-24844-2

Strohmier H (2023) The importance of cybersecurity in supply chain. https://doi.org/10.1109/ISDFS58141.2023.10131834

Subramanian N, Joshi A, Bagga D (2023) Transparent and traceable food supply chain management. arXiv.org. https://doi.org/10.48550/arXiv.2305.12188

Taschner A, Charifzadeh M (2020) Challenges in supply chain management and supply chain management accounting. https://doi.org/10.1007/978-3-658-28597-5_3

Venkatesh VG, Kang K, Wang B, Zhong RY, Zhang A (2020) System architecture for blockchain-based transparency of supply chain social sustainability. Robot Comput-Integr Manuf 63:Article 101896. https://doi.org/10.1016/j.rcim.2019.101896

Wang H (2023) Supply chain resources and economic security based on artificial intelligence and blockchain multi-channel technology. Int J Inf Technol Syst Approach. https://doi.org/10.4018/ijitsa.322385

Wang S, Tian G, Islam T, Khaskhelly FZ (2023) Effect of blockchain technology for sustainable performance in supply chain management. https://doi.org/10.35784/preko.3954

Weng CE, Becker M (2022) An implementation of trust chain framework with hierarchical content identifier mechanism by using blockchain technology. Sensors 22(13):Article 4831. https://doi.org/10.3390/s22134831

Xia L, Baghaie S, Sajadi SM (2024) The digital economy: Challenges and opportunities in the new era of technology and electronic communications. Ain Shams Eng J 15(2):102411. https://doi.org/10.1016/j.asej.2023.102411

Yavaprabhas Y, Pournader M, Seuring S (2023) Blockchain is the 'trust-building machine' for supply chain management. Ann Oper Res 327(1):49–88. https://doi.org/10.1007/s10479-022-04868-0

Yontar E (2023) The role of blockchain technology in the sustainability of supply chain management: Grey based dematel implementation. Clean Logist Supply Chain. https://doi.org/10.1016/j.clscn.2023.100113

Zhou S, Li K, Xiao L, Cai J, Liang W, Castiglione A (2023) A systematic review of consensus mechanisms in blockchain. Mathematics. https://doi.org/10.3390/math11102248

Zoughalian K, Marchand J, Ghita B (2022) A blockchain-secured pharmaceutical distribution system to fight counterfeiting. Int J Environ Res Public Health 19(7):Article 4091. https://doi.org/10.3390/ijerph19074091

Blockchain for Banking

Vijaya Kittu Manda and Akhil John Mampilly

Abstract Blockchain has several inherent features that address the various banking sector issues. Though Blockchain can disrupt the functioning of banks because of features that eliminate intermediaries, global regulations do not favor such a drastic step. This chapter discusses decentralization and distributed ledger features, immutability and tamper-resistance, security through cryptography, smart contracts and process automation, and transparency and auditability that help the current banking sector. A brief comparison of Blockchain technology and traditional banking systems is provided. The chapter discusses specific banking applications such as cross-border payments and remittances, trade financing, supply chain finance, know-your-customer (KYC), regulatory compliance, remittance, collateral management, and asset tokenization, where clear benefits can be derived. The chapter then discusses challenges and regulatory hurdles, and the future of Blockchain in banking. Overall, the technology solves existing problems and opens new opportunities for the banking sector. The discussion in the chapter benefits researchers, industry professionals, and regulators by helping them explore the technology and strengthen the technology infrastructure of the banking sector.

1 Introduction

The banking and financial services (BFS) industry is constantly changing and stands to be a beneficiary of the technological revolution that is happening simultaneously. Technology is a financial inclusion enabler that attracts banks to invest heavily in technology regularly. They are always in a race against each other because of competition in the banking sector to ensure financial transactions are done safely and securely.

V. K. Manda (✉)
PBMEIT, Visakhapatnam, Andhra Pradesh, India
e-mail: vijaykittu@hotmail.com

A. J. Mampilly
Lincoln University College (LCN), Petaling Jaya, Selangor, Malaysia

S. Krishnan et al. (eds.), *Unveiling Blockchain Enabled Business Applications*, Contributions to Management Science, https://doi.org/10.1007/978-981-95-2337-5_12

Organizations deploying the technology are found to have a strategic advantage (Mbaidin et al. 2024).

Blockchain technology has emerged as a critical technology potentially transforming banking operations and services, attracting immense interest from researchers (Saravanan et al. 2020). Blockchain originated as the underlying framework for cryptocurrencies, especially Bitcoin. However, its fundamental characteristics help several industries and sectors, including the banking sector.

Its relevance in banking stems from its unique characteristics—decentralization, transparency, immutability, and security—which collectively address longstanding industry challenges. By its basic features, Blockchain can address typical problems encountered in the banking and finance space—such as double-spending and ensuring fair, direct transactions without intermediaries, among others. Features such as a secure, transparent, and collaborative environment for financial transactions are always sought after by the Banking sector. Blockchain platforms like Ethereum, Corda, and Hyperledger Fabric are popular for implementing banking systems.

Banks must re-engineer their existing systems and workflows to capitalize on their full capabilities. Participating institutions must align technical standards and business rules to realize network effects. Regulators also play a vital role in balancing innovation against risks like consumer protection, data privacy, and anti-money laundering compliance. Consortium projects, therefore, engage regulators to address such concerns through transparent experimentation under pre-emptive guidance.

2 Features of Blockchain in Banking

Blockchain technology has core features that can bring a dramatic shift in the banking sector, thereby revolutionizing it. The banking sector is a primary beneficiary due to its inherent features. Blockchain brings a secure, transparent, decentralized approach to financial transactions (Mampilly et al. 2024). Blockchain implementations in banking operations offer several key features that can enhance efficiency, reduce costs, and mitigate risks. This sub-section explains the salient features of Blockchain in the context of banking applications.

Blockchain, alongside Big Data and eXtensible Business Reporting Language (XBRL), has all the essential features that help in accounting and auditing. These technologies, for example, can provide real-time reporting, reduce doubtful debts, reduce tax evasion, and obtain digital confirmation of transactions (Weshah 2024).

Deployment of Blockchain in Banks and Financial Services depends on four Critical Success Factors (CSFs)—Organizational Readiness, Regulatory Compliance, Technical Expertise, and Financial Resources (Mbaidin et al. 2024).

Blockchain can work together with Near-field communication (NFC) of the Internet of Things (IoT) to secure transactions in mobile banking (Chorey and Sahu 2024). Blockchain has numerous use cases for smart city infrastructure and implementation (Krishnan et al. 2021). Due to the increased importance of green technology, Green Blockchain systems have also emerged (Saravanan et al. 2023).

2.1 Decentralization and Distributed Ledger

Blockchain runs on the concept of a distributed ledger system on a decentralized network. This is in contrast to the traditional banking systems that primarily rely on central authorities or intermediaries responsible for maintaining and controlling the ledgers of transactions. This eliminates the need for banking and financial intermediaries, and the transactions are processed end-to-end. With this, the importance and role of correspondent banks come down.

The complete transaction ledger of the Blockchain is replicated and distributed across multiple nodes in the network. While this might look like wasting precious network storage resources, it ensures that no single entity has complete control over the data, reducing the risk of single points of failure and increasing system availability and resilience. Furthermore, it eliminates the need for intermediaries, potentially reducing banking institutions' operational costs and transaction fees (Attia and Abed 2024).

It facilitates peer-to-peer transactions and ensures direct interaction between parties. This translates to improved efficiency and reduced transaction costs in the banking sector. Decentralization also enhances resilience against systemic failures, as the distributed nature of Blockchain prevents a single point of failure, thereby bolstering the stability and reliability of banking operations.

2.2 Immutability and Tamper-Resistance

Blockchain technology's most significant features are its immutability and tamper-resistant nature. In the banking sector, where data integrity and record-keeping are paramount, the immutable ledger provided by Blockchain can ensure the authenticity and non-repudiation of transactions. Blocks are linked with each other using a cryptographic hashing technique. Thus, a chain of blocks comprising an unalterable chronological record is created. Attempts to modify or alter the data stored on the Blockchain would require an immense computational effort, making it impractical and infeasible. Hence, tamper-proofing records (data immutability) can bring in banking operations' integrity, trust, and reliability, allowing banks to reduce fraud, prevent data manipulation, and reduce reconciliation discrepancies. Audit processes can happen in parallel and smoothly, making Blockchain a tool that financial auditing and compliance professionals love.

2.3 Security Through Cryptography

Blockchain uses several advanced cryptographic tools and techniques to ensure the security of transactions. The list includes:

1. Hash functions (such as SHA-256 and above)
2. Public Key Cryptography (Asymmetric Cryptography)
3. Digital Signatures and Elliptic Curve Digital Signature Algorithm (ECDSA)
4. Merkle Trees (Hash Trees)
5. Peer-to-Peer Network Encryption.

Public-key cryptography generates unique digital identities for participants, enabling secure authentication and authorization processes. Cryptographic hashing algorithms validate and verify transactions and check the data integrity.

All transactional data is encrypted and stored in blocks linked (chained) with each other. This makes it virtually impossible to tamper with the data. This level of security can help the banking sector because of the need to protect sensitive financial information. These robust security mechanisms can significantly mitigate cyber threat risks, such as hacking and fraud, data breaches, and unauthorized access, which are significant concerns for banking institutions dealing with sensitive financial information. The cryptographic security features help safeguard assets and personal information much better than traditional banking systems. These considerations are becoming important as the world moves toward open banking (Jeng 2022).

2.4 Smart Contracts and Process Automation

Blockchain 2.0 introduced the concept of Smart Contracts (Ahmed et al. 2024). They are self-executing computer programs that legally implement decisions. The decisions can be predefined business rules, contract terms, and conditions coded and put onto the Blockchain.

Smart contracts can automate various processes in the banking sector. Some applications include loan origination and disbursements, interest payments, securities trading, and regulatory compliance. Because of the automation of complex financial agreements, there is less need for manual interventions and third-party verification. Associated costs that are incurred can be saved. Token contracts are smart contracts responsible for issuing tokens, such as to making payments. Additionally, smart contracts can enhance transparency by providing an auditable trail of contract execution, increasing stakeholder trust and accountability. Overall, they improve operational efficiency and reduce administrative costs.

2.5 Transparency and Auditability

Blockchain has inherent transparency and auditability features. Every transaction on the Blockchain is visible to all participants, ensuring all actions are traceable and accountable. This makes it easy for the auditors to express their opinions on the transactions following the audit reporting frameworks. The features can significantly benefit the banking sector because the industry is blamed for its opaque processes and lack of visibility. The distributed nature of the Blockchain ledger allows authorized participants to access and verify transactions, promoting transparency and enabling effective auditing (Bejjar and Siala 2024). This feature can enhance banking institutions' regulatory compliance, risk management, and governance.

Furthermore, the immutable and chronological record of transactions facilitates comprehensive auditing, enabling regulatory bodies and internal auditors to monitor and assess banking activities efficiently. The transactions can be monitored in real-time and help in compliance and anti-money laundering (AML) efforts.

3 Blockchain vs. Traditional Banking

Traditional banking systems have been functioning for centuries and comprise the well-established infrastructure of centralized institutions. These institutions, including commercial banks, central banks, and regulatory bodies, act as intermediaries, facilitating financial transactions and ensuring systemic stability. The centralized operational models of the industry led to issues related to efficiency, security, and transparency. These classical limitations led to the exploration of alternative models and technologies. Blockchain is a leading technology from the Industry 4.0 disruption technologies stack.

Figure 12.1 lists fundamental Blockchain features that contrast traditional banking systems, which are centralized, opaque, and manually intensive processes. Blockchain suggests a new way to conduct and record banking and financial transactions that challenge the traditional centralized banking model. The contrast between Blockchain and Traditional banking lies in their underlying architectures, governance structures, and operational mechanisms. The technology features promise to bring innovative updates to the banking sector.

In traditional banking systems, the centralized intermediaries facilitate transactions, maintain ledgers, and ensure participant trust. These institutions act as trusted third parties and verify and authorize transactions. They maintain control over information and fund flow in the banking system. The intermediaries bring integrity to the financial system. However, the centralized approach comes with problems—they lead to single points of failure, are prone to cyberattacks, and have loopholes that pave the way for data manipulation or censorship. Conversely, Blockchain overcomes all these using cryptography and distributed ledger technology (DLT) to establish a trustless environment.

The transaction records are concurrently maintained across multiple nodes of the network. This mitigates the risks associated with central points of failure. It also leads to enhanced transactional throughput. The transactions are cryptographically secured and validated using methods called consensus mechanisms. These are done by a network of participants instead of individuals. They thus can remove the need for central custodians. This promotes greater transparency and immutability because

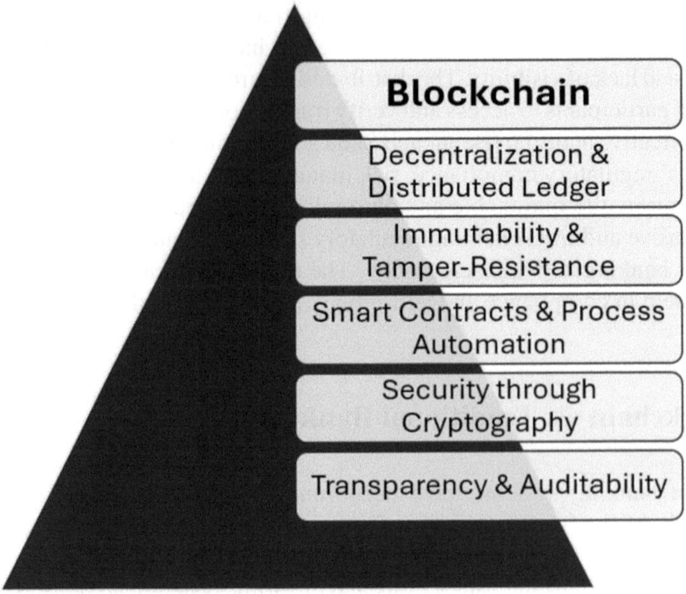

Fig. 12.1 Key features of Blockchain. *Source* Author compilation

all validated transactions are permanently recorded on the shared ledger, accessible to authorized participants.

The traditional banking systems follow a hierarchical governance structure model. The regulatory bodies and financial institutions dictate policies and procedures. Decision-making processes are often opaque. The control over financial data and transactions lies within the purview of these centralized entities. Conversely, Blockchain networks are governed by consensus protocols and cryptographic rules, which bring transparency and help in the democratic decision-making process among network participants.

Traditional banking systems depend on complex reconciliation processes and clearing mechanisms. These are considered mandatory activities to ensure the accuracy and consistency of financial records and effective financial management (Onuoha and Amponsah 2012). Reconciliation processes in the traditional systems are often cumbersome. The processes are time-consuming, prone to errors, and susceptible to fraud or manipulation. Blockchain inherently has transparency and facilitates real-time transaction visibility. Cryptographic techniques and distributed consensus mechanisms bring immutability and integrity to transaction features. So, there is no need for reconciliation between the parties (or their transactions). This further reduces the risk of fraud. Every step of a transaction is immutably recorded, creating an audit trail. The data is thus tamper-proof and auditable by authorized participants. This promotes greater trust and accountability within the financial ecosystem.

Traditional banking systems are inaccessible in underserved or remote areas. They often impose barriers to entry for unbanked or underbanked populations. Blockchain technology can help promote financial inclusion (Schuetz and Venkatesh 2020; Galanti and Özsoy 2023) using its decentralized nature and global reach. It can help build a transparent, inclusive, and accountable digital economy (Maupin 2017). So, individuals and businesses can access financial services irrespective of geographic location or socioeconomic status.

Scalability can be a potential issue in Blockchain platforms because they struggle to handle the transaction volumes. Similarly, traditional banking systems are generally designed and optimized to comply with legal and regulatory frameworks. Blockchain systems sometimes challenge the established norms, raising concerns regarding compliance, anti-money laundering (AML) regulations, and Know Your Customer (KYC) requirements. Regulatory bodies and financial institutions are looking for a way to use the technology while remaining compliant.

The differences between the two systems exist in other areas, such as fundamental architectures, governance structures (Truby et al. 2022), operational mechanisms, and regulatory implications. However, the widespread adoption of Blockchain in the banking sector will require addressing regulatory challenges, ensuring compliance, and promoting collaboration between traditional institutions and emerging Blockchain-based solutions.

Blockchain has both disruptive features that bring about a new form of banking. On the other hand, it optimizes the existing banking system. For example, a Blockchain-based electronic cheque clearance can help prevent fraudulent alterations, double spending, and counterfeit e-cheques (Singh et al. 2019). Adopting Blockchain helped commercial banks expand their credit and lending practices by reducing management costs and narrowing the difference between the market equilibrium and the bank deposit in case interest rates have not yet been marketized (Wang 2023).

4 Applications of Blockchain in Banking

Key drivers in the adoption of Blockchain technology include technical implementation challenges, customers' use cases and pilot programs, social and cultural factors, and the need for local customizations (Zhuo et al. 2023), compliance and security enhancements, industry collaboration and knowledge sharing, legacy systems integration and interoperability (Ijeh and Rahayu 2023). Table 12.1 lists the broad range of benefits that the banking sector gets from Blockchain adoption.

4.1 Cross-Border Payments and Remittances

Over the years, payment system-related technologies have improved and made lives happier (Vijaya Kittu et al. 2024). Improved international trade and financial systems,

Table 12.1 Blockchain benefits to the Banking sector

Management indicators	Business indicators	Technology indicators
• Cultural • Regulatory • Governance	• Performance • Business process	• Data quality • Distributed ledger characteristics • Infrastructure • Information Exchange & Transactions

Data Source Attia and Abed (2024)

aided by digitalization, have accelerated this (He 2021) and increased the number of cross-border payment transactions and volume. Cross-border e-commerce (CBEC), too, has witnessed a rise (Chen et al. 2023).

However, the total cost of a typical $200 bank-based cross-border payment service remains high. While the cost of remittances has come down over the years, they are still at highs, with banks charging 12.09% as a fee as of Q2 2023, and they are termed the most expensive type of service provider. The global average cost of sending remittances stands at 6.2%, with UN SDGs targeting to bring it down to 3% by 2030. The costs from G8 countries stand at 5.83%, while from G20 countries stands at 6.43%. The costs stand at 13.9% for remittances from South Africa (The World Bank 2023) (Fig. 12.2).

All cross-border transactions cost a charge involving two components—fee and foreign exchange (FX) margin. Digital channels incur lower costs compared to non-digital (cash-based) transactions. Despite the cost, the settlement time is high at close to ten days in some instances (Zhuo et al. 2023). Traditional cross-border payments are troubled by slow processing times and high fees due to the involvement of multiple intermediaries.

Blockchain-based systems are found to facilitate direct peer-to-peer transactions between banks or individuals, significantly reduce settlement times, cut costs, and help increase gross domestic product (GDP). Countries like Russia were able to use Blockchain systems to mitigate risks associated with increasing sanction pressure, including restrictions on access to international clearing services and payment systems (SWIFT, among others) (Safiullin et al. 2023).

Blockchain will help in faster and more inexpensive remittance services when universal wallet systems that can be used across homogeneous and heterogeneous Blockchains exist (Coutinho et al. 2023). The technology improves the overall efficiency within the global financial system.

IBM's Blockchain World Wire is a popular cross-border payment system.

4.2 Trade Financing

Trade finance is a financial service often provided by banks for international trade by its banking clients. Banks provide funding guarantees and credit facilities, ensuring everyone gets paid, and the goods/services are delivered on agreed terms. Supply

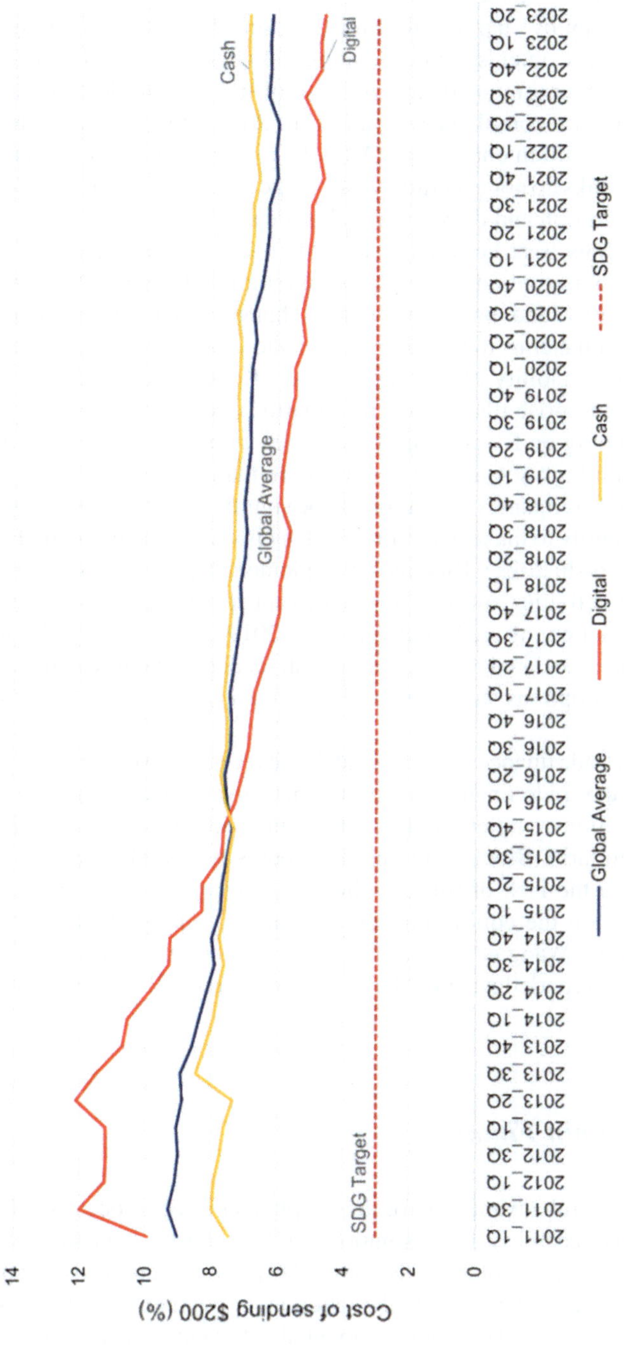

Fig. 12.2 Cost of Remittances are coming down but still are at highs. *Source* The World Bank (2023)

chain and logistics, customs clearances, insurance, and logistics are some industries that involve many trade finance activities.

Banks earn money by charging fees for their services, such as issuing Letter of Credits, managing documents, and mitigating risks. Further, international business activities come with enormous delays because of the need to do counterparty due diligence and contract compliance (Joshi et al. 2024). Small businesses, such as micro, small, and medium enterprises (MSMEs), often do not get sufficient export support because banks do not generally wish to take risks. Blockchain as a technology can bring value addition and disruption to this area.

Blockchain brings trust between stakeholders such as exporters and importers without needing to depend on a third party (Ibrahim and Truby 2022). It improves transactional security and promotes data exchanges, facilitating the expression of benevolence, enhancing the efficiency and the quality of communication, and increasing the predictability of trading partners (Kowalski et al. 2021). A significant increase in enterprise trade credit information disclosures was observed when systems were running on Blockchain. Enterprises with weak bargaining power and those in areas with high social trust benefit more than others (Luo et al. 2023). Of course, established companies were found to be significantly and positively impacted because of their ability to make better financial decisions upon deploying Blockchain (Basumatary and Joshi 2022). Blockchain implementations that use smart contracts automate certain activities and potentially disrupt traditional international finance products such as letters of credit (Chang et al. 2019) because they eliminate non-payment risks and associated delays. Smart contract deployments can help in invoice trading, but such adoptions should be evaluated on a case-by-case basis (Wang and Xu 2022).

Decentralized trade finance is considered the next evolution of the concept wherein the public owns the trade finance model, the platform provides liquidity, and upon successful transaction completion, the participants will be incentivized (Kumar and Amin 2022). Similarly, Blockchain-based accounts receivable chains are in practice where despite the lack of full coordinates of a supply chain, Blockchain was able to help increase the equilibrium order quantity, supply chain efficiency, and participants' profits (Wang et al. 2023). Barclay's Ornua L/C, HSBC—Corda L/C, MizuhoMarubeni Hyperledger, and Marco Polo are popular Blockchain trade finance platforms.

4.3 Supply Chain Finance

Blockchain has several advantages for the supply chain sector because of improved transparency and traceability. It has solutions for the industry's accounting, accountability, and assurance problems (Rijanto 2024). Smart contracts help banking and financial service providers make automatic payments and release funds when conditions are met. This reduces manual intervention and reduces payment delays. Blockchain-based payment platforms help in invoice financing, receivables discounting, reverse factoring, and payables financing by connecting buyers, suppliers, and financial institutions on a single platform. Newer forms of financial solutions have emerged, improving liquidity and transaction processing at a lower cost.

Popular supply chain finance platforms include Chainyard, TradeIX, BRD, Veem, Inception Finance, and TradeFinancer.

4.4 KYC & AML

Banks and financial institutions have to mandatorily collect, check, and periodically ensure that the KYC information of their customers is maintained in the database as per the Anti-Money Laundering (AML) regulations. Each institution has to collect the KYC information separately, and hardly any shared infrastructure has been created for this purpose. In addition, the process and requirements differ from country to country and are usually specified by the central banks of the respective countries.

Banking firms find the traditional Know-Your-Customer (KYC) system unreliable, costly, and time-consuming. The Thomson Reuters Survey says KYC updates are delayed by 3 to 4 months! The situation forces banks to explore and invest in various technologies (Mansoor et al. 2023).

Centralized KYC (CKYC) has come into force, with institutions agreeing to share the KYC details once any of the institutions capture them. This reduces the burden on the customer of sharing KYC documents repeatedly. Further, electronic KYC (eKYC) usage has become common, doing away with paperwork. While this reduces the need to repeat the submission of documents, a few of the old KYC problems persist, for which technologies such as Blockchain come into the picture.

Blockchain-based KYC, in conjunction with distributed data storage technologies like Interplanetary File Systems (IPFS) (Mamun et al. 2020), helps run a KYC system at a fractional cost, is easy to update and verify, gives secured data storage services, and, with off-chain systems, allows home-branch-based KYC update and verification. Blockchain's immutable and transparent nature facilitates strong KYC and AML compliance within the banking sector. The shared ledger provides a secure platform for storing and verifying customer identities and transaction histories. This streamlines compliance processes for banks and reduces the risk of fraud and financial crimes. Hyperledger Fabric is considered a popular platform for KYC implementation (Ullah et al. 2021). Researchers are also working on using Biometrics with Blockchain to offer superior KYC services (Sarier 2024).

4.5 Regulatory Compliance

Banks spend substantial resources on meeting regulatory compliance. While specific regulatory requirements are globally applicable, others are regional and must be maintained to meet the national banking regulator requirements. The concept is constantly evolving, the latest being the EU Anti-Money Laundering and Countering the Financing of Terrorism Framework (Kaiser 2024).

Blockchain can help banks provide a tamper-proof, immutable record of transactions for audit. The complete audit trail by regulators can be done in parallel because all the data is available on the distributed ledger and accessible from any node. Reporting can be automated and streamlined, allowing banks to provide real-time data to regulators. Smart contracts can automate specific compliance processes, such as "Know Your Customer" (KYC) and Anti-Money Laundering (AML) checks, and ensure that they are conducted accurately and consistently.

The adoption of Blockchain systems in banks requires serious consideration because of stringent regulatory requirements (Yerram et al. 2021). Regulatory technologies are vital as the European Union approaches the Digital Decade by 2030 (Valkanov 2023). Some regulatory bodies have set up Blockchain-based "regulatory sandboxes" to allow banks and fintech companies to test new financial services in a controlled environment. This helps in ensuring compliance before a full-scale launch.

4.6 Remittance

Bitcoin is one of the most popular remittance tools and is the second most popular option after traditional bank transfer (Apatu and Goudar 2024). Also, remittance is one of the biggest reasons El Salvador treats Bitcoin as a legal tender. The country gets 25% of its national income from international remittances (Ba and Şen 2024).

4.7 Credit Scoring, Disbursement, and Payment Defaults

Blockchain technology can help traditional banks that follow traditional lending practices. Technology can also bring in disruptive changes that lead to newer forms of lending, which often require regulatory approvals before such things can be implemented. Significant themes in credit management include fintech credit appraisal, credit default learning algorithms, the efficiency of credit assessment (Jain et al. 2023), and interest rate marketization (Wang 2023), among others.

The immutable system of storing credit scores and related records allows small and medium-sized enterprises (SMEs) to showcase their low-risk and high-quality, and access bank credit without many hassles, sometimes even when they do not have sufficient collateral (Wang et al. 2019; Joshi et al. 2024). Blockchain with machine learning can predict credit default probabilities, allowing institutions and regulators to prevent the misuse of financial assets (Tong et al. 2024).

Credit scoring is an essential criterion of the creditworthiness of an individual or an entity and is a crucial step that decides various aspects of a credit disbursement. The technology eliminates the need for intermediaries such as credit rating agencies (CRA) to generate and update credit scores. Several existing models come with limitations, such as excessive centralization, fake data identification, and data transmission lag (Wang 2020b). Blockchain systems propose to address the defects and make the overall process better (Wang 2020a).

The new Blockchain-based system reduces the paperwork required for credit score evaluation (Hassija et al. 2020). It can create and secure the digital identity of individuals (including underprivileged people such as those from rural areas or those who do not have proper bank accounts) (Jain et al. 2019). Securely stored transactional records can be fed to a machine learning model, such as a random forest model, for credit data feature extraction, feature selection, credit model construction, and model evaluation to arrive at efficient and accrued credit scores (Yang et al. 2022). Several creditworthiness clouds have been developed as public Blockchain clients and creditworthiness lookup services (Li et al. 2018).

The disbursement process can be authorized and automated based on the score using smart contracts (Patel et al. 2021). The technology also showed the ease of credit supervision and improved information management (Li 2023).

While the above discussion deals with credit in traditional banking, an implementation called Decentralized Finance (DeFi) has tools, techniques, and systems that allow for a new concept called Decentralized Credit (Packin and Lev Aretz 2023). As a part of this, the system determines creditworthiness using smart contracts. This is done using a more inclusive and efficient method than traditional systems. Traditional credit bureaus like TransUnion are exploring partnerships with Blockchain-based platforms to integrate decentralized credit scoring into their existing frameworks.

4.8 Collateral Management

Collateral management involves identifying, valuing, and managing assets pledged by borrowers to secure loans. Several challenges are frequently encountered in the traditional mechanisms governing these processes. There are gaps in valuation, protracted verification procedures, and the risk of fraud. It is a risk management feature that brings financial stability and ensures regulatory requirements are met (Goblet et al. 2024).

Blockchain has features that help banks ensure efficient and secure collateral management. Banks can integrate Blockchain into their existing collateral management systems and benefit from real-time transparency and immutable record-keeping. This allows for instant verification of collateral ownership and valuation. Because of this, there is no need for centralized verification bodies. Smart contracts automate the enforcement of collateral-related agreements, streamlining operations and reducing the administrative burden on banking institutions.

Blockchain provides a single source of truth regarding the status and valuation of assets. So, the risk of over-collateralization and under-collateralization is eliminated (Park and Kahn 2019). By enabling dynamic, real-time revaluation of assets, banks can more accurately manage loan-to-value ratios, enhancing the responsiveness of collateral management to market fluctuations. Furthermore, utilizing Blockchain technology in this domain inherently increases the security of collateral management processes. Cryptographic data protection and eliminating centralized points of vulnerability render the system highly resilient to tampering and fraudulent activities.

Banks can benefit from new opportunities within collateral management that emerge from Blockchain adoption. Tokenization of physical assets on Blockchain platforms, for example, can facilitate the creation of new financing models, such as asset-backed lending on decentralized finance (DeFi) platforms (Manda et al. 2024). This integration of Blockchain technology thus leads to a significant evolution in collateral management practices that align them more closely with the industry demands.

4.9 Asset Tokenization & Fractional Ownership

Tokenization is converting rights on an asset into a digital token in a Blockchain system (Thomason and Ivwurie 2023). It is a critical component in the Web 3.0 economy. It helps improve liquidity, reduce settlement times, and open new investment opportunities. Using this, customers benefit from buying, selling, and trading assets securely and transparently at affordable denominations.

Asset tokenization represents a transformative application of Blockchain technology within the banking sector (Ferrara and Santoro 2023). It entails the fractionalization (fractional ownership) of real-world assets, such as real estate, artworks, private equity, or intellectual property, into digital tokens (Lacity and Treiblmaier 2022). This allows for the disaggregation of assets to be sold in smaller denominations. Smaller denominations bring more liquidity and involvement in the previously inaccessible asset class product. Tokens are issued on a Blockchain network according to standards like ERC-20, which defines token creation and transaction rules. These tokens reside on a Blockchain ledger, representing ownership rights and facilitating secure, transparent trading on secondary markets. Purchasers can acquire and trade these tokens just like any other cryptocurrency. The underlying asset backs the tokens and provides token holders with ownership rights proportional to the number of tokens owned (Sazandrishvili 2020).

There will be high entry barriers for illiquid assets like private equity or real estate, one being the difficulty of valuing them (Brandon et al. 2022). One reason for this is the need for significant capital. Tokenization addresses this barrier by enabling fractional ownership. Investors can purchase portions of a more significant asset represented by buying tokens in the assets. Thus, tokenization democratizes access to previously exclusive investment opportunities. This brings in greater liquidity for illiquid assets and expands the investor pool. The risk of fraud and errors in asset transactions can be reduced with immutability and transparency.

Furthermore, smart contracts and self-executing code embedded within the Blockchain can automate various aspects of asset management. For instance, smart contracts can govern dividend distribution based on token ownership, streamline voting rights for token holders, and trigger automatic asset redemption upon predefined conditions. This programmable functionality enhances efficiency and transparency within asset management processes.

Implementing asset tokenization in banking also streamlines the settlement process because transactions executed on a Blockchain can be settled in real time, reducing the need for intermediaries. By doing so, the associated costs and time delays can come down, benefiting banks. Customer experience will improve by giving them faster access to funds and investment opportunities. Banks can use this new "excellent customer service" idea to build new services (Attia and Abed 2024). Moreover, the programmable nature of smart contracts on Blockchain platforms automates compliance, ensuring that all transactions adhere to regulatory requirements and contractual obligations automatically, mitigating risk and operational overhead.

Developers are experimenting with adding programmability to tokens to enable features like binding tokens to revenue or profit sharing from the asset. Additional regulations may be needed to address disclosure standards and investor protection concerns. Regulatory frameworks surrounding tokenized assets remain nascent, posing challenges for widespread adoption. Classifying these assets and developing clear regulatory guidelines are essential for mitigating bank risks and ensuring investor protection.

5 Challenges and Regulatory Hurdles

While Blockchain brings various advantages to the banking sector, several technology adoption challenges will act as obstacles and delay the broad implementation of the technology. These challenges are discussed in eight dimensions as follows:

1. **Regulatory Compliance**: Regulatory compliance is a requirement in all jurisdictions. Traditional banking systems are well-integrated and fit perfectly into the existing regulatory frameworks. They are designed and updated regularly to bring financial stability, protect consumers, and prevent financial crimes in the economy. The global and decentralized nature of Blockchain makes compliance difficult. Adapting these frameworks to accommodate the decentralized and borderless nature of Blockchain remains a challenge.

    a. **Uncertainty**: Many countries and jurisdictions still lack clear regulations and guidelines for Blockchain and cryptocurrencies. Banks feel this is a considerable uncertainty and is challenging to deal with.

b. **Anti-Money Laundering (AML) and Know Your Customer (KYC)**: Ensuring compliance with AML and KYC regulations is challenging on a decentralized network. The identity of participants can be more difficult to verify because of the usage of features such as Privacy Enhancement Technologies (PET).
c. **Cross-Border Transactions**: Cross-border transactions involve transactions done in different banking jurisdictions. Blockchain for international transactions may or may not comply with respective compliance and regulations.

2. **Privacy and Data Protection**:

a. **Data Privacy Laws**: Implementing and staying compliant with data protection regulations, such as the General Data Protection Regulation (GDPR) in the European Union, will be challenging. For example, the right to be forgotten is one such challenge because of the immutable ledger feature.
b. **Confidentiality**: Banks deal with a lot of confidential and private information. Ensuring that data on a Blockchain is secure and cannot be accessed by unauthorized parties is a significant challenge.

3. **Security Concerns**:

a. **51% Attacks**: Although rare, a 51% attack is possible. In this, a user or a group can gain control of most of the mining power and manipulate the network.
b. **Smart Contract Vulnerabilities**: Flaws in smart contract code can lead to security vulnerabilities, potentially resulting in the loss of funds or data breaches.

4. **Scalability and Performance**:

a. **Transaction Throughput**: Traditional banking systems process thousands of transactions per second, a rate that current Blockchain solutions struggle to match. This is a substantial practical constraint to the widespread adoption of Blockchain in banking.
b. **Latency**: The time taken to finalize transactions (commit) on a Blockchain can be longer than that of traditional systems. User experience can be affected, and certain transactions may time out and may not complete.

5. **Integration with Existing Systems**:

a. **Legacy Systems**: Integrating Blockchain technology with existing banking systems and infrastructure can be complex and costly.
b. **Interoperability**: Ensuring that different Blockchain systems can communicate and work together seamlessly is essential for widespread adoption.

6. **Operational Challenges**:

a. **Cost Implications**: Initial setup costs can be significant, including technology investment and training personnel.

   b. **Change Management**: Adopting Blockchain technology requires substantial changes in internal processes, which can meet resistance from staff and management.

7. **Legal and Contractual Issues**:

   a. **Legal Status of Smart Contracts**: The enforceability of smart contracts as legal contracts is still being debated in many jurisdictions.
   b. **Jurisdiction**: Determining the applicable jurisdiction and laws for transactions recorded on a global, decentralized network presents challenges.

8. **Standardization**:

   a. **Lack of Standards**: The absence of industry-wide standards for implementing Blockchain technology can hinder interoperability and the development of best practices.

6 Future of Blockchain in Banking

Research progress connecting Blockchain and Banking is happening in two directions:

1. Technological advancements are changing the way banking is done. These include census mechanisms and scalability, storage, and regulation, as well as the selection of appropriate Blockchain platforms.
2. Introducing new financial products helps banks expand their offerings in the industry. This includes products like Digital currencies (including cryptocurrencies) and peer-to-peer lending.

Blockchain technology holds immense transformative potential for the banking sector. While initial applications have centered around cross-border payments and trade financing due to these activities' laborious and costly nature, Blockchain is positioned to disrupt a far wider range of processes. Areas such as automated compliance, financial inclusion, and the future of banking infrastructure itself might witness the most dramatic shifts as regulatory frameworks accommodate distributed ledger technologies, and new permissioned Blockchain networks will offer instantaneous KYC checks across jurisdictions. This can increase the satisfaction of existing customers (Sachin et al. 2023) and potentially result in hundreds of millions entering the formal financial system. Cryptographically secure ecosystems may shift compliance duties from discrete tasks onto self-regulating code. Regulatory sandboxes will be crucial in responsibly calibrating such systems and even aid in financial inclusion (Jenik and Lauer 2017). On the infrastructure front, private Blockchains or hybrid architectures blending legacy databases with immutability layers can streamline core banking frameworks (Johansson and Charpentier 2017). Remittance networks, collateral registries, syndicated lending platforms, and financial market utilities represent operational improvements that banks can make with Blockchain.

Scalability, bandwidth, and cross-chain (Ou et al. 2022) operability challenges must be addressed. As costs are reduced through technological progress, more thoughtful implementations and applications emerge (Niranjanamurthy et al. 2019). Blockchain innovations show creative ways of reducing cyber threats and building trust and confidence when doing financial transactions (Yerram et al. 2021). Its relevance to banking will grow steadily over the long run. Prudently, progress in these areas can happen through cooperation between banking groups (Cucari et al. 2022) and overseers to use the technology to its fullest potential, creating value across the entire financial services spectrum.

7 Conclusion

The key features of Blockchain technology and how they can transform various functions within the banking sector are discussed in the chapter. By using the decentralized and immutable nature of distributed ledgers, along with features like smart contracts and cryptography, Blockchain holds the potential to streamline many banking processes like cross-border payments, trade financing, regulatory compliance, and remittances. The applications of Blockchain outlined here, from asset tokenization to credit scoring, show how this technology may shift certain activities from centralized intermediaries to peer-to-peer models. However, significant challenges around scalability, integration, regulations, and consumer adoption remain to be addressed. Banks can gain transparency, security, and efficiency benefits if Blockchain technologies are implemented successfully by continuing to explore opportunities, address hurdles, and work with regulators. Overall, Blockchain has the prospect of reshaping tomorrow's banking industry significantly.

References

Ahmed M, Ahmad A, Zeshan F, Turab H (2024) Blockchain basics: a deep dive into the blocks, chains, and consensus. In: Darwish D (ed) Advances in web technologies and engineering. IGI Global, pp 57–89. https://doi.org/10.4018/979-8-3693-1532-3.ch003

Apatu E, Goudar P (2024) Bitcoin use cases: a scoping review. Challenges 15(1):15. https://doi.org/10.3390/challe15010015

Attia M, Abed AH (2024) A comprehensive investigation for quantifying and assessing the advantages of blockchain adoption in banking industry. 2024 6th International Conference on Computing and Informatics (ICCI), 322–331. https://doi.org/10.1109/ICCI61671.2024.10485028

Ba H-L, Şen ÖF (2024) Explaining variation in national cryptocurrency regulation: implications for the global political economy. Rev Int Polit Econ: 1–24. https://doi.org/10.1080/09692290.2024.2325403

Basumatary A, Joshi S (2022) Adoption of blockchain in trade finance and its impact on financial decision making. 2022 International Conference on Decision Aid Sciences and Applications (DASA), 556–559. https://doi.org/10.1109/DASA54658.2022.9765150

Bejjar MA, Siala Y (2024) The impact of blockchain technology on the financial audit: In Alqatan A, Hussainey K, Hamed M, Benameur K (eds) Advances in finance, accounting, and economics. IGI Global, pp 272–300. https://doi.org/10.4018/979-8-3693-1678-8.ch012

Brandon RG, Hoesli M, Shan J (2022) The valuation of illiquid assets: a focus on private equity and real estate. J Altern Investments 25(1):111–128. https://doi.org/10.3905/jai.2022.1.163

Chang SE, Luo HL, Chen Y (2019) Blockchain-enabled trade finance innovation: a potential paradigm shift on using letter of credit. Sustainability 12(1):188. https://doi.org/10.3390/su12010188

Chen J, Lan Y, Chang Y (2023) Consumer behaviour in cross-border e-commerce: systematic literature review and future research agenda. Int J Consum Stud 47(6):2609–2669. https://doi.org/10.1111/ijcs.12969

Chorey P, Sahu N (2024) Framework for dynamic block sizing in secured and efficient transaction systems. J Integr Sci Technol 12(5):1–7

Coutinho K, Khairwal N, Wongthongtham P (2023) Towards a truly decentralized blockchain framework for remittance. J Risk Financ Manag 16(4):240. https://doi.org/10.3390/jrfm16040240

Cucari N, Lagasio V, Lia G, Torriero C (2022) The impact of Blockchain in banking processes: the Interbank Spunta case study. Technol Anal Strateg Manag 34(2):138–150. https://doi.org/10.1080/09537325.2021.1891217

Ferrara G, Santoro C (2023) Digital twin through physical assets tokenization in Blockchain. In: Fotia L, Messina F, Rosaci D, Sarné GML (eds) Security, trust and privacy models, and architectures in IoT environments. Springer International Publishing, pp 149–166. https://doi.org/10.1007/978-3-031-21940-5_9

Galanti S, Özsoy ÇY (2023) Can Blockchain help improve financial inclusion? A comparative study. J Econ Issues 57(2):438–449. https://doi.org/10.1080/00213624.2023.2200650

Goblet J, Willame F, Grauls J (2024) A mixed-integer linear programming approach for collateral management. Decis Anal J 10:100429. https://doi.org/10.1016/j.dajour.2024.100429

Hassija V, Bansal G, Chamola V, Kumar N, Guizani M (2020) Secure lending: blockchain and prospect theory-based decentralized credit scoring model. IEEE Trans Netw Sci Eng 7(4):2566–2575. https://doi.org/10.1109/TNSE.2020.2982488

He D (2021) Digitalization of cross-border payments. China Econ J 14(1):26–38. https://doi.org/10.1080/17538963.2020.1870272

Ibrahim IA, Truby J (2022) Governance in the era of Blockchain technology in Qatar: a roadmap and a manual for Trade Finance. J Bank Regul 23(4):419–438. https://doi.org/10.1057/s41261-021-00165-1

Ijeh D, Rahayu IP (2023) Investigating the adoption of Blockchain technology for cross-border payments in the Swedish financial industry: a diffusion theory approach through expert interviews [Graduate, Linnaeus University]. https://www.diva-portal.org/smash/get/diva2:1772069/FULLTEXT02.pdf

Jain N, Agrawal T, Goyal P, Hassija V (2019) A blockchain-based distributed network for secure credit scoring. 2019 5th International Conference on Signal Processing, Computing and Control (ISPCC), 306–312. https://doi.org/10.1109/ISPCC48220.2019.8988510

Jain V, Tiwari R, Mehrotra R, Bohra NS, Misra A, Pandey DC (2023) Role of technology for credit risk management: A bibliometric review. IEEE Int Conf Blockchain Distrib Syst Secur (ICBDS) 2023:1–6. https://doi.org/10.1109/ICBDS58040.2023.10346300

Jeng L (ed) (2022) Open Banking, 1st edn. Oxford University Press. https://doi.org/10.1093/oso/9780197582879.001.0001

Jenik I, Lauer K (2017) Regulatory Sandboxes and Financial Inclusion. https://www.cgap.org/sites/default/files/researches/documents/Working-Paper-Regulatory-Sandboxes-Oct-2017.pdf

Johansson T, Charpentier V (2017) Blockchain database; technical background and a reconnaissance on an implementation within the banking industry [Independent thesis Basic level (degree of Bachelor), KTH Skolan för datavetenskap och kommunikation]. https://www.diva-portal.org/smash/get/diva2:1127299/FULLTEXT01.pdf
Joshi A, Ranade M, Patvardhan N, Sharma P (2024) An enquiry into the adoption and effectiveness of blockchain technology in trade finance. 2024 IEEE International Students' Conference on Electrical, Electronics and Computer Science (SCEECS), 1–7. https://doi.org/10.1109/SCEECS61402.2024.10482370
Kaiser R (2024) Compliance requirements in the future EU anti-money laundering and countering the financing of terrorism framework. J Payments Strategy Syst 18(1):20–29
Kowalski M, Lee ZWY, Chan TKH (2021) Blockchain technology and trust relationships in trade finance. Technol Forecast Soc Chang 166:120641. https://doi.org/10.1016/j.techfore.2021.120641
Krishnan S, Balas VE, Golden EJ, Robinson YH, Kumar R (eds) (2021) Blockchain for smart cities. Elsevier
Kumar S, Amin R (2022) Decentralized trade finance using Blockchain and lightning network. Secur Priv 5(6):e260. https://doi.org/10.1002/spy2.260
Lacity MC, Treiblmaier H (eds) (2022) Blockchains and the token economy: theory and practice. Springer International Publishing. https://doi.org/10.1007/978-3-030-95108-5
Li G (2023) Evaluation of credit bank information management system model based on blockchain technology. 2023 IEEE World Conference on Applied Intelligence and Computing (AIC), 995–1000. https://doi.org/10.1109/AIC57670.2023.10263885
Li Y, Liang X, Zhu X, Wu B (2018) A blockchain-based autonomous credit system. 2018 IEEE 15th International Conference on E-Business Engineering (ICEBE), 178–186. https://doi.org/10.1109/ICEBE.2018.00036
Luo Y, Duan J, Li A, Shao J (2023) Blockchain information disclosure and trade credit. Emerg Mark Financ Trade 59(9):3036–3059. https://doi.org/10.1080/1540496X.2023.2203809
Mamun AA, Al Mamun A, Hasan SR, Hasan SR, Bhuiyan MS, Bhuiyan MS, Kaiser MS, Kaiser MS, Abu Yousuf M, Yousuf MA (2020) Secure and Transparent KYC for Banking System Using IPFS and Blockchain Technology. 2020 IEEE Region 10 Symposium (TENSYMP), 348–351. https://doi.org/10.1109/TENSYMP50017.2020.9230987
Manda VK, Abukari AM, Gupta V, Bharathi MJ (2024) Revolutionizing finance with decentralized finance (DeFi): In: Darwish D (ed) Advances in web technologies and engineering. IGI Global, pp 127–150. https://doi.org/10.4018/979-8-3693-1532-3.ch006
Mansoor N, Antora KF, Deb P, Arman TA, Manaf AA, Zareei M (2023) A review of blockchain approaches for KYC. IEEE Access 11:121013–121042. https://doi.org/10.1109/ACCESS.2023.3328536
Mampilly AJ, Manda VK, Perumal CPSM (2024) Blockchain in cybersecurity. In: Chaubey NK, Chaubey N (eds) Advances in information security, privacy, and ethics. IGI Global, pp 49–88. https://doi.org/10.4018/979-8-3693-9225-6.ch003
Maupin JA (2017) Blockchains and the G20: building an inclusive, transparent and accountable digital economy. JSTOR. http://www.jstor.org/stable/resrep05191
Mbaidin HO, Alomari KM, AlMubydeen IO, Sbaee NQ (2024) The critical success factors (CSF) of blockchain technology effecting excel performance of banking sector: case of UAE Islamic Banks. Int J Data Netw Sci 8(1):289–306. https://doi.org/10.5267/j.ijdns.2023.9.024
Niranjanamurthy M, Nithya BN, Jagannatha S (2019) Analysis of blockchain technology: pros, cons and SWOT. Clust Comput 22(S6):14743–14757. https://doi.org/10.1007/s10586-018-2387-5
Onuoha LN, Amponsah EB (2012) Bank reconciliation as a due process imperative for effective financial management. https://www.researchgate.net/profile/Luke-Onuoha/publication/266281448
Ou W, Huang S, Zheng J, Zhang Q, Zeng G, Han W (2022) An overview on cross-chain: mechanism, platforms, challenges and advances. Comput Netw 218:109378. https://doi.org/10.1016/j.comnet.2022.109378

Packin NG, Lev Aretz Y (2023) decentralized credit scoring: black box 3.0. SSRN Electronic Journal. https://doi.org/10.2139/ssrn.4375920

Park H, Kahn CM (2019) Collateral, rehypothecation, and efficiency. J Financ Intermediation 39:34–46. https://doi.org/10.1016/j.jfi.2018.06.001

Patel SB, Bhattacharya P, Tanwar S, Kumar N (2021) KiRTi: a blockchain-based credit recommender system for financial institutions. IEEE Trans Netw Sci Eng 8(2):1044–1054. https://doi.org/10.1109/TNSE.2020.3005678

Rijanto A (2024) Blockchain technology roles to overcome accounting, accountability and assurance barriers in supply chain finance. Asian Rev Account. https://doi.org/10.1108/ARA-03-2023-0090

Sachin P, Hari Prasad J, Latha TR (2023) Digital identity verification: transforming KYC processes in banking through advanced technology and enhanced security measures. International Research Journal of Modernization in Engineering Technology and Science. https://doi.org/10.56726/IRJMETS44476

Safiullin M, Elshin L, Burganov R (2023) Blockchain as a promising technology for international settlements (on the example of the Russian economy). Int J Prof Bus Rev 8(9):e03218. https://doi.org/10.26668/businessreview/2023.v8i9.3218

Saravanan K, Balas VE, Julie G, Robinson YH, Balaji S, Kumar R (eds) (2020) Handbook of research on blockchain technology. Academic Press, an Imprint of Elsevier

Saravanan K, Kumar R, Balas VE (eds) (2023) Green blockchain technology for sustainable smart cities. Elsevier

Sarier ND (2024) Best of two worlds: efficient, usable and auditable biometric ABC on the blockchain. https://eprint.iacr.org/2024/502.pdf

Sazandrishvili G (2020) Asset tokenization in plain English. J Corp Account Finance 31(2):68–73. https://doi.org/10.1002/jcaf.22432

Schuetz S, Venkatesh V (2020) Blockchain, adoption, and financial inclusion in India: research opportunities. Int J Inf Manage 52:101936. https://doi.org/10.1016/j.ijinfomgt.2019.04.009

Singh N, Kumar T, Vardhan M (2019) Blockchain based e-Cheque clearing framework. Scalable Comput Pract Experience 20(3):511–526. https://doi.org/10.12694/scpe.v20i3.1506

The World Bank (2023) Remittance prices worldwide (Issue 39). The World Bank. https://remittanceprices.worldbank.org/sites/default/files/rpw_main_report_and_annex_q223.pdf

Thomason J, Ivwurie E (2023) Advancements in the new world of web 3: A look toward the decentralized future. IGI Global. https://doi.org/10.4018/978-1-6684-6658-2

Tong Z, Hu Y, Jiang C, Zhang Y (2024) User financial credit analysis for blockchain regulation. Comput Electr Eng 113:109008. https://doi.org/10.1016/j.compeleceng.2023.109008

Truby J, Dahdal A, Caudevilla O (2022) Global blockchain-based trade finance solutions: analysis of governance models and impact on local laws in six jurisdictions. Global J Comp Law 11(2):167–196. https://doi.org/10.1163/2211906X-11020001

Ullah N, Al-Dhlan KA, Al-Rahmi WM (2021) KYC optimization by blockchain based hyperledger fabric network. 2021 4th International Conference on Advanced Electronic Materials, Computers and Software Engineering (AEMCSE), 1294–1299. https://doi.org/10.1109/AEMCSE51986.2021.00264

Valkanov N (2023) Financial regulatory technologization in the context of europe's digital decade. Izvestia Journal of the Union of Scientists - Varna. Econ Sci Ser 12(1):85–94. https://doi.org/10.56065/IJUSV-ESS/2023.12.1.85

Vijaya Kittu M, Yadav A, Khaliq LN (2024) Role of payment systems in financial well-being: an examination from India. In: Emerging perspectives on financial well-being. IGI Global, p 24. https://www.igi-global.com/gateway/chapter/347786

Wang W (2020a) A SME credit evaluation system based on blockchain. 2020 International Conference on E-Commerce and Internet Technology (ECIT), 248–251. https://doi.org/10.1109/ECIT50008.2020.00064

Wang W (2020b) Exploring personal credit evaluation model based on blockchain. 2020 International Conference on E-Commerce and Internet Technology (ECIT), 163–166. https://doi.org/10.1109/ECIT50008.2020.00043
Wang R (2023) Blockchain and bank lending behavior: a theoretical analysis. SAGE Open 13(1):215824402311645. https://doi.org/10.1177/21582440231164597
Wang X, Xu F (2022) The value of smart contract in trade finance. Manufacturing & Service Operations Management, msom.2022.1126. https://doi.org/10.1287/msom.2022.1126
Wang R, Lin Z, Luo H (2019) Blockchain, bank credit and SME financing. Qual Quant 53(3):1127–1140. https://doi.org/10.1007/s11135-018-0806-6
Wang C, Chen X, Xu X, Jin W (2023) Financing and operating strategies for blockchain technology-driven accounts receivable chains. Eur J Oper Res 304(3):1279–1295. https://doi.org/10.1016/j.ejor.2022.05.013
Weshah S (2024) Business digitalization in accounting and auditing fields. In: Alzoubi HM, Alshurideh MT, Vasudevan S (eds) Technology innovation for business intelligence and analytics (TIBIA), vol 147. Springer Nature Switzerland, pp 111–123. https://doi.org/10.1007/978-3-031-55221-2_6
Yang F, Qiao Y, Qi Y, Bo J, Wang X (2022) BACS: Blockchain and AutoML-based technology for efficient credit scoring classification. Ann Oper Res. https://doi.org/10.1007/s10479-022-04531-8
Yerram SR, Goda DR, Mahadasa R, Mallipeddi SR, Varghese A, Ande JRPK, Surarapu P, Dekkati S (2021) The role of blockchain technology in enhancing financial security amidst digital transformation. Asian Bus Rev 11(3):125–134. https://doi.org/10.18034/abr.v11i3.694
Zhuo X, Irresberger F, Bostandzic D (2023) Building a connected future: the role of blockchain in cross-border payments and financial inclusion in Africa. SSRN Electronic Journal. https://doi.org/10.2139/ssrn.4550837

Investigating the Applications of Blockchain in Banking: A Hybrid MADM Approach

Hamidreza Fallah Lajimi, Alireza Arab, and Mojtaba Fallahpour

Abstract Nowadays, businesses face challenges like managing processes, ensuring effectiveness, maintaining information transparency, and safeguarding data security. Blockchain technology has emerged as a solution to bolste transparency and trust. Within the banking sector, blockchain plays a role in enhancing risk control and enabling real-time information analysis. While this technology is leveraged to improve performance, efficiency, and security, banks must optimize their applications within constraints. Hence, this study aims to explore the applications of blockchain in banking. In this way, at first, the applications of blockchain in banking and also the attributes for evaluating them were identified through a literature review. Subsequently, the Bayesian Best Worst Method was utilized to obtain the importance weight of each attribute, followed by employing the VIKOR technique to rank these applications. The findings reveal that security, privacy, and transparency are deemed as the most important attributes, respectively. Also, the most significant application pertains to financial supply chain management. Drawing from these research outcomes, practical and managerial suggestions have been provided.

1 Introduction

In today's business environment, organizations and companies encounter a range of barriers that differ based on their industry. These barriers encompass management processes, the pursuit of enhanced effectiveness (Noy and Zhang 2023), assurance of information transparency (Chen et al. 2023), safeguarding data and information,

H. F. Lajimi · A. Arab (✉) · M. Fallahpour
Department of Industrial Management, Faculty of Economics and Administrative Sciences, University of Mazandaran, Babolsar, Iran
e-mail: alireza.arab@umz.ac.ir

H. F. Lajimi
e-mail: h.fallah@umz.ac.ir

M. Fallahpour
e-mail: m.fallahpour03@umail.umz.ac.ir

S. Krishnan et al. (eds.), *Unveiling Blockchain Enabled Business Applications*,
Contributions to Management Science, https://doi.org/10.1007/978-981-95-2337-5_13

as well as meeting the escalating expectations for promptness and responsibility (Cooper et al. 2022). Over the years, as technology has progressed, numerous new tools and technologies have made their way into the business world, offering promises of enhancing and transforming existing processes. One such technology that stands out is blockchain, which has gained recognition for its capabilities in areas like currencies, such as Bitcoin. It holds the potential to address critical business challenges (Swan 2015). Transparency is an aspect of technology, and it plays a significant role in building trust between business partners and customers (Sun et al. 2022). Blockchain ensures that all transactions and data can be traced and audited by creating a distributed and unchangeable ledger (Chen et al. 2023). Blockchain technology has a range of applications across sectors and industries (Krishnan et al. 2023, 2021, 2020). One of the most important applications of this technology is in the banking industry.

The banking industry did not undergo an immediate transformation; rather, it underwent a lengthy process of transitioning from conventional banking to banking 4.0. This transition brought about a revolutionary change in the financial sector, offering enhanced decision-making, risk management, future predictability, and customer satisfaction (Nirvan et al. 2023). Traditional financial institutions typically employ mortgage guarantees as their primary method of risk reduction in the process of borrowing funds due to their limited access to customer data. Blockchain technology, which ensures the veracity and dependability of crucial resources like customer information and transaction records through non-alterable modifications to the system, enables banks to shift from singular to comprehensive risk control (He 2021). In the banking industry, various applications of these technologies, such as cryptocurrency (Song et al. 2023), business and process innovation (Sharma et al. 2023), etc., are employed to enhance performance, increase efficiency, and elevate process security.

The concept of this technology originated from the implementation of the Bitcoin protocol and network. Satoshi Nakamoto described this in 2008 (Lai and Chuen 2018). Over time, it has grown into an idea that includes various distributed ledger technologies. Blockchain technology utilizes techniques, like hashing and asymmetric to ensure the security and integrity of transactions (Chen et al. 2021). The introduction of Ethereum, a network that supports contracts, has expanded the functional possibilities of this technology. With its ledger system, proof of work, and cryptography, blockchain finds applications in domains (Sherman et al. 2019).

Until now, the development of blockchain computing has gone through four stages. The first stage, known as Blockchain 1.0, focused on the aspect of programming currency. This stage began with the emergence of Bitcoin, which symbolized the rise of blockchain-based cryptocurrency (Sun et al. 2022). Next came Blockchain 2.0, which introduced contracts. This platform allowed people to upload applications and execute them independently by incorporating security measures and encryption. It ensured execution based on predetermined logic (Leng et al. 2020). Building upon these advancements, Blockchain 3.0 brought about governance or digital society. This generation of blockchain has the potential to accelerate the development of societies by enabling synchronized data tracking and reducing surveillance costs for communities. In the future, it's possible that blockchain's decentralized computing model could revolutionize societies through improved collaboration and governance (Su et al. 2023). Currently we are in the phase of Blockchain 4.0, which focuses on integrating with emerging technologies and industries. Its main goal is to align itself with areas like the Internet of Things (IoT), large-scale data sets, and Artificial Intelligence (AI). The essence of Blockchain 4.0 lies in its vision of a decentralized, secure, and interconnected ecosystem, leveraging blockchain's trust and transparency to induce transformative changes across various sectors (Mukherjee and Pradhan 2021).

Blockchain 3.0, with its focus on governance and its application in society, brings forth a range of capabilities that can have a significant impact on the banking sector. In this context, the use of Blockchain 3.0 shows promise to bring about transformations within this industry. This phase of technology emphasizes data transparency to an extent, as highlighted by Su et al. (2023), and plays a crucial role in establishing a well-coordinated and highly secure banking system. By adopting Blockchain 3.0, financial institutions can enjoy benefits, such as launching a synchronized and securely managed data tracking system that effectively reduces costs associated with traditional surveillance methods. The decentralized nature of Blockchain 3.0 is a factor that ensures there is no authority controlling the data, thus fostering collaboration and increased transparency within the financial ecosystem (Mukherjee and Pradhan 2021).

Banks have started to realize the benefits that blockchain offers, including security protocols and more efficient transaction processes, but it's important to recognize that banks operate within limitations, such as resources and technological infrastructure. In this way, they need to choose the most effective and efficient applications of the blockchain in their operation that have the maximum possible amount of benefits of using this technology. Since this selection process is an expert-oriented task that needs the subjective judgment of experts, and also, several attributes that often conflict with each other are involved in evaluating and choosing these applications, this problem is a multi-attribute decision-making (MADM) problem. Hence, the current research aims to provide an approach for the investigation and analysis of the applications of blockchain in the banking industry by using a hybrid MADM approach, and a case study in Iran's banking industry is utilized to show the proposed approach. The proposed approach leads to the selection of the most optimal applications of blockchain for implementation in banking, and as a result, it leads to the optimal use of resources and the achievement of competitive advantage, and it can be used according to any specific cases (banks) with their specialized characteristics. For this purpose, the applications of blockchain in banking and the attributes for evaluating them are identified through a literature review. After that, by using the Bayesian Best Worst Method (BBWM), the importance weights of the attributes were obtained. Then, for the final evaluation of the applications, the VIKOR method was utilized, and finally, some practical and future research suggestions were provided based on this result.

2 Attributes and Applications

Different attributes have been considered for blockchain technology. Through reviewing the articles and books in the literature, the most important attributes can be considered as follows:

C1. Privacy: Companies' handling of user data has raised privacy concerns. Blockchain technology minimizes breach risks (Suma 2019; Wang et al. 2020). In this technology-based network, no single entity holds the power to control or manipulate data (Soni and Bhushan 2019). In a blockchain network, data is recorded as encrypted transactions on a distributed ledger instead of being stored in a file on a server. This ensures data privacy (Zhang et al. 2020). Ensuring data privacy and refraining from trading user information may gain significance in scenarios where both regulatory authorities and customers emphasize privacy (Tapscott and Tapscott 2016).

C2. Lower fraud risk: Blockchain is known for its nature, making it extremely difficult to alter its content. The data is stored on a ledger, which is replicated across nodes (Gan and Lau 2024). In the realm of technology, there exists no database that can be tampered with (Dai et al. 2017; Mansoor et al. 2023). This particular feature has the potential to help companies minimize instances of fraud, as altering records on blockchains is extremely difficult. As a result, it can significantly decrease business risks (Kamal and Tayyab 2017).

C3. Transparency: The blockchain ledger is capable of storing various types of information through coding. This includes important documents such as birth, death, and marriage documents, property deeds, and ownership titles (Tapscott and Tapscott 2016). Blockchain enables greater transparency among companies and stakeholders by providing the possibility for data auditing (Gan and Lau 2024; Yele and Litoriya 2024). The sharing and exchange of data among companies can enable decision-making and foster collaboration (Albeshr and Nobanee 2020). This will be particularly relevant to industries where real-time data sharing is essential but challenging due to system limitations or a lack of integration (Khadka 2020).

C4. Security: The internet poses various risks to individuals and businesses, such as hacking, phishing, malware, and ransomware attempts. Therefore, it is crucial to remain cautious (Tapscott and Tapscott 2016). It is important to prioritize the security of our systems in light of the increasing sophistication of cyber-attacks. The prevalence of cybercrime poses a threat to businesses, underscoring the necessity for us to actively tackle this matter. The use of technology can provide benefits such as encryption and decentralization, making it more difficult for cyber theft and data corruption to occur. In such a setup, hacking databases simultaneously becomes a challenging task (Gervais et al. 2016; Mainelli and Smith 2015; Tapscott and Tapscott 2016). In blockchain, there isn't a component, and the network doesn't rely on centralized servers, like banks or cloud services that have experienced downtime due to targeted attacks in the past (Kamal and Tayyab 2017).

C5. Speed: Remittance transactions, stock, and bank loan trades usually take 3–7, 2–3, and 23 days to process, respectively. Also, SWIFT manages 15 million orders daily and requires several days to clear them. In contrast, blockchain transactions are instant or completed within minutes (Hassani et al. 2018; Khadka 2020). Businesses can process transactions, assets, or information faster, leading to better service levels (Kamal and Tayyab 2017; Tapscott and Tapscott 2016).

C6. Quality: The data stored in a blockchain system reflects the information that exists in reality because of its mechanism for creating distributed consensus (Hashem et al. 2023; Rekha and Resmi 2021; Tapscott and Tapscott 2016). This ensures that a transaction is declined if the majority of nodes can't validate it, improving data quality. By incorporating technology, businesses can enhance their data (Benton and Radziwill 2017; Kamal and Tayyab 2017; Koteska et al. 2017).

C7. Traceability: The implementation of technology enhances the traceability and mobility of assets, including goods and raw materials, thereby providing more flexibility to supply chains. Moreover, it effectively mitigates the risks associated with fraud and counterfeiting (Agbezoutsi et al. 2021; Mitani and Otsuka 2020; Pradhan et al. 2021). Every transaction made on the blockchain can be traced back to its source. This ability to trace is vital in ensuring the legitimacy of products in industries such as pharmaceuticals, luxury goods, and organic products where the origin is significant (Agbezoutsi et al. 2019; Biswas et al. 2023; Brandín and Abrishami 2024; Li et al. 2024).

The banking sector can greatly benefit from the applications of blockchain technology. It is possible to eliminate intermediaries, enhance transaction security and trust and improve the efficiency of operations (Bhowmick and Kundu 2022). Banks have already started using blockchain for purposes like promoting inclusion, negotiating value, reducing transaction costs, and exploring business models (Dewangan et al. 2023). Moreover, blockchain technology can facilitate insurance innovation in banks by detecting fraud, improving pricing strategies, and cutting costs in the insurance sector (Kour 2023). Various applications of these technologies (Table 13.1) are being implemented in the banking industry to enhance performance, increase efficiency, and ensure process security.

The research conducted by various scholars provides valuable insights into this field. Griffiths and Baudier (2023) emphasize promoting responsible banking practices through technology, aligning with UN principles. Nirvan et al. (2023) focus on enhancing banking services' efficiency and decision-making with blockchain and AI. Iacoviello and Bruno (2023) highlight blockchain's role in lending processes, while Sharma et al. (2023) stress its importance in streamlining banking operations. Song et al. (2023) review blockchain's use in enterprise financing, discussing advancements and challenges. Gan et al. (2021) examine applications in banking and finance, suggesting areas for further research. Huyut and Temur (2021) analyze implementation decisions in banking using decision analysis methods. He (2021) explores blockchain's impact on banks, positioning it as a driver of digital economy

Table 13.1 Applications of the blockchain technology in banking

Application	Definition	Reference
Cryptocurrencies *(A1)*	The concept of technology initially explored in a research paper on Bitcoin's cryptography has led to the emergence of various cryptocurrencies like Bitcoin, Ethereum, and Litecoin. Cryptocurrency, considered Blockchain 1.0, has gained research attention. Non-Fungible Tokens (NFTs), a recent development, stand out from traditional cryptocurrencies as unique digital assets stored on a blockchain. Unlike fungible cryptocurrencies, NFTs cannot be replicated or exchanged on a one-to-one basis, each representing distinct items with proof of ownership and authenticity	W. Chen et al. (2018b), Gan et al. (2021), Gao and Su (2020), Ghosh et al. (2020), Iacoviello and Bruno (2023), Miraz and Ali (2018), Poongodi et al. (2020), Sethaput and Innet (2023), Sharma et al. (2023), Song et al. (2023), and Tasatanattakool and Techapanupreeda (2018)
Smart contract *(A2)*	Smart contracts and software programs executing agreements on platforms face inefficiencies due to a lack of sharing and management methods, hindering reuse. A proposed remedy involves employing tags to organize and enhance contract reusability and efficiency. Additionally, a framework for recommendations and a warning algorithm aim to enhance contract development and implementation. For instance, categorizing contracts with tags and implementing recommendation frameworks and alert algorithms can boost the effectiveness, efficiency, and security of smart contracts across platforms	Alaba et al. (2024), Almahirah and Salameh (2021), Alsalim and Ucan (2023), Hamilton (2020), John et al. (2023), Khan et al. (2021), Kongmanee et al. (2019), Mbaidin et al. (2024), Mohamed and Faisal (2024), Peters and Panayi (2016), and Xie et al. (2024)
Financial digitalization and disintermediation *(A3)*	Blockchain technology's trust element enables non-interference in certain banking and financial sectors, permitting unauthorized modifications. Financial product information management can be facilitated through a blockchain-based platform, characterized by a distributed network architecture ensuring tamper-resistant features, information transparency, and secure sharing environments. For example, if a customer disputes an unauthorized payment deduction, banks can promptly investigate by accessing the blockchain system for transparent transaction histories. This ensures secure and immutable storage of financial data, bolstering trust in banking transactions	Cai (2018), B. Chen et al. (2018a), Gan et al. (2021), Ikeda and Hamid (2018), Kabra et al. (2020), and Sharma et al. (2023)

(continued)

Table 13.1 (continued)

Application	Definition	Reference
Financial regulation *(A4)*	Blockchain technology enhances the flow of information between government agencies and stakeholders by promoting transparency and transaction tracking. For instance, the Ministry of Economy can utilize blockchain to monitor financial transaction settlements among banks. Each bank would transmit its transactions to a blockchain network, where they are securely and transparently recorded on the blockchain ledger. This allows the Ministry of Economy to monitor transactions in real-time, swiftly identifying discrepancies or issues as they arise	Collomb et al. (2019), Dashkevich et al. (2020), Ducas and Wilner (2017), Gan et al. (2021), Hyvärinen et al. (2017), and Sherman et al. (2019)
Financial inclusion *(A5)*	Financial inclusion seeks to offer accessible and affordable financial services to those with limited income. Around two billion adults worldwide lack access to banks and financial services. Blockchain technology has the potential to create a transparent financial system, crucial for poverty reduction and local economic development. Mobile-based digital wallets enable secure, low-cost transactions outside traditional banking systems, reducing fraud and fostering economic inclusion and growth through facilitating payments, credit access, and building credit history	Gan et al. (2021), Kshetri (2020), Nirvan et al. (2023), Norta et al. (2019), Schuetz and Venkatesh (2020), and Sharma et al. (2023)
Financial Supply Chain *(A6)*	Traditional supply chain financing faces challenges like information dispersion and credit transfer difficulties. However, blockchain technology addresses these issues by ensuring transaction accuracy, alleviating concerns about information manipulation, and reducing financing costs for SMEs. Smart contracts play a crucial role by enabling banks to automate repayment processes based on predetermined conditions, enhancing efficiency and trust in the financing ecosystem	Guo and Liang (2016), Sharma et al. (2023), Song et al. (2023), and Tasatanattakool and Techapanupreeda (2018)
Bank credit information systems *(A7)*	Factors contributing to the ineffectiveness of bank credit information systems include data scarcity, poor quality, and challenges in data sharing between institutions, alongside unclear ownership of user data, leading to privacy and security concerns. Collaboration among stakeholders is crucial to resolving these issues, with blockchain technology offering potential support. For example, small business owners like Sarah may struggle to access loans due to these challenges. However, blockchain provides a secure and decentralized approach to data storage and sharing, potentially improving credit assessment processes and addressing privacy concerns	Iacoviello and Bruno (2023), Nirvan et al. (2023), Wang and Lin (2020), and Zou and Xue (2020)

(continued)

Table 13.1 (continued)

Application	Definition	Reference
Hyperledger *(A8)*	Open-source tools and several subprojects are offered by Hyperledger, which is a collaborative blockchain framework project sponsored by the Linux Foundation. This license is based on blockchain and does not require anonymity miner verification, nor does not require the relevant currency to be used as a promotional measure. Blockchain can be expanded with less delay. For example, consider two different banks in a country that have shared current accounts and need to carry out frequent transactions between these accounts. Using Hyperledger Fabric, both banks can create a private blockchain infrastructure that offers high security and efficiency. Then, transactions such as transferring funds between the shared accounts can be conducted quickly and with minimal costs because there is no need for intermediaries or validation by mining nodes. This reduction in time and costs ultimately adds value to the banks' customers and improves the performance of the banking system	Sakho et al. (2019), Tasatanattakool and Techapanupreeda (2018), and Zou and Xue (2020)
Payment clearing system: distributed cleanup mechanism *(A9)*	Hyperledger, a collaborative blockchain framework backed by the Linux Foundation, offers various open-source tools and subprojects. Unlike conventional blockchain systems, Hyperledger doesn't rely on anonymous miners for transaction verification or require a native currency. This setup facilitates rapid scalability. For instance, imagine two banks with shared accounts requiring frequent transactions. Through Hyperledger Fabric, they can establish a private blockchain infrastructure, ensuring security and efficiency. Transactions like fund transfers between shared accounts can then occur swiftly and cost-effectively, without intermediaries or mining node validation. This efficiency cuts time and costs, benefiting customers and boosting overall banking system performance	Dashkevich et al. (2020), Guo and Liang (2016), and Sharma et al. (2023)

(continued)

Table 13.1 (continued)

Application	Definition	Reference
Create data ownership *(A10)*	In the digital age, individuals generate valuable data online but lack ownership or control over it as it's typically held by internet corporations. Privacy concerns and fragmented data hinder its effective utilization. Technology solutions, like encryption, can empower individuals to control their data. This allows credit agencies to maintain accuracy and trustworthiness while reducing data collection costs. For instance, encrypted financial transaction data can be securely stored and selectively shared with banks, enabling informed decisions while preserving privacy	Dashkevich et al. (2020), Griffiths and Baudier (2022), and Guo and Liang (2016)
Promote data sharing *(A11)*	Blockchain technology enables credit agencies to automate big data registration, securely storing and sharing encrypted customer credit status across institutions. In a proposed blockchain credit solution, banks upload encrypted summary data onto the blockchain as part of the KYC process. When a credit check is requested, the blockchain notifies the bank, allowing them to share encrypted summary data with the requester, ensuring security and privacy in credit data sharing among financial institutions	Griffiths and Baudier (2022), Guo and Liang (2016), Iacoviello and Bruno (2023), Nirvan et al. (2023), and Sharma et al. (2023)

growth. Zou and Xue (2020) propose technology enhancements for credit management systems to bolster trust. Dashkevich et al. (2020) outline technology applications for central banks, addressing research gaps. Tasatanattakool and Techapanupreeda (2018) and Miraz and Ali (2018) discuss the challenges and applications of blockchain beyond cryptocurrencies. Guo and Liang (2016) focus on blockchain's applications in banking and the hurdles to its effectiveness. These studies collectively shape our understanding of blockchain's impact on banking operations and industry streamlining.

3 Methodology

In this section, some preliminaries on the research framework (Fig. 13.1) are presented. In the first step, the most frequent applications of blockchain in banking, and also the attributes for evaluating them, are extracted from a literature review. After that, the weight of the attributes (stated in Sect. 13.2) is calculated by using the BBWM. In the next step, the applications of the blockchain in banking (Table 13.1) are prioritized by using the VIKOR method and the opinion of experts. Finally, some practical and future research suggestions were provided based on the results. The

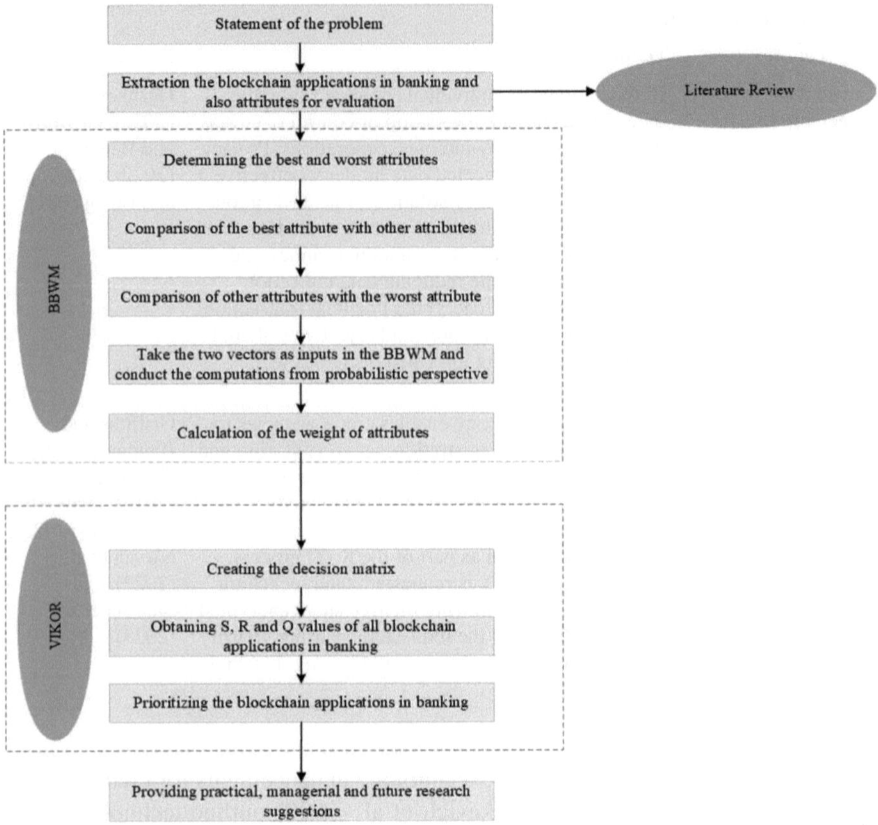

Fig. 13.1 The executive framework of research. *Source* Author's own work

expert team for this research comprises five individuals with the following qualifications: extensive experience in the banking sector (minimum of 10 years), possession of an advanced academic degree, familiarity with blockchain technology and its applications in banking, and a keen interest in contributing to this research endeavor.

3.1 Bayesian Best-Worst Method (BBWM)

Rezaei (2015) introduced the Best-Worst Method (BWM) as a MADM method to obtain the importance weight of attributes in a decision-making problem. This method has efficient pairwise comparisons in comparison to other famous pairwise comparison methods, such as AHP, making it a time-efficient method that mitigates cognitive biases such as equalizing bias (Rezaei et al. 2022) and anchoring bias (Rezaei et al. 2024), and is a simple and practical approach. For the latest information about this method, see Qun et al. (2024).

Although this method offers benefits, it's limited in its capacity to facilitate group decision-making. It's primarily suited for individual decision-making. In this way, Mohammadi and Rezaei (2020) introduced the BBWM. The steps of this method are as follows (Mohammadi and Rezaei 2020):

1. Determining the attributes to be evaluated: The attributes $c_j = \{c_1, c_2, \ldots, c_n\}$ is determined within the scope of expert evaluation and literature research.
2. Creating a special form survey: Each expert determines the most and the least important attributes from the set c_j. Then, Experts should make pairwise comparisons between the best attribute and all other attributes (BO), and also, between the other attributes and the worst attribute (OW) by using the numbers between 1 and 9 (1 is equally important and 9 is absolutely important).
3. Calculating the weights by using the probabilistic perspective provides input to calculate the aggregate weight in a way that $\sum_j w_j = 1$ and $w_j \geq 0, \forall j$. These input vectors are modeled as a multinomial distribution (13.1):

$$\begin{aligned} &\left(A_B^k | w^k\right) \sim \text{multinomial}\left(\frac{1}{W^k}\right); \\ &\left(A_W^k | w^k\right) \sim \text{multinomial}\left(w^k\right), \ \forall_k = 1, 2, \ldots, K \end{aligned} \tag{13.1}$$

Also, the importance weight of the attributes (w^*) can be obtained based on the following probabilistic models (13.2–13.4)

$$w^k | w^* \sim \text{Dir}\left(\gamma \times w^*\right), \quad \forall_k = 1, 2, \ldots, K \tag{13.2}$$

$$\gamma \sim \text{gamma}(0.1, 0.1), \tag{13.3}$$

$$w^* \sim \text{Dir}(1) \tag{13.4}$$

4. In the end, the importance weights of attributes and also, the confidence level and related graph of the calculated weights, can be adopted by solving the model proposed by Mohammadi and Rezaei (2020).

3.2 VIKOR Method

Opricovic (1998) developed the VIKOR method. It provides a procedure for ranking the decision-making alternatives by identifying cost and benefit criteria. It has been applied in different sectors in recent years (Ak et al. 2022; Saner et al. 2022). The procedure of this method is described as follows (Opricovic and Tzeng 2004):

1. Construct the decision matrix: In this matrix (13.5), $j = 1, 2, \ldots, M; i = 1, 2, \ldots, N$, shows the real performance value of the jth alternative in terms of ith attribute.

$$\begin{bmatrix} f_{11} & \cdots & f_{1N} \\ \vdots & \ddots & \vdots \\ f_{M1} & \cdots & f_{MN} \end{bmatrix} \tag{13.5}$$

2. Determining the benefit and cost values in a decision matrix as (13.6) or (13.7):

$$f_i^* = \max_j f_{ij}, \quad f_i^- = \min_j f_{ij}, \quad \text{if the } i\text{th attribute is a benefit} \tag{13.6}$$

$$f_i^* = \min_j f_{ij}, \quad f_i^- = \max_j f_{ij}, \quad \text{if the } i\text{th attribute is a cost} \tag{13.7}$$

3. Calculate the values S_j, R_j and Q_j as (13.8–13.10):

$$S_j = \sum_{i=1}^{n} w_i\left(f_i^* - f_{ij}\right)/\left(f_i^* - f_i^-\right) \tag{13.8}$$

$$R_j = \max\left[w_i\left(f_i^* - f_{ij}\right)/\left(f_i^* - f_i^-\right)\right] \tag{13.9}$$

$$Q_j = \upsilon\frac{\left(S_j - S^*\right)}{\left(S^- - S^*\right)} + (1 - \upsilon)\frac{\left(R_j - R^*\right)}{\left(R^- - R^*\right)} \tag{13.10}$$

   Where w_i is the importance weight of the attribute ith, which is calculated by BBWM, $S^* = \min S_j$, $S^- = \max S_j$, $R^* = \min R_j$, $R^- = \max R_j$ and $\upsilon \in [0, 1]$ is the group utility, so we assumed this utility $\upsilon = 0.5$.
4. Ranking the alternatives can be calculated by sorting the values S_j, R_j and Q_j in decreasing order.
5. Provide a compromise solution if the following two conditions (condition 1: Acceptable advantage (13.11) and condition 2: Acceptable stability) are satisfied. In this equation, J is the number of alternatives.

Condition 1

$$Q(A_2) - Q(A_1) \geq \left(\frac{1}{J - 1}\right) \tag{13.11}$$

Condition 2

(A_1) should have the top rank by S or/and R.

In case any one of the conditions is not fulfilled, then the solutions provided as follows are proposed:

- If only condition 2 is not fulfilled: Alternatives (A_1) and (A_2);

- If condition 1 is not fulfilled: Alternatives $(A_1, A_2, \ldots, A_m)$. that A_m is calculated by using (13.12).

$$Q(A_m) - Q(A_1) < \left(\frac{1}{J-1}\right) \tag{13.12}$$

4 Data Analysis

In this section, following the executive framework of research (Fig. 13.1), at first, the weight of each of the blockchain applications evaluation attributes (stated in Sect. 13.2) was calculated by using the BBWM. For this purpose, Experts conducted pairwise comparisons between the best attribute and all other attributes, presented in Table 13.2. Additionally, experts performed pairwise comparisons between the other attributes and the worst attribute, shown in Table 13.3.

In the end, the final weights were obtained by using MATLAB R2016a software (Table 13.4).

Table 13.2 Pairwise comparisons between the best attribute and all other attributes

Expert number	1	2	3	4	5
The Best attribute	C4	C5	C1	C3	C4
C1	2	4	1	3	3
C2	3	8	3	2	4
C3	4	4	5	1	6
C4	1	3	2	3	1
C5	5	1	6	8	7
C6	7	2	7	6	5
C7	5	2	3	2	9

Table 13.3 Pairwise comparisons between the other attributes and the worst attribute

Expert number	1	2	3	4	5
The Worst attribute	C6	C2	C6	C5	C7
C1	4	2	7	6	7
C2	4	1	5	5	5
C3	5	3	3	8	3
C4	7	3	4	5	9
C5	2	8	3	1	3
C6	1	6	1	3	4
C7	2	2	2	4	1

Table 13.4 The results of the Bayesian BWM

Attribute	Weight
C1	0.1963
C2	0.1351
C3	0.1419
C4	0.2088
C5	0.1102
C6	0.1011
C7	0.1066

According to the results (Table 13.4 and Fig. 13.2), *Security* emerged as the top-ranked attribute in terms of importance, followed by *Privacy* in second place and *Transparency* in third. Additionally, Fig. 13.3 shows the credal graph for the results. For example, *Security* (C1) is the most important and *Quality* (C6) is the least important. Also, *Security* is certainly more important than *Quality*, *Speed*, *Traceability*, *Lower Fraud Risk*, and *Transparency* attributes, with a confidence greater than 0.92, but *Security* is more desirable than *Privacy* with a confidence of 0.6.

After that, following the executive framework of research (Fig. 13.1), the ranking of the blockchain applications in banking (provided in Table 13.1) was calculated by using the VIKOR method. Table 13.5 demonstrates the intended S, R, and Q values of the applications of blockchain in banking.

Based on the results of VIKOR (Table 13.5 and Fig. 13.4), the alternative (*A6: Financial supply chain*) is the best ranked by the Q, if the two considered conditions are fulfilled. The results show that both conditions were satisfied for the A6. The final ranking of applications calculated as $A6 > A2 > A1 \approx A3 > A5 \approx A10 > A11 \approx A7 \approx A9 > A4 > A8$.

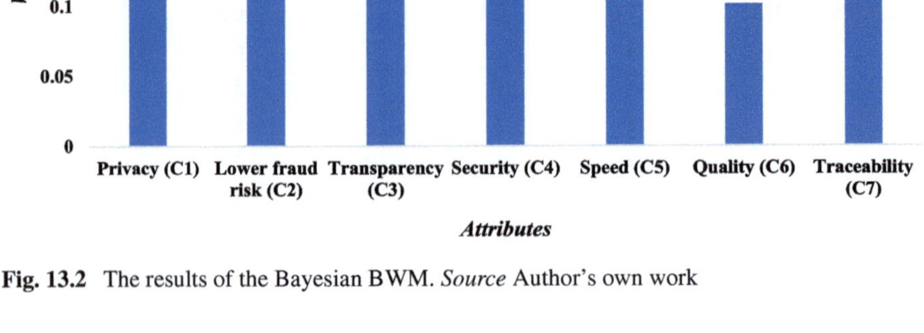

Fig. 13.2 The results of the Bayesian BWM. *Source* Author's own work

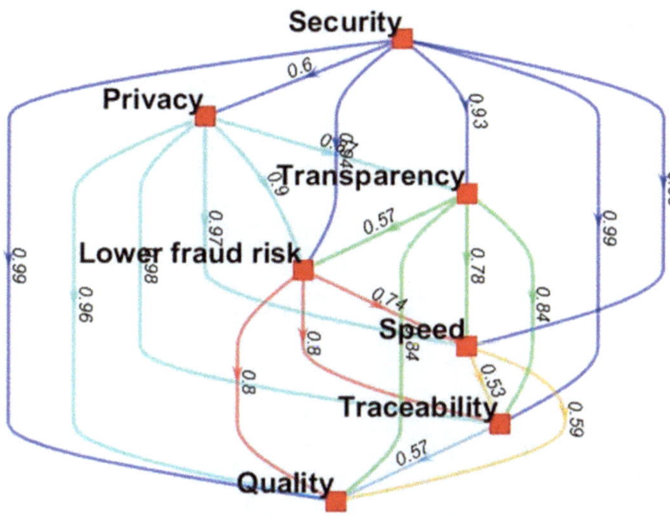

Fig. 13.3 The credal graph. *Source* Author's own work

Table 13.5 The results of the VIKOR method

Applications	S	Rank S	R	Rank R	Q	Rank Q
Cryptocurrencies *(A1)*	0.380	3	0.101	3	0.278	3
Smart contract *(A2)*	0.232	1	0.123	6	0.215	2
Financial digitalization and disintermediation *(A3)*	0.402	4	0.097	2	0.283	4
Financial regulation *(A4)*)	0.592	9	0.209	11	0.829	10
Financial inclusion *(A5)*	0.482	6	0.104	4	0.383	5
Financial supply chain *(A6)*	0.313	2	0.058	1	0.073	1
Bank credit information systems *(A7)*	0.517	7	0.174	8	0.645	8
Hyperledger *(A8)*	0.779	11	0.196	10	0.959	11
Distributed cleanup mechanism *(A9)*	0.562	8	0.174	8	0.686	9
Create data ownership *(A10)*	0.482	5	0.107	5	0.389	6
Promote data sharing *(A11)*	0.625	10	0.139	7	0.628	7

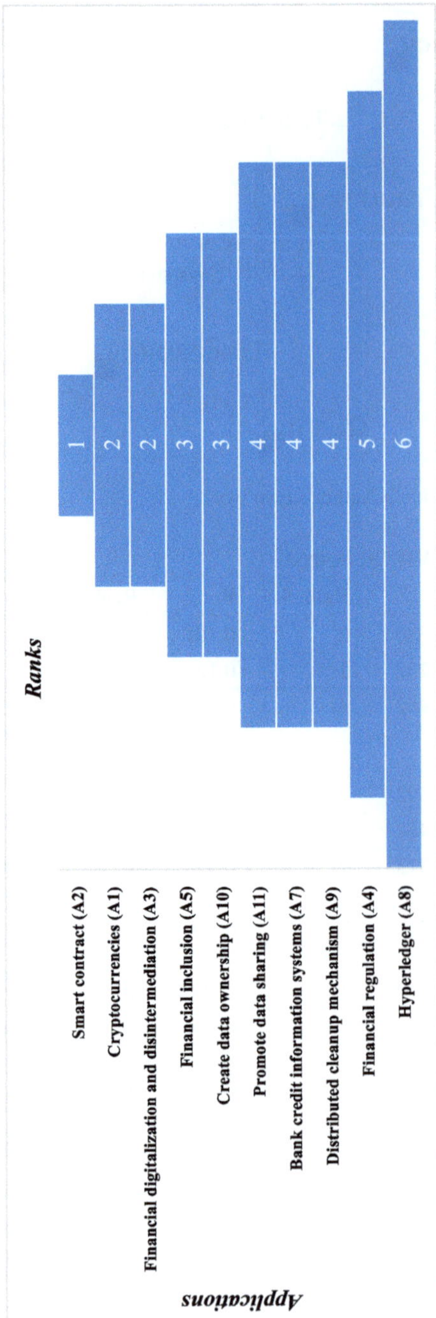

Fig. 13.4 Ranking of the applications. *Source* Author's own work

5 Discussion

According to the results of the BBWM method, security, privacy, and transparency are the most important attributes in the mentioned case study. These attributes are also mentioned as important ones in the research literature. For privacy (Soni and Bhushan 2019; Suma 2019; Wang et al. 2020), for security (Dasgupta et al. 2019; Gervais et al. 2016; Mainelli and Smith 2015; Sakho et al. 2019) and for transparency (Al Mamun et al. 2020; Albeshr and Nobanee 2020; Gan and Lau 2024) pointed out that these attributes have an important role in the blockchain.

Blockchain security involves a set of characteristics and approaches specifically designed to prevent changes, unauthorized entries, and cyber intrusions on the network. It is achieved through the use of consensus mechanisms among a network of interconnected computing devices (Kamal and Tayyab 2017; Mainelli and Smith 2015; Tapscott and Tapscott 2016). In the world of banking, keeping an environment is crucial because the industry constantly faces cyber attacks. The blockchain offers a level of security with its ledger system and advanced encryption methods. This is extremely important as it protects both assets and valuable customer information.

The concept of privacy relates to the ability to protect financial information from being disclosed, even though transactions are recorded on a shared ledger. This is accomplished by utilizing addresses and employing techniques such as signature loops (Soni and Bhushan 2019; Wang et al. 2020; Zhang et al. 2020). Maintaining privacy is extremely important for individuals involved in banking transactions, as their personal and financial information must remain confidential unless they explicitly agree to share it. The use of technology ensures the handling of privacy concerns by incorporating anonymous addresses and other innovative methods.

Transparency in the realm of blockchain pertains to the inherent quality whereby all transactions are meticulously documented and easily traceable in the central register (Gan and Lau 2024; Yele and Litoriya 2024). The primary objective of this characteristic is to facilitate the scrutiny of any transaction by any individual, fostering heightened confidence in the overall system and diminishing the likelihood of clandestine activities and deceitful practices (Khadka 2020; Tapscott and Tapscott 2016). Transparency is crucial in building trust and preventing corruption. In the banking industry, being transparent can give users an understanding of how a financial institution is performing, ensuring that their transactions are accurate and trustworthy.

These three metrics work together within the framework to create a trading environment that's resistant to security breaches. By preventing manipulation and fraudulent activities, the security aspect ensures that transactions are trustworthy. Additionally, measures are in place to protect data while transparency provisions promote trust and accountability. Given the banking sector's reliance on trust and security, combining these three metrics through technology could have an impact on how financial transactions are carried out, resulting in cost savings and improved customer service.

According to the results of the ranking, the application of the financial supply chain has a higher rank than the rest of the applications. This application is also an important one in research literature, such as (Choi 2023; Du et al. 2020; Gong et al. 2022; Liu 2020; Wang and Wang 2022). Within the banking sector, blockchain technology is instrumental in safeguarding the security and privacy of transactions. Privacy within the financial supply chain refers to safeguarding an individual's financial information while recording transactions on a shared ledger. Blockchain achieves this by using addresses and advanced encryption techniques. Security, on the other hand, involves protecting transactions and financial data through encryption methods and consensus mechanisms. This ensures that genuine and secure transactions remain intact while deterring activities. Another vital aspect is transparency, which entails recording all transactions in a ledger that can be easily traced. This feature allows any party to monitor transactions, thereby enhancing trust in the system and reducing corruption risks. By integrating security, privacy, and transparency, blockchain technology establishes a trading environment within the financial supply chain. It assures customers of transaction validity and integrity while also contributing to cost-reduction efforts. This technology has the potential to ensure the credibility and reliability of transactions, which can greatly benefit customers and lead to cost savings.

6 Conclusion

In today's business world, businesses are grappling with a range of issues surrounding managing operations, improving productivity, safeguarding data, and meeting rising demands. Blockchain technology has emerged as a solution in the banking industry. It brings advantages in terms of openness, risk control, and operational efficiency. Nevertheless, banks face obstacles when it comes to integrating this technology, due to resources and their current tech setup. Hence, it is crucial to choose the most optimal applications of the blockchain in their operations. In this way, this research provided an approach to the investigation of the applications of blockchain in the banking industry by using a hybrid MADM approach, and a case study in Iran's banking industry.

By using the BBWM, key attributes revolved around privacy, security, and transparency. Likewise, with the VIKOR method, the financial supply chain operation was introduced as the most important application. In light of these insights, some practical and managerial recommendations are provided as follows:

The results of this study emphasize the significance of incorporating technology into financial supply chains. Establishing an unchangeable ledger blockchain offers the ability to ensure security and transparency within this sector. Employing a blockchain can securely document all transactions in a reliable environment with advanced encryption and safeguard sensitive data from unauthorized access. This innovation not only minimizes the risk of fraud and security breaches but also thwarts unauthorized entry by enhancing identity verification procedures, establishing a robust defense against potential manipulation of financial information. Additionally, it is imperative to conduct information security-oriented training programs to equip all supply chain participants with defenses against cyber threats and maintain awareness. This is crucial not only for safeguarding data but also for upholding the trustworthiness and stability of the entire supply chain ecosystem. Utilizing intelligence capabilities for data analysis plays a role in predicting and preventing suspicious activities while aiding in identifying transaction trends that could signal illicit practices like tax evasion or money laundering. Establishing transparent reporting standards fosters trust among supply chain stakeholders, promoting data-informed decision-making, and more effective management of risks. Furthermore, fostering collaboration and interaction among supply chain members through shared platforms and collaborative initiatives enhances coordination and synergy in transactions and financial operations. In the realm of banking and finance, all sectors must work together and align their efforts when incorporating technology. This collaboration is important to ensuring consistency and trustworthiness across areas of the economy.

The following suggestions are being considered for future research:

- The proposed approach of this study can be used in different cases to make a comparison of the results.
- This research focused on conventional banking. Different paradigms, such as resilience, sustainability, and others, can be utilized to investigate this problem in banking.
- The issue has been resolved with an export-oriented approach. In future research, statistical methods can be employed to examine the proposed framework.
- The cause-and-effect relationship between attributes can be analyzed and dissected using techniques such as DANP, ISM, and FCM.
- The existing uncertainty inherent in evaluating this issue can be addressed by using fuzzy, grey, or neutrosophic approaches.

References

Agbezoutsi KE, Urien P, Dandjinou TM (2019) Towards blockchain services for mobile money traceability and federation. In: 2019 3rd Cyber Security in Networking Conference (CSNet)

Agbezoutsi KE, Urien P, Dandjinou TM (2021) Mobile money traceability and federation using blockchain services. Ann Telecommun 76:223–233

Ak MF, Yucesan M, Gul M (2022) Occupational health, safety and environmental risk assessment in textile production industry through a Bayesian BWM-VIKOR approach. Stoch Env Res Risk Assess 36(2):629–642

Alaba FA, Sulaimon HA, Marisa MI, Najeem O (2024) Smart contracts security application and challenges: a review. Cloud Comput Data Sci 5(1):15–41

Albeshr S, Nobanee H (2020) Blockchain applications in banking industry: a mini-review. Available at SSRN 3539152

Almahirah Z, Salameh M (2021) The effect of smart blockchain contracts on the financial services industry in the banking sector in Jordan. Ilkogretim Online 20(5)

Al Mamun A, Hasan SR, Bhuiyan MS, Kaiser MS, Yousuf MA (2020) Secure and transparent KYC for banking system using IPFS and blockchain technology. In: 2020 IEEE region 10 symposium (TENSYMP)
Alsalim MSH, Ucan ON (2023) Secure banking and international trade digitization using blockchain. Optik 272:170269
Benton MC, Radziwill NM (2017) Quality and innovation with blockchain technology. arXiv preprint arXiv:1710.04130
Bhowmick D, Kundu P (2022) Application of Blockchain in the World of Finance & Business. Int J Sci Res Eng Manag 6(03):1–6
Biswas D, Jalali H, Ansaripoor AH, De Giovanni P (2023) Traceability vs. sustainability in supply chains: the implications of blockchain. Eur J Oper Res 305(1):128–147
Brandín R, Abrishami S (2024) IoT-BIM and blockchain integration for enhanced data traceability in offsite manufacturing. Autom Constr 159:105266
Cai CW (2018) Disruption of financial intermediation by FinTech: a review on crowdfunding and blockchain. Account Finance 58(4):965–992
Chen B, Tan Z, Fang W (2018a) Blockchain-based implementation for financial product management. In: 2018 28th international telecommunication networks and applications conference (ITNAC)
Chen W, Xu Z, Shi S, Zhao Y, Zhao J (2018b) A survey of blockchain applications in different domains. In: Proceedings of the 2018 International Conference on Blockchain Technology and Application
Chen L, Cong LW, Xiao Y (2021) A brief introduction to blockchain economics. In: Information for efficient decision making: big data, blockchain and relevance. World Scientific, pp 1–40
Chen Z, Ji X, Li M, Li J (2023) How corporate social responsibility auditing interacts with supply chain information transparency. Ann Oper Res 329(1):1221–1240
Choi T-M (2023) Supply chain financing using blockchain: impacts on supply chains selling fashionable products. Ann Oper Res 331(1):393–415
Collomb A, De Filippi P, Klara S (2019) Blockchain technology and financial regulation: a risk-based approach to the regulation of ICOs. Eur J Risk Regul 10(2):263–314
Cooper AF, Moss E, Laufer B, Nissenbaum H (2022) Accountability in an algorithmic society: relationality, responsibility, and robustness in machine learning. In: Proceedings of the 2022 ACM Conference on Fairness, Accountability, and Transparency
Dai J, Wang Y, Vasarhelyi MA (2017) Blockchain: an emerging solution for fraud prevention. CPA J 87(6):12–14
Dasgupta D, Shrein JM, Gupta KD (2019) A survey of blockchain from security perspective. J Bank Financ Technol 3:1–17
Dashkevich N, Counsell S, Destefanis G (2020) Blockchain application for central banks: a systematic mapping study. IEEE Access 8:139918–139952
Dewangan S, Verma SK, Parganiha B, Dewangan S (2023) Applications and implementations of blockchain technology across the various sectors. In: Building secure business models through blockchain technology: tactics, methods, limitations, and performance. IGI Global, pp 1–19
Ducas E, Wilner A (2017) The security and financial implications of blockchain technologies: regulating emerging technologies in Canada. Int J 72(4):538–562
Du M, Chen Q, Xiao J, Yang H, Ma X (2020) Supply chain finance innovation using blockchain. IEEE Trans Eng Manage 67(4):1045–1058
Gan Q, Lau RYK (2024) Trust in a 'trust-free'system: blockchain acceptance in the banking and finance sector. Technol Forecast Soc Chang 199:123050
Gan Q, Lau RYK, Hong J (2021) A critical review of blockchain applications to banking and finance: a qualitative thematic analysis approach. Technol Anal Strateg Manag: 1–17
Gao W, Su C (2020) Analysis of earnings forecast of blockchain financial products based on particle swarm optimization. J Comput Appl Math 372:112724

Gervais A, Karame GO, Wüst K, Glykantzis V, Ritzdorf H, Capkun S (2016) On the security and performance of proof of work blockchains. In: Proceedings of the 2016 ACM SIGSAC conference on computer and communications security

Ghosh A, Gupta S, Dua A, Kumar N (2020) Security of cryptocurrencies in blockchain technology: state-of-art, challenges and future prospects. J Netw Comput Appl 163:102635

Gong Q, Ban M, Zhang Y (2022) Blockchain, enterprise digitalization, and supply chain finance innovation. China Econ Transit= Dangdai Zhongguo Jingji Zhuanxing Yanjiu 5(2):131–158

Griffiths PDR, Baudier P (2022) Enabling responsible banking through the application of Blockchain. J Innov Econ Manag: I–XXXIII

Griffiths PDR, Baudier P (2023) Enabling responsible banking through the application of Blockchain. J Innov Econ Manag 41(2): 17–49

Guo Y, Liang C (2016) Blockchain application and outlook in the banking industry. Financ Innov 2:1–12

Hamilton M (2020) Blockchain distributed ledger technology: an introduction and focus on smart contracts. J Corp Account Finance 31(2):7–12

Hashem R, Mubarak A-RI, Abu-Musa A (2023) The impact of blockchain technology on audit process quality: an empirical study on the banking sector. Int J Audit Account Stud 5(1):87–118

Hassani H, Huang X, Silva E (2018) Banking with blockchain-ed big data. J Manag Analytics 5(4):256–275

He Q (2021) Application of blockchain technology in commercial banks. In: E3S Web of Conferences, Vol. 235, p 03070. EDP Sciences

Huyut Ü, Temur GT (2021) Blockchain application decision analysis with dematel and vikor methods: a case in the turkish banking sector. Proc Book 56

Hyvärinen H, Risius M, Friis G (2017) A blockchain-based approach towards overcoming financial fraud in public sector services. Bus Inf Syst Eng 59:441–456

Iacoviello G, Bruno E (2023) Exploring a new business model for lending processes in the banking sector using Blockchain technology: an Italian case study

Ikeda K, Hamid M-N (2018) Applications of blockchain in the financial sector and a peer-to-peer global barter web. In: Advances in computers, vol 111. Elsevier, pp 99–120

John K, Kogan L, Saleh F (2023) Smart contracts and decentralized finance. Annu Rev Financ Econ 15:523–542

Kabra N, Bhattacharya P, Tanwar S, Tyagi S (2020) MudraChain: Blockchain-based framework for automated cheque clearance in financial institutions. Futur Gener Comput Syst 102:574–587

Kamal HT, Tayyab S (2017) The impact of blockchain on business models: a study on how the attributes of blockchain affect the elements of business model

Khadka R (2020) The impact of blockchain technology in banking: how can blockchain revolutionize the banking industry?

Khan SN, Loukil F, Ghedira-Guegan C, Benkhelifa E, Bani-Hani A (2021) Blockchain smart contracts: applications, challenges, and future trends. Peer-to-Peer Netw Appl 14:2901–2925

Kongmanee J, Kijsanayothin P, Hewett R (2019) Securing smart contracts in blockchain. In: 2019 34th IEEE/ACM International Conference on Automated Software Engineering Workshop (ASEW)

Koteska B, Karafiloski E, Mishev A (2017) Blockchain implementation quality challenges: a literature. In: SQAMIA 2017: 6th workshop of software quality, analysis, monitoring, improvement, and applications

Kour M (2023) Blockchain technology changing landscape of banking industry. In: 2023 2nd International Conference on Applied Artificial Intelligence and Computing (ICAAIC)

Krishnan S, Balas VE, Julie EG, Yesudhas HR, Balaji S, Kumar R (eds) (2020) Handbook of research on blockchain technology. Academic Press

Krishnan S, Balas VE, Julie EG, Yesudhas HR, Kumar R (eds) (2021) Blockchain for smart cities. Elsevier

Krishnan S, Balas VE, Kumar R (eds) (2023) Green Blockchain technology for sustainable smart cities. Elsevier

Kshetri N (2020) Blockchain-based financial technologies and cryptocurrencies for low-income people: technical potential versus practical reality. Computer 53(1):18–29

Lai R, Chuen DLK (2018) Blockchain–from public to private. In: Handbook of blockchain, digital finance, and inclusion, vol 2. Elsevier, pp 145–177

Leng J, Ruan G, Jiang P, Xu K, Liu Q, Zhou X, Liu C (2020) Blockchain-empowered sustainable manufacturing and product lifecycle management in industry 4.0: a survey. Renew Sustain Energy Rev 132:110112

Li J, Wang Z, Guan S, Cao Y (2024) ProChain: a privacy-preserving blockchain-based supply chain traceability system model. Comput Ind Eng 187:109831

Liu Z (2020) Literature review of supply chain finance based on blockchain perspective. Open J Bus Manag 9(1):419–429

Mainelli M, Smith M (2015) Sharing ledgers for sharing economies: an exploration of mutual distributed ledgers (aka blockchain technology). J Financ Perspect 3(3)

Mansoor N, Antora KF, Deb P, Arman TA, Manaf AA, Zareei M (2023) A review of blockchain approaches for KYC. IEEE Access

Mbaidin H, Sbaee N, AlMubydeen I, Alomari K (2024) Key success drivers for implementation blockchain technology in UAE Islamic banking. Uncertain Supply Chain Manag 12(2):1175–1188

Miraz MH, Ali M (2018) Applications of blockchain technology beyond cryptocurrency. arXiv preprint arXiv:1801.03528

Mitani T, Otsuka A (2020) Traceability in permissioned blockchain. IEEE Access 8:21573–21588

Mohamed A, Faisal R (2024) Exploring metaverse-enabled innovation in banking: leveraging NFTS, blockchain, and smart contracts for transformative business opportunities. Int J Data Netw Sci 8(1):35–44

Mohammadi M, Rezaei J (2020) Bayesian best-worst method: a probabilistic group decision making model. Omega 96:102075

Mukherjee P, Pradhan C (2021) Blockchain 1.0 to blockchain 4.0—the evolutionary transformation of blockchain technology. In Blockchain technology: applications and challenges. Springer, pp 29–49

Nirvan S, Verma S, Kathuria S, Singh R, Akram SV (2023) Enhanced Banking Services using Blockchain and Artificial Intelligence. In: 2023 Second International Conference on Augmented Intelligence and Sustainable Systems (ICAISS)

Norta A, Leiding B, Lane A (2019) Lowering financial inclusion barriers with a blockchain-based capital transfer system. In: IEEE INFOCOM 2019-IEEE Conference on Computer Communications Workshops (INFOCOM WKSHPS)

Noy S, Zhang W (2023) Experimental evidence on the productivity effects of generative artificial intelligence. Available at SSRN 4375283

Opricovic S (1998) Multicriteria optimization of civil engineering systems. Fac Civil Eng Belgrade 2(1):5–21

Opricovic S, Tzeng G-H (2004) Compromise solution by MCDM methods: a comparative analysis of VIKOR and TOPSIS. Eur J Oper Res 156(2):445–455

Peters GW, Panayi E (2016) Understanding modern banking ledgers through blockchain technologies: future of transaction processing and smart contracts on the internet of money. Springer

Poongodi M, Sharma A, Vijayakumar V, Bhardwaj V, Sharma AP, Iqbal R, Kumar R (2020) Prediction of the price of Ethereum blockchain cryptocurrency in an industrial finance system. Comput Electr Eng 81:106527

Pradhan NR, Singh AP, Kumar V (2021) Blockchain-enabled traceable, transparent transportation system for blood bank. In: Advances in VLSI, Communication, and Signal Processing: Select Proceedings of VCAS 2019

Qun W, Xinwang L, Ligang Z, Jindong Q, Jafar R (2024) An analytical framework for the best–worst method. Omega 123:102974

Rekha A, Resmi A (2021) An empirical study of blockchain technology, innovation, service quality and firm performance in the banking industry. In: Eurasian Economic Perspectives: Proceedings of the 29th Eurasia Business and Economics Society Conference
Rezaei J (2015) Best-worst multi-criteria decision-making method. Omega 53:49–57
Rezaei J, Arab A, Mehregan M (2022) Equalizing bias in eliciting attribute weights in multiattribute decision-making: experimental research. J Behav Decis Mak 35(2):e2262
Rezaei J, Arab A, Mehregan M (2024) Analyzing anchoring bias in attribute weight elicitation of SMART, Swing, and best-worst method. Int Trans Oper Res 31(2):918–948
Sakho S, Jianbiao Z, Essaf F, Badiss K (2019) Improving banking transactions using blockchain technology. In: 2019 IEEE 5th International Conference on Computer and Communications (ICCC)
Saner HS, Yucesan M, Gul M (2022) A Bayesian BWM and VIKOR-based model for assessing hospital preparedness in the face of disasters. Nat Hazards 111(2):1603–1635
Schuetz S, Venkatesh V (2020) Blockchain, adoption, and financial inclusion in India: research opportunities. Int J Inf Manage 52:101936
Sethaput V, Innet S (2023) Blockchain application for central bank digital currencies (CBDC). Cluster Comput 26(4):1–15
Sharma A, Sharma D, Bansal R (2023) Emerging role of blockchain in banking operations: an overview. In: Grima S, Sood K, Özen E (eds) Contemporary studies of risks in emerging technology, Part A. Emerald Publishing Limited, pp 1–12
Sherman AT, Javani F, Zhang H, Golaszewski E (2019) On the origins and variations of blockchain technologies. IEEE Secur Priv 17(1):72–77
Song H, Yang Y, Tao Z (2023) Application of blockchain in enterprise financing: literature review and knowledge framework. Nankai Bus Rev Int 14(3):373–399
Soni S, Bhushan B (2019) A comprehensive survey on blockchain: working, security analysis, privacy threats and potential applications. In: 2019 2nd international conference on intelligent computing, instrumentation and control technologies (ICICICT)
Su D, Zhang L, Peng H, Saeidi P, Tirkolaee EB (2023) Technical challenges of blockchain technology for sustainable manufacturing paradigm in Industry 4.0 era using a fuzzy decision support system. Technol Forecast Soc Change 188:122275
Suma V (2019) Security and privacy mechanism using blockchain. J Ubiquitous Comput Commun Technol (UCCT) 1(01):45–54
Sun Y, Shahzad M, Razzaq A (2022) Sustainable organizational performance through blockchain technology adoption and knowledge management in China. J Innov Knowl 7(4):100247
Swan M (2015) Blockchain: blueprint for a new economy. O'Reilly Media, Inc
Tapscott D, Tapscott A (2016) Blockchain revolution: how the technology behind bitcoin is changing money, business, and the world. Penguin
Tasatanattakool P, Techapanupreeda C (2018) Blockchain: challenges and applications. In: 2018 International Conference on Information Networking (ICOIN)
Wang Y, Lin C (2020) Research on the application of blockchain in credit bank. In: 2020 International Conference on Information Science and Education (ICISE-IE)
Wang L, Wang Y (2022) Supply chain financial service management system based on block chain IoT data sharing and edge computing. Alex Eng J 61(1):147–158
Wang H, Ma S, Dai H-N, Imran M, Wang T (2020) Blockchain-based data privacy management with nudge theory in open banking. Futur Gener Comput Syst 110:812–823
Xie P, Kassim AAM, Wei M, Helmi RAA (2024) Comprehensive review of blockchain applications in fintech companies. J Syst Manag Sci 14(1):98–119
Yele S, Litoriya R (2024) Blockchain-based secure dining: enhancing safety, transparency, and traceability in food consumption environment. Blockchain Res Appl 5(2):100187
Zhang Q, Zhu J, Ding Q (2020) OBBC: a blockchain-based data sharing scheme for open banking. In: Blockchain Technology and Application: Second CCF China Blockchain Conference, CBCC 2019, Chengdu, China, October 11–13, 2019, Revised Selected Papers 2

Zou G, Xue Y (2020) Application of blockchain technology in credit management for credit bank system. In: Proceedings of the 2020 4th international conference on electronic information technology and computer engineering

Smart Governance in Vietnam—A Case Study on How Blockchain Enables Zero Trust Cyber Forensic Accounting

Pham Quang Huy and Vu Kien Phuc

Abstract The present study aims to provide a comprehensive understanding of how the implementation of blockchain technology (BT) contributes to the advancement of smart government establishment (SG). Additionally, it sheds light on the mediating function of zero trust cyber forensic accounting (ZTCFA) in the relationship between BT and SG. This research employed a quantitative methodology and utilized a deductive strategy. The hypothesized model, which was based on statistical data obtained from a survey administered to a cross-sectional sample of 812 accountants employed in public sector organizations (PSOs), was examined using structural equation modelling. Positive interrelationships between BT, ZTCFA, and SG are validated by the outcome analyses in terms of significance and effect magnitude. As a result of these observations, the leaders of PSOs are able to discern and capitalize on opportunities to become effective in the digital transformation journey, in addition to formulating robust strategies to effectively direct their SG process. Insights with greater specificity that are provided by this work facilitated the enactment of regulations and laws by policymakers with regard to digitalization and the formulation and operationalization of SG.

1 Introduction

Worldwide, governments strive to meet the demands of their citizens by providing high-quality services that are also efficient, affordable, and socially fair. According to Alqaryouti et al. (2024), one of the most essential ways to get consumer input is to look at digital sources, such reviews of smart applications and information found online. The fast expansion of worldwide government services and applications has been propelled by emerging information communication technologies (Almuraqab

P. Q. Huy (✉) · V. K. Phuc
School of Accounting, University of Economics Ho Chi Minh City (UEH), Ho Chi Minh City, Vietnam
e-mail: pquanghuy@ueh.edu.vn

V. K. Phuc
e-mail: phucvk@ueh.edu.vn

S. Krishnan et al. (eds.), *Unveiling Blockchain Enabled Business Applications*, Contributions to Management Science, https://doi.org/10.1007/978-981-95-2337-5_14

et al. 2021; Hujran et al. 2023). According to Gil-Garcia et al. (2014) and Lemke et al. (2020), current e-government maturity models are unable to account for the lightning-fast pace of technological development in the field, which will drastically alter how future governments provide public services and engage with the public. By empowering governments to deliver intelligent, integrated, personalized, and interactive public services to stakeholders—including citizens, businesses, employees, and other public organizations—smart government (SG) is seen as the pinnacle of public agency modernization (Althunibat et al. 2021; Hujran et al. 2023). Making ensuring that SG systems are secure, private, secret, intact, and available is, thus, very important. In the absence of future improvements in security technology and countermeasures, even the most advanced government systems will be susceptible to privacy and security breaches.

Numerous publications have covered the various blockchain algorithms and kinds (Assiri and Alnami 2024). A thorough investigation of every kind is presented by Krishnan et al. (2020). There has been much anticipation that blockchain technology (BT) will transform public agencies and services ever since its inception (Warkentin and Orgeron 2021). Yli-Huumo et al. (2016) found that BT can help government organizations with transactions and information exchanges that require trust and authentication. Due to its unique properties, such as trustworthy recordkeeping, shared trustworthy business processes, decreased running cost, and better discoverability concerning audit trials, government organizations can utilize BT to develop e-government systems. According to Shan et al. (2021), the implementation of a collaborative governance model of SG using BT has significant implications for enhancing governance capabilities and system. BT offers a solution to the security issues faced by internet of things services. It ensures secure data sharing and maintains data integrity (Alketbi et al. 2018). In the research on the potential use of the BT for smarter governments, Ølnes (2016) argued that bitcoin could become an emerging technology with great potential for authenticating numerous kinds of documents that are persistent in public sector organizations (PSOs). In their study, Mohamed et al. (2022) discovered that the primary attributes of BT that are anticipated to offer the greatest benefit to customers are decentralization and the ability to be shared and made public. On the other hand, the most sought-after attributes for the effective provision of SG services are speed, trust, and participation (Mohamed et al. 2022). Moreover, several academic notes underline the potential of BT for smart cities (i.e., Krishnan et al. 2021, 2023).

The primary issue surrounding the implementation of BT in government organizations pertains to the security of public data. Government authorities are actively seeking secure and reliable solutions to enhance e-government services for citizens (Warkentin and Orgeron 2021). Besides, there has been a lot of appreciations about zero-trust networks because they challenge the conventional wisdom about security-based boundary protection. This network successfully prevented large-scale information security incidents and maintained the network's healthy operation. From an implementation standpoint, the only way to make a system more resistant to intrusion attacks is to regularly upgrade it with new methodologies and technologies that keep up with the rate of research and innovation, such as BT, zero trust, and other advanced techniques (Awan et al. 2023).

Given the preceding, the objective of this research is to analyze the ways in which BT enhances and optimizes the functioning of SG. Furthermore, this study examines the role of zero trust cyber forensic accounting (ZTCFA) as a mediator in the connection between SG and BT. Therefore, the current study sought to investigate the following research questions (RQs):

RQ1. *To what extent does BT affect SG?*
RQ2. *Does ZTCFA mediate the association between BT and SG?*

SG has many advantages, but it is still a relatively new concept that is gaining the attention of scholars in the domains of public administration, information systems, and political science (Anthopoulos et al. 2022). Although many governments have yet to completely implement BT, some nations have launched BT initiatives to investigate the possibility of using it to provide citizens with public services (Jun 2018). Remarkably, limited research has been conducted on the application of BT in government, despite the fact that few scholars have examined its potential to address societal requirements (Olnes et al. 2017). In this regard, this study's contribution to the area is the new insights it offers for continuing this line of investigation, backed up by strong empirical data. Furthermore, the research's culmination would serve as an indispensable cornerstone, providing research hints for subsequent investigations concerning the function of ZTCFA during the SG era. The results of this study contributed to the expanding corpus of literature concerning the incorporation of zero trust principles into the field of digital forensics.

Researchers and practitioners seeking smart solutions for the government establishment could use the study's findings and lessons learned before making any decisions. The research also suggested necessary policy steps for decision-making in the context of SG establishment when planning future actions, from a policymaker's perspective. Alternatively, the findings may provide practitioners and lawmakers with fresh perspectives to develop focused methods to boost BT implementation and ZTCFA, both of which have the potential to bring about beneficial outcomes down the road.

The rest of this manuscript is structured as follows. Section 2 gives a synopsis of the essential theoretical and conceptual elements. The development of research hypotheses and the research model are both covered in Sect. 3. This study's methodology is detailed in Sect. 4. Section 5 offers the analyses and results of the empirical study. Several implications and extensive suggestions for future study directions round out the report.

2 Literature Review

2.1 *Contingency Theory*

Numerous insights have been generated by the influential organizational theory of contingency (Abboh et al. 2024). At present, a significant framework for organizational design is provided by contingency theory (Donaldson 2006). Contingencies such as the environment and organizational strategy influence the achievement of an organization. Theoretically, there is a broad consensus that the optimal organizational structure is contingent upon various elements, such as personnel characteristics, technological implementation, and market conditions (Nasrallah and Qawasmeh 2009). Due to its prevalence, contingency theory has attracted considerable interest from scholars in the domains of auditing and accounting (Badara 2015). As a result, contingency theory may serve as an appropriate conceptual framework to illuminate the field of forensic accounting research. Pham and Vu (2024) employed this theoretical framework to explicate the interconnectedness that exists between accelerated internationalization, circular metaverse business model innovation, and digital forensic accounting.

2.2 *Blockchain*

BT, which was initially unveiled in 2008, is a distributed database designed to uphold a continuously expanding inventory of data records while ensuring protection against tampering and modification (Zhang et al. 2021). The data is accessible to all participants, which is why BT is referred to as a distributed ledger (Ølnes et al. 2017). The BT is a decentralized system that is accessible from any location on the planet. Working without the intervention of a third party in a highly uncertain environment is advantageous when this technology is utilized. BT ensures complete security and integrity by generating a hash of the preceding transaction. These interconnected nodes are referred to as the "blockchain." As stated by Narayanan et al. (2016), mining refers to the procedure by which a new block is generated.

2.3 *Zero Trust Cyber Forensic Accounting*

Zero trust digital forensics is an investigative method in which every part of an inquiry is considered untrustworthy unless it is confirmed to be reliable (Neale et al. 2022). Cyber forensic accounting, as defined by Hossain et al. (2023), involves the utilization of forensic accounting principles to investigate and avert cybercrimes. In the context of this study, ZTCFA pertains to the investigation and prevention of cybercrimes by employing a method in which every facet of an investigation is presumed to be unreliable until it is corroborated.

2.4 Smart Government

Governments worldwide strive to provide high-quality government services that are efficient, cost-effective, and equitable, with a focus on meeting the demands of the people. The concept of SG emerged as a progression from open government, with the aim of enhancing the accessibility, efficiency, and attractiveness of governments to their citizens (Anthopoulos et al. 2022). According to Alqaryouti et al. (2024), digital sources such as smart application reviews and online feedback information are regarded as significant sources for gathering customers' feedback and input. The notion of "smart government" is based on the principles of open data, sharing, service integration, innovation, citizen involvement, information openness, and collaborative policy creation (Younus et al. 2024).

3 Hypothesis Development and Research Model

3.1 Hypothesis Development

The operations of government are anticipated to be impacted by BT. Contracts between parties have historically been instrumental in fostering confidence and coordination. Within the BT, this contract's digital manifestation is referred to as a "smart contract" (Verma and Sheel 2022). One domain in which these smart contracts have the potential to be implemented is the transfer of title for immovable property, such as houses and land. A smart contract is a BT-based software application that implements contract conditions in accordance with a consensus mechanism. All nodes provide their endorsement of contract conditions. The conditions and regulations that have been reached in advance are embedded within an automated smart contract (Olnes et al. 2017). Parties to a transaction reach an agreement on contract terms, which are then enforced by the BT. A smart contract is a mechanism that facilitates the exchange of digital assets in accordance with a formula based on mutual agreement among the involved parties. Moreover, BT has the potential to be implemented in any transaction or exchange of information that involves government intervention—such as inventory and asset registry, health data, ideas, patents, votes, and health information (Swan 2015). Additionally, BT facilitates the exchange of assets whose ownership is ambiguous, such as land or automobiles. Occasionally, this can make locating the true proprietor of the asset challenging. The BT enables buyers to conduct transactions and identify legitimate proprietors (Verma and Sheel 2022). When transactions involve more than two parties, such as obtaining government authorization to organize large-scale events (i.e., concerts and demonstrations) in which a large number of people are anticipated to partake, BT can also be advantageous. Numerous entities are engaged in such occurrences, including the police department, health department, and municipality. The coordination of the event is contingent upon the consent of all relevant government departments to manage substantial public gatherings (Verma and Sheel 2022). To that end, the first research hypothesis of this study is formulated in light of the aforementioned considerations.

Hypothesis 1 (H1). BT significantly and positively induces an effect on SG.

The potential for a paradigm shift in financial record-keeping exists with the incorporation of BT into accounting (Bonsón and Bednárová 2019). This technological advancement provides a transparent and immutable ledger, which has a profound impact on conventional accounting procedures. According to Benedetti and Nikbakht (2021), the risk of fraudulent activities can be reduced through the utilization of decentralized and secure data storage (Tang et al. 2022; Iftikhar et al. 2021). The data integrity of BT is guaranteed by its cryptographic characteristics, which significantly diminish the probability of unauthorized modifications or interference. In addition, the integrity of financial transactions is bolstered by the transparency that BT inherently provides (Centobelli et al. 2022; Han et al. 2023). A shared real-time ledger is accessible to stakeholders, which promotes accountability and diminishes the inherent opaqueness of conventional accounting systems. The profound potential of BT highlights its wider impacts in reshaping the fundamental aspects of financial record-keeping, thereby facilitating the implementation of accounting practices that are more secure, streamlined, and transparent. The adoption of this technology has the potential to bolster the precision and safeguarding of financial data (Demirkan et al. 2020; Han et al. 2023). Consequently, this could result in a diminished probability of accounting inaccuracies and an enhancement of information protection (Kolisnyk et al. 2023; Matskiv et al. 2023).

Further elaborating on its implications, BT documents transactions across multiple workstations using an immutable alphanumeric recording method that prevents modifications to subsequent blocks. The utilization of an immutable alphanumeric process in this decentralized and tamper-resistant method of recording transactions across multiple workstations carries significant implications. It ensures that transactions recorded in blocks cannot be modified or removed without the agreement of all network participants. The integrity and reliability of the recorded information are greatly augmented by its immutable quality. It eliminates the need for an intermediary in record-keeping based on consensus, thereby reducing fraud and errors (Bonsón and Bednárová 2019). Accounting on the BT necessitates smart contracts (Smith and Castonguay 2020) that implement automatically when specific conditions are met. A distributed peer-to-peer system is utilized to store data in multiple copies, thereby reducing the likelihood of manipulation (Centobelli et al. 2022). To that end, the second research hypothesis of this study is formulated in light of the aforementioned considerations.

Hypothesis 2 (H2). BT significantly and positively induces an effect on ZTCFA.

Cyber forensic accounting, as defined by Hossain et al. (2023), involves the utilization of forensic accounting principles to investigate and deter cybercrimes. Cyber forensic accounting is a branch of accounting that focuses on identifying and preventing financial fraud and misbehavior using investigative and analytical techniques. Cyber forensic has the potential to enhance the ability to detect and prevent financial fraud, losses, and risks through technology advancements. Data analytics technologies play a crucial role in cyber forensic accounting by identifying trends, irregularities, and potential financial frauds. Data analysis is a fundamental aspect of this process. Common data analytics techniques employed by digital forensic accountants encompass data mining, machine learning, and predictive modelling. These techniques can analyze vast quantities of data to detect correlations between data points and discover signs of fraud. Cyber forensic accountants have the ability to assess the effectiveness of internal controls connected to cybersecurity and provide suggestions for improvements to prevent fraud and risks in the functioning of PSOs.

The deployment of zero trust ensures secure access to the company's resources and enhances security across the whole digital attack surface (Bertino 2021). The concept of zero trust is founded on the principles of complete distrust, a risk-focused approach to trust, and ongoing verification of trust, drawing on the perspectives of Wylde (2021). It imposes particular limitations on each subject's capacity to access the entire network (Bertino 2021). Implementing a zero-trust approach enhances the level of network visibility within an organization. Consequently, it promotes the act of observing and identifying potentially suspect traffic, as opposed to conventional security protocols. To that end, the third research hypothesis of this study is formulated in light of the aforementioned considerations.

Hypothesis 3 (H3). ZTCFA significantly and positively induces an effect on SG.

3.2 Research Model

The hypothesized model which stresses on the impact which BT and ZTCFA had on SG is demonstrated in Fig. 1.

4 Research Methodology

4.1 Operationalization of Variables for Measurement

In order to authenticate the model, a survey was disseminated to staff members working in the accounting departments. The survey has two distinct phases: questionnaire development and data collection. The survey approach has been considered

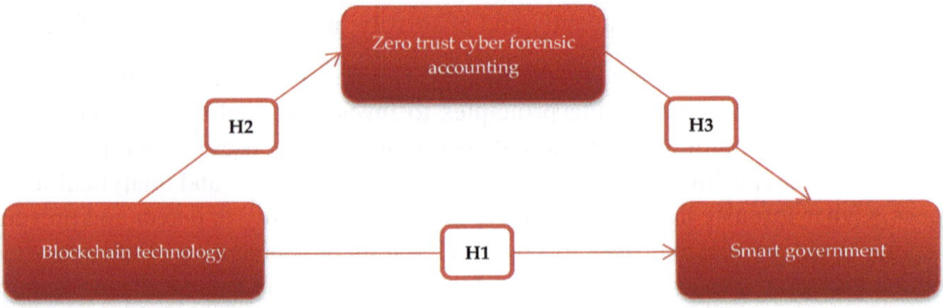

Fig. 1 Hypothesized model

appropriate for its intended aims of assessing hypotheses, constructing measuring scales, and characterizing the population (Wamba et al. 2020). The development of a series of questions was achieved by combining pertinent measurements with recognized literature and hypotheses. The question recitals were modified to accommodate the specific circumstances pertaining to this investigation. Initially, the authors carried out a pilot study with 30 respondents from Vietnamese PSOs. Cronbach's alpha was used to evaluate the internal consistency of each construct in the questionnaire. The Cronbach's alpha coefficient for the small-scale pilot test exceeded 0.7, indicating that the questionnaire consistently generated valid and reliable replies. By incorporating the results obtained from the pilot study into the theoretical framework of the official questionnaire survey. The assessment of the research construct was performed by employing a Likert scale consisting of five discrete levels: 1 denoted substantial disagreement; 5 represented substantial agreement.

Blockchain technology. The BT construct in the current investigation was determined as a higher-level combination of four core constructs—Decentralization; Persistency; Anonymity; Auditability which were modified from the findings of Zarrin et al. (2021) and were pre-tested with practitioners.

Zero trust cyber forensic accounting. The ZTCFA construct in the current investigation was determined as a higher-level combination of four core constructs—cyber anti-fraud policies, zero-trust governance, digitally designed forensic procedures, and management control systems. More specifically, the measurement scale for cyber anti-fraud policies was drawn on the criteria derived from the findings of Hossain et al. (2023). The measurement scale for zero-trust governance was drawn on the criteria derived from the findings of Al-Taee and Flayyih (2023). The measurement scale for digitally designed forensic procedures was drawn on the criteria derived by Ogundele et al. (2023). The management control systems measurement scale was drawn on the criteria derived from the findings of Almbaidin (2014).

Smart government. The SG construct in the current investigation was determined as a higher-level combination of four core constructs—Smart governance; Smart resources; Smart services; Smart infrastructure which were modified from the findings of Anthopoulos et al. (2022), Younus et al. (2024), Alqaryouti et al. (2024), Koman et al. (2024) and were pre-tested with practitioners.

4.2 *Sampling Procedures and Data Collection*

Vietnam was chosen as the site for data acquisition in the course of this investigation. There is a notable surge in the inclination of PSOs in Vietnam to participate in sustainable development initiatives through the adoption and implementation of a wide range of emergent technologies. In light of the fact that the organization serves as the unit of analysis in this research, each of these organizations was invited to send a representative from their respective accounting departments to participate in the survey. This investigation utilized a combination of convenience and snowball sampling techniques to collect its data. Each accountant of the PSO was duly informed that their participation was entirely voluntary and intended solely for academic purposes. Furthermore, they were guaranteed the utmost confidentiality regarding their identities. The participants were approached via email and telephone on multiple occasions in an effort to elicit their participation in the survey. Each participant was given a document containing their response. A clarification was provided to each accountant regarding the procedure for completing the response document. They were granted a response period of nine months (from June 2023 to March 2024) within which they were obligated to sufficiently complete the response documents. Following an exhaustive examination of all 910 responses, it was determined that 98 were incomplete, representing a 10.77% loss of data rate. These factors were not taken into account. The 812 responses that were gathered and utilized as representative samples of the 812 organizations were subjected to analysis. The information pertaining to the 812 organizations is presented in Table 1. The statistical data analysis was conducted using IBM SPSS Statistics version 29 and AMOS version 28 as supporting software.

5 Result Analysis and Discussion

5.1 *Evaluation on Measurement Model*

Construct reliability and Convergent validity. All factor loadings in this investigation surpassed the threshold of 0.6, signifying that the factors exerted a substantial influence on the model (Ide et al. 2024). The composite reliability, which measure the internal consistency of latent constructs, exceeded the threshold of 0.70 suggested by Rashid et al. (2024), indicating a satisfactory level of internal consistency. Furthermore, the Cronbach's alpha values surpassed the 0.70 criterion set by Kline (2023),

Table 1 Distribution of participants based on demographic characteristics

Item	Contents	Frequency	Valid (%)
Gender			
	Male	531	65.39
	Female	281	34.61
Age (years)			
	Under 30	22	2.71
	30–Under 40	683	84.11
	40–Under 50	64	7.88
	Above 50	43	5.30
Education			
	Undergraduate	783	96.43
	Postgraduate	29	3.57
Experience (years)			
	Under 10	27	3.33
	10–Under 20	513	63.18
	20–Under 30	233	28.69
	Over 30	39	4.80

which provided further support for the internal consistency and reliability of the latent constructs. The average variance extracted (AVE) values exceeded the lowest limit of 0.50 recommended by Khan et al. (2023), indicating strong convergent validity. The result analyses shown in Table 2 demonstrated that this research has achieved convergent validity and internal reliability.

Discriminant validity. Discriminant validity was assessed through a comparison between the correlations among the proposed constructs and the square root of the AVE. Each latent variable's AVE exceeded the values for off-diagonal correlation (Fornell and Larcker 1981). The discriminant validity was demonstrated by the results presented in Table 3.

According to Henseler et al. (2015), discriminant validity among latent variables was evaluated by using heterotrait-monotrait (HTMT) values. Discriminant validity was confirmed when the HTMT values were below 0.90. As shown in Table 4, the HTMT values were less than 0.90, confirming the discriminant validity of all constructs investigated in this study.

Goodness-of-fit index evaluation. The goodness-of-fit metric, as suggested by Kang and Ahn (2021), assessed the degree to which the predicted model accurately reflected the observed phenomenon in the data. In the absence of a singular metric to quantify the adequacy of a model's fit, a number of fit indices were chosen to evaluate its suitability (Hair et al. 2014). Five parameters were utilized to assess the model's fit: Chi-square/df, comparative fit index (CFI), Tucker Lewis index (TLI), goodness of

Table 2 Construct reliability and validity

Constructs and operationalization		Convergent validity		Construct reliability		Result
		Factor loadings	AVE	Cronbach's Alpha	Composite reliability	
Blockchain technology	BT					
Decentralization	DE	0.721–0.835	0.633	0.875	0.878	Retained
Persistency	PE	0.693–0.760	0.520	0.816	0.816	Retained
Anonymity	AN	0.724–0.807	0.590	0.853	0.857	Retained
Auditability	AU	0.683–0.803	0.576	0.847	0.849	Retained
Zero trust cyber forensic accounting	ZTCFA					
Cyber anti-fraud policies	CAP	0.783–0.815	0.661	0.799	0.799	Retained
Zero trust governance	ZTG	0.652–0.784	0.533	0.855	0.856	Retained
Digitally designed forensic procedures	DDFP	0.730–0.801	0.602	0.863	0.863	Retained
Management control systems	MCS	0.718–0.877	0.644	0.782	0.786	Retained
Smart government	SG					
Smart governance	SGO	0.668–0.748	0.522	0.768	0.769	Retained
Smart resources	SR	0.676–0.794	0.579	0.803	0.807	Retained
Smart services	SS	0.671–0.834	0.578	0.849	0.850	Retained
Smart infrastructure	SI	0.676–0.781	0.520	0.848	0.849	Retained

fit index (GFI), and root mean square error of approximation (RMSEA). Additionally, the p-values associated with each element variable utilized in these constructs remained intrinsically zero, indicating that there was no degradation in the process of translating the constructs into their decomposed forms (Ellis et al. 2021). The results of each parameter in Table 5 demonstrated that the overall model fit was satisfactory.

5.2 *Evaluation on Structural Model*

Direct effect. Building on the result analysis gleaned from Table 6, BT demonstrated vigorous influences on SG ($\beta = 0.220$; $p = 0.000$) and ZTCFA ($\beta = 0.287$; $p =$

Table 3 Results summary of Discriminant validity using Fornell–Larcker process

	PE	SS	AN	ZTG	DDFP	MCS	AU	SGO	DE	SR	CAP	SI
PE	1											
SS	0.220	1										
AN	0.104	0.250	1									
ZTG	0.204	0.192	0.157	1								
DDFP	0.148	0.156	0.149	0.078	1							
MCS	0.249	0.195	0.105	0.157	0.082	1						
AU	0.190	0.317	0.114	0.134	0.137	0.165	1					
SGO	0.254	0.269	0.155	0.198	0.125	0.135	0.226	1				
DE	0.272	0.331	0.027	0.149	0.051	0.275	0.151	0.050	1			
SR	0.111	0.047	0.172	0.382	0.110	0.066	0.120	0.133	0.073	1		
CAP	0.321	0.074	0.157	0.352	0.073	0.061	0.114	0.110	0.045	0.223	1	
SI	0.386	0.286	0.151	0.373	0.121	0.135	0.073	0.145	0.428	0.233	0.113	1

Table 4 Results summary for discriminant validity on Heterotrait–Monotrait ratio

	PE	SS	AN	ZTG	DDFP	MCS	AU	SGO	DE	SR	CAP	SI
PE												
SS	0.185											
AN	0.087	0.209										
ZTG	0.173	0.161	0.132									
DDFP	0.124	0.132	0.131	0.072								
MCS	0.215	0.163	0.095	0.126	0.068							
AU	0.163	0.266	0.103	0.119	0.118	0.146						
SGO	0.213	0.220	0.128	0.165	0.099	0.106	0.191					
DE	0.227	0.277	0.019	0.116	0.037	0.225	0.132	0.051				
SR	0.091	0.042	0.130	0.314	0.095	0.056	0.098	0.108	0.057			
CAP	0.272	0.065	0.125	0.285	0.072	0.053	0.092	0.097	0.038	0.173		
SI	0.320	0.234	0.127	0.305	0.102	0.113	0.062	0.114	0.334	0.181	0.098	

0.000). In the same vein, ZTCFA ($\beta = 0.259$; $p = 0.000$) was also found to be positively correlated with SG.

Indirect effect. The present study employed the AMOS plugin, which was the main subject of investigation in a study conducted by Gaskin and Lim (2018), to examine the mediation effect. According to the study conducted by Baron and Kenny (1986), the results presented in Table 6 showed that all the necessary conditions for mediation were met. BT, as an independent variable, demonstrated a positive indirect correlation with SG, which was a dependent variable, when ZTCFA, acting as a mediator, was

Table 5 Results of measurement and structural model analysis

The goodness of fit measures	Minimum cutoff	Parameter estimates of measurement model	Parameter estimates of structural model	Recommended by
Chi-square/df	< 3	1.248	1.397	Hair et al. (2014)
TLI	≥ 0.9	0.933	0.923	Little (2013)
CFI	≥ 0.9	0.940	0.927	Hair et al. (2014)
GFI	$0.774 \leq x \leq 0.923$	0.890	0.879	Motawa and Oladokun (2015)
RMSEA	< 0.07	0.019	0.023	Guo et al. (2019)

Table 6 Structural relationships and hypothesis testing

Hypothesis No	Hypothesized path			Standardized coefficients Beta	S.E	C.R	P	Status
H1	BT	→	SG	0.220	3.896	3.060	0.000	Accepted
H2	BT	→	ZTCFA	0.287	3.771	3.757	0.000	Accepted
H3	ZTCFA	→	SG	0.259	3.995	3.511	0.000	Accepted

Table 7 Research findings the mediating role of ZTCFA

Indirect path	Standardized	Lower	Upper	P
BT → ZTCFA → SG	0.074	0.048	0.119	0.008

present. The statistical analysis shown in Table 7 indicated that the ZTCFA functioned as a partial mediator in the association between BT and SG.

6 Concluding Remark

6.1 Practical Implication

Expanding on the managerial perspectives, the current research's acquired observations furnished a plethora of actionable takeaways. In regard to SG, given the significant role it has assumed in contemporary society, it is imperative that all executives within PSOs acquire managerial cognitive capacity and prioritize this domain. Additionally, PSOs ought to guarantee that the BT implementation has transcended its technical nature and become intertwined with an organization-wide transformation endeavor, necessitating the utilization of analytics strategies, managerial backing, and proactive, meticulous, and laborious administration in order to fully capitalize on BT's potential. Thus, emphasis has been placed on concentrating tangible resources, including BT infrastructure and digital platform, in addition to other essential resources required for the implementation of BT initiatives. Besides, the results emphasized the significance of ZTCFA in preventing and detecting various categories of fraud. As a result, organizations are compelled to implement ZTCFA techniques to reduce the recurrence of cyber fraud. Importantly, it has been proposed that accounting personnel participate in a series of training and development programs designed to enhance their knowledge, abilities, and skills in order to better carry out their duties.

It should be emphasized by legislators and government agencies that SG formulation and operationalization are crucial. In order to make use of emergent trends, legislators and practitioners must collaborate to establish standards and best practices for the implementation of digital technologies. In the realm of forensic accounting, policymakers and influential members of government ought to accord priority to cyber anti-fraud policies that pertain to data analytics, digital forensics, and other applications of digital technology. This will guarantee that the investigation outcomes are both dependable and precise.

6.2 Limitation and Orientations for Further Research

The current academic endeavor in a novel field of study, while emphasizing the significant findings, encountered a number of unavoidable constraints. Nonetheless, the outlined constraints may serve as launching points for the development of novel approaches in subsequent endeavors. To begin with, the main limitation of this study was that it was conducted in a specific region of Vietnam. Due to the lack of external validity in this study, it is advisable to exercise caution when considering any generalizations that may result from the research. Subsequent research endeavors might therefore be focused on alternative national and cultural contexts in order to validate the findings of the current study. Secondly, due to financial and schedule constraints, the informants included in the present study were selected using convenience and snowball sampling methods. Therefore, additional research employing systematic sampling techniques to obtain a large sample size may be able to extrapolate these findings and offer global resolutions for all the PSOs in order to gain a more comprehensive understanding of the SG enhancement. Furthermore, it is important to note that the generalizability of the research findings regarding the organizations comprising the population from which the current study sample was derived is limited, the data utilized for the deployment of the analysis was of a declarative nature. While the accountant survey responses were highly valuable due to the informants' exceptional knowledge and insight regarding ZTCFA and BT implementation issues within their respective organizations, it is important to note that the data they provided may only reflect their personal perspectives. Unknown were the degree to which the informants trusted their abilities to provide the information and the degree to which the dependability of their responses in comparison to reality was assessed. By conducting comparable research with personnel and departments from various types of organizations, these issues might be resolved. In the fourth place, cross-sectional data analysis rendered the parameters static as opposed to dynamic. The interrelationships among the variables under investigation in this study were fluid and susceptible to modification as time progressed. Building upon these analyses, comprehensive longitudinal studies could undoubtedly delve deeper and provide additional theoretical elaboration.

Acknowledgements This research was funded by University of Economics Ho Chi Minh City (UEH) in Vietnam.

References

Abboh UA, Majid AHA, Fareed M, Abdussalaam II (2024) High-performance work practices lecturers' performance connection: Does working condition matter? Manag Educ 38(2):79–90. https://doi.org/10.1177/08920206211051468

Alketbi A, Nasir Q, Talib MA (2018) Blockchain for government services—use cases, security benefits and challenges. In: 2018 15th learning and technology conference (L&T), Jeddah, Saudi Arabia, 25–26 February 2018, pp 112–119.

Almbaidin TH (2014) The effectiveness of accounting information system in Jordanian banks: from the management perspective. Int Bull Bus Adm 14(14):1451–1243. https://doi.org/10.5296/ijafr.v4i1.5323

Almuraqab NAS, Jasimuddin SM, Mansoor W (2021) An empirical study of perception of the end-user on the acceptance of smart government service in the UAE. J Glob Inf Manag 29(6):1–29. https://doi.org/10.4018/JGIM.20211101.oa11

Alqaryouti O, Siyam N, Abdel Monem A, Shaalan K (2024) Aspect-based sentiment analysis using smart government review data. Appl Comput Inform 20(1/2):142–161. https://doi.org/10.1016/j.aci.2019.11.003

Al-Taee SHH, Flayyih HH (2023) Impact of the electronic internal auditing based on it governance to reduce auditing risk. Corp GovAnce Organ Behav Rev 7(1):94–100. https://doi.org/10.22495/cgobrv7i1p9

Althunibat A, Binsawad M, Almaiah MA, Almomani O, Alsaaidah A, Al-Rahmi W, Seliaman ME (2021) Sustainable applications of smart-government services: a model to understand smart-government adoption. Sustain 13(6):1–28. https://doi.org/10.3390/su13063028

Anthopoulos L, Sirakoulis K, Reddick CG (2022) Conceptualizing smart government: interrelations and reciprocities with smart city. Digit GovMent: Res Pract 2(4):1–28. https://doi.org/10.1145/3465061

Assiri B, Alnami H (2024) Smart partitioned blockchain. SensS 24(13):1–13. https://doi.org/10.3390/s24134033

Awan SM, Azad MA, Arshad J, Waheed U, Sharif T (2023) A blockchain-inspired attribute-based zero-trust access control model for IoT. Inf 14:1–26. https://doi.org/10.3390/info14020129

Badara MS (2015) Empirical evidence of performance measurement of internal audit function on its effectiveness. Acad J Manag Sci Res 1(1):1–10

Baron RM, Kenny DA (1986) The moderator–mediator variable distinction in social psychological research: conceptual, strategic, and statistical considerations. J Pers Soc Psychol 51(6):1173–1182. https://doi.org/10.1037/0022-3514.51.6.1173

Benedetti H, Nikbakht E (2021) Returns and network growth of digital tokens after cross-listings. J Corp Finan 66:1–18. https://doi.org/10.1016/j.jcorpfin.2020.101853
Bertino E (2021) Zero trust architecture: does it help? IEEE Secur Priv 19(05):95–96. https://doi.org/10.1109/MSEC.2021.3091195
Bonsón E, Bednárová M (2019) Blockchain and its implications for accounting and auditing. Meditari Account Res 27(5):725–740. https://doi.org/10.1108/MEDAR-11-2018-0406
Centobelli P, Cerchione R, Del Vecchio P, Oropallo E, Secundo G (2022) Blockchain technology design in accounting: game changer to tackle fraud or technological fairy tale? Account, Audit & Account J 35(7):1566–1597. https://doi.org/10.1108/AAAJ-10-2020-4994
Demirkan S, Demirkan I, McKee A (2020) Blockchain technology in the future of business cyber security and accounting. J Manag Anal 7(2):189–208. https://doi.org/10.1080/23270012.2020.1731721
Donaldson L (2006) The contingency theory of organizational design: challenges and opportunities. In: Burton, RM, Håkonsson DD, Eriksen B., et al. (eds) Organization design. Information and organization design series, vol 6. Springer, Boston, MA, pp 19–40.
Ellis ME, Downey JP, Chen AN, Lu H-K (2021) Why Taiwanese seniors use technology. Asia Pac Manag Rev 26(3):149–159. https://doi.org/10.1016/j.apmrv.2021.01.001
Fornell C, Larcker DF (1981) Evaluating structural equation models with unobservable variables and measurement error. J Mark Res 18(1):39–50. https://doi.org/10.2307/3151312
Gaskin J, Lim J (2018) Indirect effects. AMOS plugin. Gaskination's StatWiki. Retrieved from http://statwiki.kolobkreations.com/index.php?title=Plugins.
Gil-Garcia JR, Helbig N, Ojo A (2014) Being smart: emerging technologies and innovation in the public sector. Gov Inf Q 31:I1–I8. https://doi.org/10.1016/j.giq.2014.09.001
Guo J, Marsh HW, Parker PD, Dicke T, Lüdtke O, Diallo TMO (2019) A systematic evaluation and comparison between exploratory structural equation modeling and Bayesian structural equation modeling. Struct Equ Model: Multidiscip J:1–28. https://doi.org/10.1080/10705511.2018.1554999
Hair JF, Black WC, Babin BJ, Anderson RE (2014) Multivariate data analysis, 7th edn. Pearson Education, Essex, UK
Han H, Shiwakoti RK, Jarvis R, Mordi C, Botchie D (2023) Accounting and auditing with blockchain technology and artificial intelligence: a literature review. Int J Account Inf Syst 48:1–16. https://doi.org/10.1016/j.accinf.2022.100598
Henseler J, Ringle CM, Sarstedt M (2015) A new criterion for assessing discriminant validity in variance-based structural equation modeling. J Acad Mark Sci 43(1):115–135. https://doi.org/10.1007/s11747-014-0403-8
Hossain MZ (2023) Emerging Trends in forensic accounting: data analytics, cyber forensic accounting, cryptocurrencies, and blockchain technology for fraud investigation and prevention. SSRN Electron J:1–30. https://doi.org/10.2139/ssrn.4450488
Hujran O, Al-Debei MM, Al-Adwan AS, Alarabiat A, Altarawneh N (2023) Examining the antecedents and outcomes of smart government usage: an integrated model. Gov Inf Q 40(1):1–22. https://doi.org/10.1016/j.giq.2022.101783
Ide H, Suwa S, Akuta Y, Kodate N, Tsujimura M, Ishimaru M, Shimamura A, Kitinoja H, Donnelly S, Hallila J, Toivonen M, Camilla Bergman-Kärpijoki C, Takahashi E, Yu W (2024) Developing a model to explain users' ethical perceptions regarding the use of care robots in home care: A cross-sectional study in Ireland, Finland, and Japan. Arch Gerontol Geriatr 116:1–10. https://doi.org/10.1016/j.archger.2023.105137
Iftikhar Z, Javed Y, Zaidi SYA, Shah MA, Iqbal Khan Z, Mussadiq S, Abbasi K (2021) Privacy preservation in resource-constrained IoT devices using blockchain—a survey. Electron 10(14):1–26. https://doi.org/10.3390/electronics10141732
Jun M (2018) Blockchain government-a next form of infrastructure for the twenty-first century. J Open Innov: Technol, Mark, ComplexS 4(1):1–12. https://doi.org/10.1186/s40852-018-0086-3

Kang H, Ahn J-W (2021) Model setting and interpretation of results in research using structural equation modeling: a checklist with guiding questions for reporting. Asian Nurs Res 15(3):157–162. https://doi.org/10.1016/j.anr.2021.06.001

Khan S, Rashid A, Rasheed R, Amirah NA (2023) Designing a knowledge-based system (KBS) to study consumer purchase intention: the impact of digital influencers in Pakistan. Kybernetes 52(5):1720–1744. https://doi.org/10.1108/K-06-2021-0497

Kline RB (2023) Principles and practice of structural equation modeling. Guilford publications, New York, NY

Kolisnyk O, Hurina N, Druzhynska N (2023) Innovative technologies in accounting and auditing: the use of Blockchain technology. Financ Credit Act-Probl Theory Pract 3:24–41. https://doi.org/10.55643/fcaptp.3.50.2023.4082

Koman G, Toman D, Jankal R, Boršoš P (2024) The importance of e-recruitment within a smart government framework. Syst 12(3):1–28. https://doi.org/10.3390/systems12030071

Krishnan S, Balas VE, Julie EG, Yesudhas HR, Balaji S, Kumar R (eds) (2020) Handbook of research on blockchain technology. Academic Press, Cambridge, MA

Krishnan S, Balas VE, Julie EG, Yesudhas HR, Kumar R (eds) (2021) Blockchain for smart cities. Elsevier.

Krishnan S, Balas VE, Kumar R (eds) (2023) Green blockchain technology for sustainable smart cities. Elsevier.

Lemke F, Taveter K, Erlenheim R, Pappel I, Draheim D, Janssen M (2020) Stage models for moving from e-government to smart government. In: Chugunov A, Khodachek I, Misnikov Y, Trutnev D (eds) Electronic governance and open society: challenges in Eurasia. EGOSE 2019. Communications in computer and information science, vol 1135. Springer, Cham, pp 152–164.

Little TD (2013) Longitudinal structural equation modeling. The Guilford Press, New York, NY

Matskiv H, Smirnova I, Malikova A (2023) The application of Blockchain technology in accounting and auditing: experience of Ukraine And Kazakhstan. Financ Credit Act-Probl Theory Pract 1:180–191. https://doi.org/10.55643/fcaptp.1.48.2023.3955

Mohamed AAD, Alkhateeb YM, Agarwal P, Abdelwahab AR, Alrababah O, Salim TA (2022) Characteristics of blockchain and smart services, for smart governments: a systematic review of the literature. Int J Inf Syst Proj Manag 10(3):30–55. https://doi.org/10.12821/ijispm100302

Motawa I, Oladokun MG (2015) Structural equation modelling of energy consumption in buildings. Int J Energy Sect Manage 9(4):435–450. https://doi.org/10.1108/ijesm-11-2014-0004

Narayanan A, Bonneau J, Felten E, Miller A, Goldfeder S (2016) Bitcoin and cryptocurrency technologies: a comprehensive introduction. Princeton University Press.

Nasrallah WF, Qawasmeh SJ (2009) Comparing multi-dimensional contingency fit to financial performance of organizations. Eur J Oper Res 194(3):911–921. https://doi.org/10.1016/j.ejor.2008.01.011

Neale C, Kennedy I, Price B, Yu Y, Nuseibeh B (2022) The case for zero trust digital forensics. Forensic Sci Int: Digit Investig 40:1–13. https://doi.org/10.1016/j.fsidi.2022.301352

Ogundele AT, Awodiran MA, Idem UJ, Anwana, Emem O (2023) Digitally designed forensic procedure a panacea to cyber fraud control in Nigeria. In: 2023 international conference on cyber management and engineering (CyMaEn). Bangkok, Thailand, pp 223–229. https://doi.org/10.1109/CyMaEn57228.2023.10050964

Ølnes S (2016) Beyond bitcoin enabling smart government using Blockchain technology. In: Scholl, HJ, et al (ed) Electronic government. EGOV 2016. Lecture notes in computer science, vol 9820. Springer, Cham, pp 253–264.

Ølnes S, Ubacht J, Janssen M (2017) Blockchain in government: benefits and implications of distributed ledger technology for information sharing. Gov Inf Q 34(3):355–364. https://doi.org/10.1016/j.giq.2017.09.007

Pham QH, Vu KP (2024) Insight into how digital forensic accounting and metaverse circular business model innovation contribute to accelerated internationalization: evidence from Vietnam-based SMEs. Cogent Bus & Manag 11(1):1–24. https://doi.org/10.1080/23311975.2024.2320203

Rashid A, Rasheed R, Ngah AH (2024) Achieving sustainability through multifaceted green functions in manufacturing. J Glob Oper Strat Sourc 17(2):402–428. https://doi.org/10.1108/JGOSS-06-2023-0054

Shan S, Duan X, Zhang Y, Zhang TT, Li H (2021) Research on collaborative governance of smart government based on blockchain technology: an evolutionary approach. Discret Dyn Nat Soc:1–23. https://doi.org/10.1155/2021/6634386

Smith SS, Castonguay JJ (2020) Blockchain and accounting governance: emerging issues and considerations for accounting and assurance professionals. J Emerg Technol Account 17(1):119–131. https://doi.org/10.2308/jeta-52686

Swan M (2015) Blockchain: blueprint for a new economy. O'Reilly Media Inc., Sebastopol

Tang X, Zhu L, Shen M, Peng J, Kang J, Niyato D, Abd El-Latif AA (2022) Secure and trusted collaborative learning based on Blockchain for artificial intelligence of things. IEEE Wirel Commun 29(3):14–22. https://doi.org/10.1109/MWC.003.2100598

Verma S, Sheel A (2022) Blockchain for government organizations: past, present and future. J Glob Oper Strat Sourc 15(3):406–430. https://doi.org/10.1108/JGOSS-08-2021-0063

Wamba SF, Gunasekaran A, Akter S, Dubey R (2020) The performance effects of big data analytics and supply chain ambidexterity: The moderating effect of environmental dynamism. Int J Prod Econ 222:1–14. https://doi.org/10.1016/j.ijpe.2019.09.019

Warkentin M, Orgeron C (2021) Using the security triad to assess blockchain technology in public sector applications. Int J Inf Manage 1(2):18–41. https://doi.org/10.1016/j.ijinfomgt.2020.102090

Wylde A (2021) Zero trust: never trust, always verify. In: 2021 international conference on cyber situational awareness, data analytics and assessment (CyberSA). IEEE, Dublin, Ireland, pp 1–4.

Yli-Huumo J, Ko D, Choi S, Park S, Smolander K (2016) Where is current research on blockchain technology?—a systematic review. PLoS ONE 11(10):1–27. https://doi.org/10.1371/journal.pone.0163477

Younus M, Nurmandi A, Suswanta S, Suardi W (2024) Conceptualizing smart citizen: relationships and reciprocity with smart government. Transform GovMent: People, Process Policy 18(1):71–85. https://doi.org/10.1108/TG-02-2023-0016

Zarrin J, Wen Phang H, Babu Saheer L, Zarrin B (2021) Blockchain for decentralization of internet: prospects, trends, and challenges. Clust Comput 24:2841–2866. https://doi.org/10.1007/s10586-021-03301-8

Zhang T, Li J, Jiang X (2021) Analysis of supply chain finance based on blockchain. Procedia Comput Sci 187:1–6. https://doi.org/10.1016/j.procs.2021.04.025

Sending Money at the Speed of Light? Cryptocurrency and the Future of Money Transfer

Rintu Augustine, A. Krishnaveni, and Sunil Uthup

Abstract Imagine sending money abroad instantly and practically for free, disrupting the slow and expensive world of traditional transfers. Cryptocurrency promises this revolution, but is it too good to be true? This chapter dives into the exciting potential and hidden complexities of using crypto for money transfer, backed by real-world financial reports. This chapter explores how blockchain-powered crypto could reshape the financial landscape, offering faster, cheaper, and more accessible options globally. A 2023 World Bank report reveals over 2 billion unbanked adults, highlighting the potential for crypto to promote financial inclusion as an alternative. However, we don't shy away from the challenges. We unveil the underlying technology, acknowledging its strengths and weaknesses based on data. A 2022 Chainalysis report clearly demonstrates trends in criminal abuse of cryptocurrency during 2021. This chapter empowers you to navigate these concerns. We analyze a 2023 Global Fintech Report showing 41% consumer interest in crypto payments, and explore expert opinions on its transformative potential. Will it replace traditional methods, or will they coexist? Join us on this data-driven exploration to understand the future of cryptocurrency for money transfer. Get informed, navigate the hype, and make empowered choices in this rapidly evolving financial landscape.

R. Augustine (✉) · A. Krishnaveni
Department of Computer Science, Karpagam Academy of Higher Education, Coimbatore, India
e-mail: rintuaugustine07@gmail.com; rintu.cs@adishankara.ac.in

A. Krishnaveni
e-mail: krishnaveni.arumugam@kahedu.edu.in

R. Augustine
Department of Computer Applications, Adi Shankara Institute of Engineering and Technology, Kalady, India

S. Uthup
HSBC, Hong Kong, China

HSBC, Singapore, Singapore

S. Krishnan et al. (eds.), *Unveiling Blockchain Enabled Business Applications*, Contributions to Management Science, https://doi.org/10.1007/978-981-95-2337-5_15

1 Introduction

In today's globalized world, the smooth and efficient movement of money across borders forms the cornerstone of international trade, personal connections, and supporting loved ones abroad. However, the current landscape of international money transfer is often marred by inefficiencies, hindering individuals, families, and businesses. Traditional systems, heavily reliant on a network of correspondent banks, can be frustratingly slow, taking days to finalize transactions, even for relatively small amounts (World Bank 2023b). This delay can be extremely inconvenient for individuals sending urgent funds to family members abroad or businesses engaging in international trade, where time is often of the essence (International Monetary Fund 2022).

Furthermore, the exorbitant fees associated with these traditional methods can be a significant burden, especially for smaller transfers. This disproportionately impacts individuals with limited financial resources, further amplifying existing economic inequalities. Moreover, the current system disproportionately excludes a significant portion of the global population. According to a 2023 World Bank report, over 2 billion adults remain unbanked, lacking access to formal financial channels, forcing them to rely on expensive and inefficient informal remittance methods (World Bank 2023a).

This not only hinders financial inclusion but also perpetuates economic disparities, limiting opportunities for individuals and hindering economic growth in developing nations (Fig. 15.1).

As we strive for a future, where sending and receiving money across borders is faster, cheaper, and more accessible, it is crucial to acknowledge the shortcomings of the existing system and actively seek innovative solutions. This chapter delves into the potential of cryptocurrency, a nascent technology with the potential to revolutionize the landscape of international money transfer. By analyzing its ability to facilitate faster, cheaper, and more accessible cross-border transactions, we can explore whether cryptocurrency can serve as a viable alternative to the current system and contribute to greater financial inclusion globally.

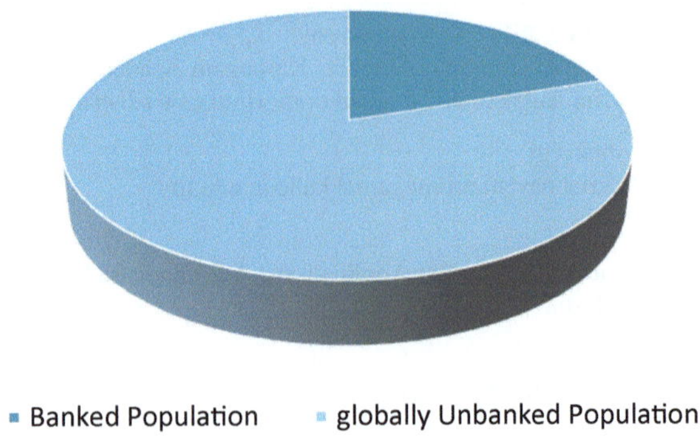

Fig. 15.1 Global distribution of banked and unbanked population (World Bank 2023a)

This exploration will involve a nuanced examination of cryptocurrency's potential benefits, including its inherent speed and transparency enabled by blockchain technology, along with a critical evaluation of the challenges it presents, such as market volatility and evolving regulatory frameworks. By understanding both its strengths and weaknesses, we can foster informed discussions and potentially pave the way for a more efficient, inclusive, and secure future for international money transfer.

2 Cryptocurrency for Money Transfer

The realm of international money transfer is ripe for innovation, and cryptocurrency, a digital or virtual asset secured by cryptography, has emerged as a potentially disruptive force. At its core lies the revolutionary blockchain technology, a decentralized, distributed ledger that records transactions across a network of computers (Nakamoto 2008). Let's explore how the fundamental principles of cryptocurrency and blockchain can transform the landscape of cross-border payments (Fig. 15.2).

2.1 *Decentralization: Transforming Transactional Dynamics*

Unlike traditional financial systems that rely on intermediaries like banks to facilitate transfers, cryptocurrency operates on a decentralized network. This means that transactions occur directly between the sender and receiver, eliminating the need for central authorities (Swan 2015). This paradigm shift offers several potential benefits:

Faster Transactions: By circumventing intermediaries, cryptocurrency transactions can potentially bypass lengthy approval processes and interbank procedures, leading

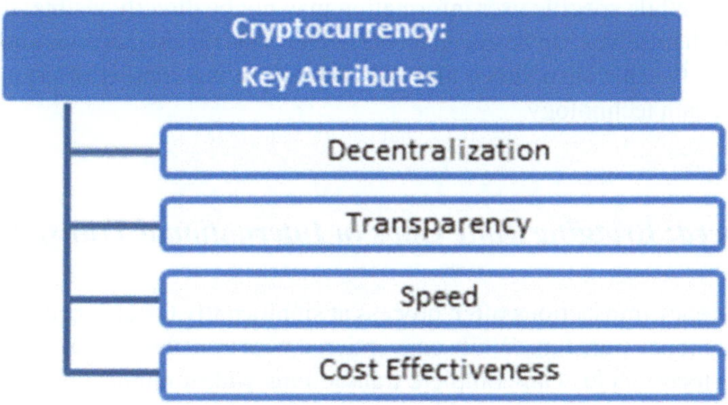

Fig. 15.2 Cryptocurrency: Key attributes (Swan 2015)

to faster completion times (Swan 2015). This can be particularly beneficial for urgent transfers, such as remittances or time-sensitive business transactions.

Reduced Fees: The absence of intermediaries also translates to lower transaction fees. Instead of paying hefty charges to banks and transfer services, users often pay minimal network fees to the miners or validators who confirm their transactions on the blockchain (Yermack 2017). This can result in significant cost savings, especially for individuals and businesses frequently sending smaller amounts of money internationally.

However, it's important to acknowledge that decentralization also presents challenges. The lack of a central authority can raise concerns about regulatory oversight and potential misuse for illegal activities. Additionally, resolving disputes or recovering funds in case of errors or fraud can be complex in decentralized systems.

2.2 Transparency: Building Trust Through Immutability

The distributed nature of blockchain technology ensures that every transaction is recorded on a shared, immutable ledger—a digital record accessible to all network participants. This inherent transparency offers several advantages:

Enhanced Security: The immutability of the ledger makes it highly resistant to tampering or fraud, as any attempt to alter the record would be readily detectable by the network (Kiayias et al. 2017). This fosters greater confidence and security in the overall transaction process.

Increased Accountability: The transparency of the ledger promotes accountability for all participants, discouraging fraudulent activities or errors. This can lead to a more secure and reliable environment for conducting financial transactions.

However, concerns around privacy can arise due to the public nature of the blockchain. While specific user information may not be directly visible, the transaction details themselves can be traced on the ledger. This raises questions about striking a balance between transparency and user privacy in the context of cryptocurrency and blockchain technology.

2.3 Speed: Bridging Time Gaps in International Transfers

Cryptocurrency transactions often process at significantly faster speeds compared to traditional money transfers. While conventional wire transfers can take days to settle, many cryptocurrencies can complete transactions within minutes or even seconds, particularly those utilizing faster blockchains. This speed offers several advantages:

Reduced Waiting Times: Faster transaction processing times mean reduced waiting periods for both the sender and the receiver. This can be crucial for individuals or businesses relying on swift transfers to meet financial obligations or seize time-sensitive opportunities.

Improved Efficiency: By accelerating transaction times, cryptocurrency can potentially streamline international payments and increase overall efficiency in global financial processes. This can benefit individuals, businesses, and entire economies through enhanced speed and convenience.

However, it's essential to acknowledge that transaction speed can vary depending on the specific cryptocurrency network and its underlying technology. Additionally, network congestion can lead to slower processing times, requiring further refinement and scaling solutions for blockchain technology.

2.4 Cost-Effectiveness: Rethinking Traditional Fees

By eliminating the need for multiple intermediaries, cryptocurrency transactions can be significantly cheaper than traditional cross-border transfers. Instead of paying hefty fees to banks and transfer services, users often pay minimal network fees to the miners or validators who confirm their transactions on the blockchain (Yermack 2017). This can result in substantial cost savings, especially for the following segments:

Individuals Sending Remittances: Migrant workers or individuals supporting families abroad can benefit greatly from lower transaction fees associated with cryptocurrency compared to traditional remittance services known for their high charges.

Small Businesses Engaging in International Trade: Businesses involved in international trade can potentially reduce their overall transaction costs by utilizing cryptocurrency, improving their financial margins, and potentially leading to more competitive pricing for their goods and services.

However, volatility in cryptocurrency prices can negate some of the cost-saving benefits. The fluctuating value of cryptocurrencies can introduce uncertainty and potential losses for users, requiring careful consideration and risk management strategies.

3 Financial Inclusion Potential

The global landscape of financial access is marred by a stark reality—an estimated 2 billion adults remain unbanked, lacking access to formal financial services (World Bank 2023a). This exclusion not only limits individual opportunities but also hinders economic growth and perpetuates inequalities. While traditional systems have served some well, they often face limitations that leave a large portion of the world's population behind. This is where cryptocurrency emerges as a potential game-changer, holding the promise of revolutionizing financial inclusion and bringing millions into the fold of the formal financial ecosystem. Let's delve deeper into how cryptocurrency can bridge this critical gap.

3.1 Accessibility for All

Traditional banking systems often erect barriers to entry, demanding extensive documentation, physical branch presence, and minimum balance requirements. These hurdles disproportionately exclude individuals in remote areas, marginalized communities, and those without formal identification, leaving them vulnerable and financially excluded (International Monetary Fund 2022). Cryptocurrency, in contrast, offers a paradigm shift in accessibility. It primarily requires an internet connection and a digital wallet, empowering individuals regardless of location or formal documentation to participate in the financial realm (Ozili 2022). Imagine individuals in remote villages, refugees fleeing conflict, or those facing challenges obtaining official identification—they can now potentially access and manage their finances through cryptocurrency, unlocking a world of possibilities.

3.2 Lowering the Entry Barrier

Opening a traditional bank account can be a cumbersome and costly process, often involving fees, lengthy procedures, and stringent eligibility criteria. Cryptocurrency offers a simpler and more inclusive alternative. Setting up a cryptocurrency wallet often requires only basic verification and entails minimal or no associated fees, making it a viable option for those who may be excluded by the traditional system due to financial constraints or limited resources (Ozili 2022). This lower barrier to entry can empower individuals to manage their finances independently, fostering financial inclusion and promoting economic participation.

3.3 Revolutionizing Remittances

For many families in developing economies, remittances, the transfer of money by migrants to their home countries, are vital lifelines. However, traditional remittance channels are often slow, expensive, and fraught with hidden fees. Cryptocurrency-based remittance platforms offer a more efficient and cost-effective solution. Transactions can be completed faster and at significantly lower costs, ensuring that a larger portion of the funds reaches its intended recipient, directly benefiting families and communities (Deloitte 2021). This increased financial flow can boost economic development, stimulate local businesses, and improve living standards in areas where remittances play a crucial role.

3.4 *Unlocking Financial Empowerment*

Beyond facilitating transactions, cryptocurrency can empower the unbanked to engage in activities previously inaccessible. Access to financial services enables individuals to save, borrow, and invest, paving the way for greater financial stability and economic opportunities. With a digital wallet, individuals can securely save their earnings, plan for the future, and potentially invest in ventures that can generate income and improve their livelihoods (World Economic Forum 2023a; World Economic Forum 2023b). This financial empowerment can lead to a cycle of growth, enabling individuals to invest in themselves and their communities, fostering greater financial inclusion and contributing to a more equitable and sustainable future.

4 The Cryptocurrency Adoption by International Banks

International banks are increasingly looking into the use of cryptocurrency as a payment option. The current section focuses on foreign banks' strategies for adopting cryptocurrencies, as well as the expanding importance of digital assets in cross-border payments and financial transactions (Table 15.1).

International banks, who have typically been conservative in their approach to innovation, are now seeing the potential benefits of cryptocurrencies in improving the efficiency, speed, and cost-effectiveness of cross-border transactions. While banks were previously hesitant to adopt cryptocurrencies due to legal uncertainties and security concerns, an increasing number of financial institutions are now exploring

Table 15.1 Financial inclusion potential of cryptocurrency-summarized (International Monetary Fund 2022; Ozili 2022; Deloitte 2021; World Economic Forum 2023a)

Accessibility for All	• Overcomes traditional banking barriers • Internet connection and digital wallet
Lowering the Entry Barrier	• Minimal setup • Low fees • Inclusive
Revolutionizing Remittances	• Cost-effective • Reduces hidden fees • Faster transaction
Unlocking Financial Empowerment	• Enables saving, borrowing, and investing • Promotes financial stability

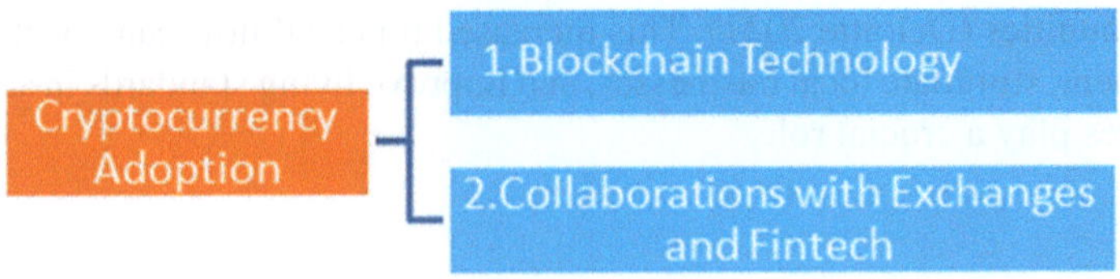

Fig. 15.3 Cryptocurrency adoption methods used by various international banks (Vigna & Casey, 2015)

various techniques to incorporate digital assets into their payment infrastructure (Vigna & Casey, 2015).

One approach taken by multinational banks is to use blockchain technology, the underlying technology behind cryptocurrencies, to streamline cross-border payment operations. In contrast to traditional correspondent banking systems, blockchain-based payment networks provide real-time settlement, increased transparency, and lower intermediation costs (Swan 2015). By incorporating blockchain into their payment infrastructure, banks can enable faster and more secure cross-border transactions, improving consumer experience and lowering operational inefficiencies. For example, HSBC, one of the world's largest banks, implemented a blockchain-based platform called HSBC FX Everywhere to manage foreign exchange transactions worth billions of dollars. By utilizing blockchain technology, HSBC achieved faster transaction settlement, improved transparency, and reduced operational costs compared to traditional payment methods (HSBC 2018) (Fig. 15.3).

Another tactic used by multinational banks is to collaborate with cryptocurrency exchanges and fintech companies to provide cryptocurrency-related services to customers. Some banks have begun to provide cryptocurrency custody, trading, and investment solutions in response to the increased demand for digital assets among consumers and institutional investors (Vigna & Casey, 2015). For instance, Swiss-based bank Sygnum partnered with cryptocurrency exchange Kraken to offer institutional clients access to cryptocurrency custody, trading, and liquidity services. This collaboration enabled Sygnum's clients to seamlessly integrate cryptocurrencies into their investment portfolios while leveraging Kraken's expertise in cryptocurrency trading and custody solutions (Sygnum 2021). By cooperating with established participants in the cryptocurrency ecosystem, banks can generate new revenue streams and expand their service offerings to meet their clients' changing needs.

Regulatory uncertainties surrounding cryptocurrencies, such as anti-money laundering (AML) and know-your-customer (KYC) compliance requirements, provide substantial challenges for banks seeking to integrate digital assets into their payment infrastructure (Swan 2015). For example, in 2019, the Financial Action Task Force (FATF) introduced guidance on virtual asset service providers (VASPs), requiring banks to implement stringent anti-money laundering (AML) and know-your-customer (KYC) procedures when dealing with cryptocurrencies (FATF 2019). The absence of clear regulatory guidance and harmonization across jurisdictions complicates compliance operations while increasing legal and reputational risks for banks.

While decentralized blockchain networks provide increased security and resilience, they also make it difficult for banks to retrieve lost or stolen cash in the event of a security breach (Swan 2015). As a result, banks must build strong cybersecurity and risk management frameworks to effectively minimize these risks.

Furthermore, the volatility and lack of market stability inherent in cryptocurrencies create new hurdles for banks when employing digital assets as a payment method. The shifting value of cryptocurrencies can cause major exchange rate and liquidity concerns, making it difficult for banks to efficiently manage their exposure to digital assets (Vigna & Casey, 2015). Furthermore, the lack of a central authority or regulatory control in the cryptocurrency market heightens these risks, demanding thorough risk assessment and mitigation techniques by institutions.

Moreover, security risks, such as cyberattacks and theft, pose significant challenges for banks when handling cryptocurrencies. In 2020, Japan-based cryptocurrency exchange Coincheck suffered a major security breach, resulting in the loss of over $500 million worth of digital assets. The incident highlighted the vulnerabilities inherent in cryptocurrency exchanges and underscored the importance of robust cybersecurity measures and risk management frameworks for banks dealing with digital assets (Coincheck 2020).

The adoption of cryptocurrencies by multinational banks is projected to accelerate as regulatory frameworks improve, technology advances, and client preferences shift toward digital assets. With developments in blockchain technology and more collaboration among banks, regulators, and industry stakeholders, the integration of cryptocurrencies into mainstream financial infrastructure is expected to accelerate in the coming years (Swan 2015).

5 Current Worldwide Cryptocurrency Scenario

The usage of cryptocurrencies has increased significantly across the globe, driven by factors such as increased digitization, efforts toward financial inclusion, and growing skepticism in existing financial systems (Chiu and Koeppl 2017). Bitcoin, the first cryptocurrency, continues to dominate the market, followed by numerous altcoins and tokens catering to a variety of use cases and functionality (Hileman and Rauchs 2020). This surge is evidenced by a myriad of real-world use cases that showcase the transformative potential of digital assets.

North America: The United States is a leading hub for cryptocurrency trading and innovation. Major companies like Tesla and MicroStrategy have made significant investments in Bitcoin, while financial institutions are increasingly offering crypto-related services (Gensler 2018). Canada has also been progressive, being one of the first countries to approve Bitcoin exchange-traded funds (ETFs).

Europe: The European Union has been proactive in developing regulatory frameworks for cryptocurrencies, with countries like Germany and Switzerland becoming key players in the crypto space. The EU's Markets in Crypto-Assets (MiCA) regulation aims to create a harmonized regulatory framework across member states, fostering innovation while ensuring consumer protection (European Commission 2020).

Asia: Asia is a major center for cryptocurrency activity. Japan has been a pioneer in cryptocurrency regulation, recognizing Bitcoin as a legal form of payment as early as 2017. South Korea is another significant market, with high levels of crypto adoption and stringent regulatory oversight. China, despite its crackdown on crypto mining and trading, continues to influence the market significantly (Chainalysis 2020).

Africa: Cryptocurrencies are gaining traction as a solution for financial inclusion and cross-border remittances. Nigeria and South Africa are leading the continent in crypto adoption, with a growing number of users leveraging Bitcoin and other cryptocurrencies for transactions and savings amidst volatile local currencies (PWC 2020).

Latin America: Countries like Argentina and Venezuela, facing economic instability and hyperinflation, have seen increased cryptocurrency adoption as a hedge against currency devaluation. El Salvador made headlines by becoming the first country to adopt Bitcoin as legal tender in 2021, a bold move aimed at boosting financial inclusion and attracting foreign investment (BBC 2021).

Oceania: Australia has been active in the cryptocurrency space, with clear regulatory guidelines and a growing number of crypto exchanges and startups. New Zealand is also emerging as a favorable environment for crypto innovation and investment (Australian Government 2019).

Despite legal and regulatory challenges, the cryptocurrency market capitalization has reached trillion-dollar thresholds, indicating increased investor confidence and institutional interest in digital assets (Hileman and Rauchs 2020). The introduction of cryptocurrency exchanges, custodial services, and investment products demonstrates the ecosystem's sophistication. Platforms like Coinbase, Binance, and Bitfinex offer a wide range of cryptocurrencies for trading, allowing users globally to speculate on price movements and diversify their investment portfolios. The proliferation of cryptocurrency investment products, such as exchange-traded funds (ETFs) and futures contracts, further underscores the growing interest in digital assets among institutional and retail investors worldwide (Vigna & Casey, 2015).

The cryptocurrency environment is characterized by ongoing innovation and experimentation, with new initiatives and technology constantly redefining the industry. Decentralized finance (DeFi) has arisen as a game-changing idea, using blockchain technology to establish an open and permissionless financial environment (Buterin 2013). DeFi protocols allow users globally to access a variety of financial services, such as lending, borrowing, trading, and yield farming, without intermediaries.

Blockchain technology is revolutionizing supply chain management by enhancing transparency, traceability, and efficiency. Companies like IBM and Walmart have collaborated on blockchain-based supply chain projects to track the provenance of products from manufacturer to consumer. In Europe, Carrefour uses blockchain to provide transparency in the supply chain of its products, enhancing consumer trust (IBM, n.d.). By leveraging blockchain technology, these companies can ensure the authenticity and integrity of their supply chains, reduce the risk of counterfeiting and fraud, and improve overall supply chain efficiency.

In the healthcare sector, blockchain technology is being used to securely manage medical records, streamline administrative processes, and facilitate interoperability between healthcare providers (Krishnan et al. 2020). Projects like MedRec and Medicalchain aim to create decentralized platforms where patients can securely access and control their medical data, ensuring privacy and data integrity. In countries like Estonia, blockchain technology is already being used to secure medical records and improve health services (Hassan et al. 2020).

Real estate tokenization platforms are leveraging blockchain technology to fractionalize ownership of real estate assets, allowing investors to purchase and trade fractional ownership tokens representing underlying real estate assets. Non-fungible tokens (NFTs) in the cryptocurrency sector allow for the tokenization and ownership of digital assets such as art, collectibles, and virtual real estate. The proliferation of NFT marketplaces and platforms has expanded opportunities for producers and collectors to monetize and sell digital content. Platforms like RealtyShares and RealT enable investors globally to diversify their investment portfolios by gaining exposure to real estate assets without the need for large capital outlays. By democratizing access to real estate investment opportunities, these platforms are transforming the way people invest in real estate worldwide (Ravindranath 2021).

6 Challenges and Risks

While the potential of cryptocurrency to facilitate faster, cheaper, and more accessible cross-border money transfers has garnered significant attention, its impact extends far beyond this realm. Beyond acting as a remittance tool, cryptocurrency holds the potential to revolutionize financial inclusion by unlocking a broader spectrum of financial services for the unbanked population—the estimated 2 billion adults globally excluded from formal financial systems.

While cryptocurrency holds immense potential for financial inclusion, it's crucial to acknowledge the challenges that need to be addressed. Price volatility can pose a risk for users, and limited technological literacy can hinder widespread adoption. Additionally, evolving regulatory frameworks need to be established to ensure responsible development and mitigate potential risks (Congressional Research Service 2021b).

Let's delve deeper into this transformative potential, exploring how cryptocurrency can empower individuals and communities beyond simple money transfers.

6.1 Fostering Financial Identity and Creditworthiness

Traditional financial systems often rely on credit histories and formal documentation to assess an individual's financial standing. This system inherently marginalizes individuals lacking access to these formal channels, perpetuating their exclusion. Blockchain technology, the underlying infrastructure of most cryptocurrencies, can potentially offer a solution through decentralized identity (DID) solutions. DIDs offer individuals greater control over their personal data and enable them to create a self-sovereign identity that can be used across various financial platforms. This digital identity record can capture an individual's financial behavior and transactions on the blockchain, building a credit history independent of traditional systems. This can empower them to access financial products and services like loans, insurance, and microfinancing, even without a formal bank account (World Economic Forum 2023a). In Sub-Saharan Africa, where millions of people lack access to formal banking services, a blockchain-based decentralized identity project called "MyIdentity" was implemented by a consortium of financial institutions and technology companies (World Economic Forum 2023a). MyIdentity allowed individuals to create digital identities on the blockchain, linked to their financial behavior and transactions.

6.2 Empowering Micro-Entrepreneurs and Small Businesses

In developing economies, small businesses and micro-entrepreneurs often face significant challenges in accessing traditional financing due to stringent eligibility criteria and high-interest rates. Cryptocurrency can offer a more accessible alternative. Decentralized Finance (DeFi) platforms built on blockchain technology enable individuals to lend, borrow, and invest directly with each other, bypassing traditional financial institutions. This facilitates peer-to-peer lending and micro-investing, potentially providing micro-entrepreneurs with the capital they need to start or expand their businesses, fostering economic growth and job creation at the grassroots level (Congressional Research Service 2021a). In a rural area of Southeast Asia, where traditional banking services are scarce and financial inclusion is limited, a cryptocurrency-based DeFi platform called "MicroFin" was launched (Congressional Research Service 2021b). MicroFin aimed to provide small business owners and micro-entrepreneurs with access to financing opportunities through peer-to-peer lending and micro-investing.

6.3 Facilitating Secure and Transparent Savings

Lack of access to secure saving mechanisms is another significant barrier to financial inclusion. Traditional saving methods in some regions may be vulnerable to theft or inflation. Cryptocurrency can offer an alternative through digital wallets that provide secure storage for individuals' savings. These wallets, often protected by strong cryptography, can empower individuals to save their earnings safely, regardless of their location or access to formal banking infrastructure (International Monetary Fund 2022). Additionally, the immutable transaction records on the blockchain offer transparency and accountability, allowing individuals to track their savings and fostering trust in the system.

6.4 Enabling Cross-Border Payments for Individuals and Businesses

While facilitating remittances is often highlighted, cryptocurrency's potential for cross-border transactions extends beyond personal transfers. Businesses, especially small and medium-sized enterprises (SMEs), can benefit from faster and cheaper international payments facilitated by cryptocurrency compared to traditional channels. This can enhance their ability to engage in international trade and reach new markets, further promoting global economic integration and fostering economic growth (Deloitte 2021).

6.5 Fostering Financial Literacy and Education

The successful and responsible integration of cryptocurrency into the financial landscape hinges on widespread financial literacy and education. Initiatives focusing on educating individuals and communities about cryptocurrencies, their risks and benefits, and responsible usage are crucial. This includes educating them on how to navigate cryptocurrency wallets, secure their digital assets, and avoid potential scams. By promoting financial literacy, we can empower individuals to make informed decisions and participate in this evolving financial ecosystem responsibly, fostering a more inclusive and sustainable financial future for all. In a hypothetical suburban community in Latin America, where access to formal financial education is limited and cryptocurrency adoption is on the rise, a non-profit organization called "CryptoEdu" could serve as a model for grassroots financial literacy initiatives.

6.6 *Bridging the Gap Between Potential and Reality*

While the potential of cryptocurrency to revolutionize financial inclusion is undeniable, it's important to acknowledge the challenges that need to be addressed. Price volatility, security concerns, and evolving regulatory frameworks remain areas that require careful consideration and ongoing development. Nevertheless, by addressing these concerns through collaborative efforts and responsible implementation, we can leverage the transformative potential of cryptocurrency to empower individuals and communities, ultimately leading to a more inclusive and equitable financial landscape for all.

However, these challenges shouldn't overshadow the transformational potential of cryptocurrency. With focused initiatives on financial education, promoting technological literacy, and fostering responsible regulatory frameworks, we can harness the power of cryptocurrency to bridge the financial gap and build a more inclusive future where everyone has the opportunity to participate in the global economy. As technology evolves and regulations adapt, cryptocurrency can become a powerful tool for financial inclusion, empowering individuals and communities to build a brighter future for themselves.

7 Navigating the Nexus: Cryptocurrency and Criminal Activities

The 2022 Crypto Crime Report finds that crypto-enabled crime reached an all-time high of $14 billion in 2021, a 79% increase from 2020 (Chainalysis 2022).

7.1 *Cryptocurrency and Money Laundering*

One of the main worries about cryptocurrencies is its ability to facilitate money laundering and other illegal financial activity. The anonymity of transactions and the lack of centralized monitoring make it an appealing choice for individuals looking to conceal the source of their funds. According to a report by the Financial Action Task Force (FATF), criminals are increasingly using cryptocurrency to launder money gained illegally (FATF 2020).

Case studies have shown how bitcoin exchanges and mixing services have been used to conceal the trail of criminal payments. For example, the renowned Silk Road marketplace, which operated on the dark web, sold drugs and other illegal products using Bitcoin as its primary currency (Christin 2013). While law enforcement organizations have attempted to counteract such acts through increased regulation and enforcement, the decentralized structure of cryptocurrencies creates unique obstacles in locating and punishing perpetrators.

7.2 Cryptocurrency and Cybercrime

In addition to money laundering, Bitcoin has been linked to a number of cybercrimes, including hacking, ransomware attacks, and fraud. The anonymity of cryptocurrency transactions makes them a tempting target for hackers looking to extort funds or steal critical information. Ransomware attacks against organizations, governments, and individuals have increased in recent years, with attackers demanding payment in cryptocurrencies like Bitcoin or Monero (Böhme et al. 2015). Ransomware payments in bitcoin increased dramatically in 2021, reaching $595 million and breaking the previous high of $416 million in 2020. The rise was fueled by the development of ransomware-as-a-service operations and a surge in victims willing to pay ransoms(Chainalysis 2022).

Furthermore, the profusion of initial coin offerings (ICOs) has created fertile ground for fraudulent schemes, in which scammers exploit investors' enthusiasm for new blockchain companies to carry out Ponzi schemes or exit scams (Hassan et al. 2018). While regulatory agencies have taken attempts to address these vulnerabilities through enhanced monitoring and enforcement activities, cryptocurrency's borderless nature makes it difficult to hold perpetrators accountable across borders.

Despite the difficulties created by cryptocurrency-related crimes, cutting-edge technology such as blockchain analytics provide intriguing answers for law enforcement and regulatory bodies. Forensic investigators can better track the movement of funds and spot suspicious transactions by using blockchain ledgers' transparency and immutability (Ron and Shamir 2013).

8 The Future of Money Transfer

As the world shrinks thanks to advancements in technology and globalization, the efficiency and accessibility of cross-border money transfers become more critical than ever. In this evolving landscape, cryptocurrency and its underpinning blockchain technology are emerging as potential game-changers, sparking debate among experts and consumers alike. Some envision a revolutionary paradigm shift, while others predict a more collaborative evolution with traditional financial systems.

8.1 Revolution or Evolution: The Experts Weigh In

Proponents of a cryptocurrency revolution highlight its potential to streamline remittances and international payments. A 2023 World Bank report suggests that the faster settlement times and lower transaction fees associated with many cryptocurrencies could offer a lifeline to migrants and businesses, paving the way for a more inclusive and efficient payments ecosystem. This is particularly appealing in regions with slow and expensive traditional transfer options.

However, other experts predict a more measured disruption. They envision cryptocurrency initially complementing traditional banking systems, especially in regions where slow and expensive transfers are commonplace (Deloitte 2022). This perspective suggests a gradual integration of blockchain technology into existing financial systems, leveraging its benefits without a complete overhaul.

8.2 Consumer Perspectives: A Mix of Curiosity and Caution

Consumer surveys reveal a growing interest and cautious optimism about cryptocurrency's role in the future of finance. A 2022 research study by Pew Research Center shows that while public awareness of cryptocurrencies has increased significantly, mainstream adoption for daily transactions remains limited (Pew Research Center 2022). Despite the cautious approach, a significant portion of respondents acknowledge potential benefits, including lower fees and ease of use, leaving room for future growth in user confidence and adoption.

Bitcoin occupies a prominent position within the cryptocurrency landscape. Its market capitalization, representing the total value of all circulating coins, ranks among the top 10 globally (CoinMarketCap 2024). Notably, India has embraced cryptocurrencies with remarkable awareness, reaching an impressive 83% of urban residents (KPMG 2023). This awareness has translated into action, with 16% of this segment actively participating by owning cryptocurrencies (KPMG 2023). Interestingly, Bitcoin dominates the Indian market with a 75% ownership rate among those invested in digital assets (KPMG 2023). This dominance extends to the global stage, where Bitcoin accounts for nearly 40% of the total cryptocurrency market value (CoinMarketCap 2024) (Fig. 15.4).

Several factors likely contribute to Bitcoin's widespread appeal. Its pioneering status as the first and most well-known cryptocurrency may have established greater trust and recognition compared to its emerging counterparts (Nakamoto 2008). Additionally, Bitcoin operates on blockchain technology, the fundamental technology underlying the entire cryptocurrency ecosystem (Nakamoto 2008). As the market evolves, it will be fascinating to witness how Bitcoin maintains its position and how other cryptocurrencies strive to capture a larger market share.

A 2023 Forbes Advisor survey conducted in the UK found that nearly two-thirds (67%) of respondents agreed (either strongly or somewhat) that cryptocurrency is a legitimate investment, while almost one-fifth (17%) were neutral and another 17% disagreed. Following diagram clearly depicts the result (Forbes Advisor Australia 2024) (Fig. 15.5).

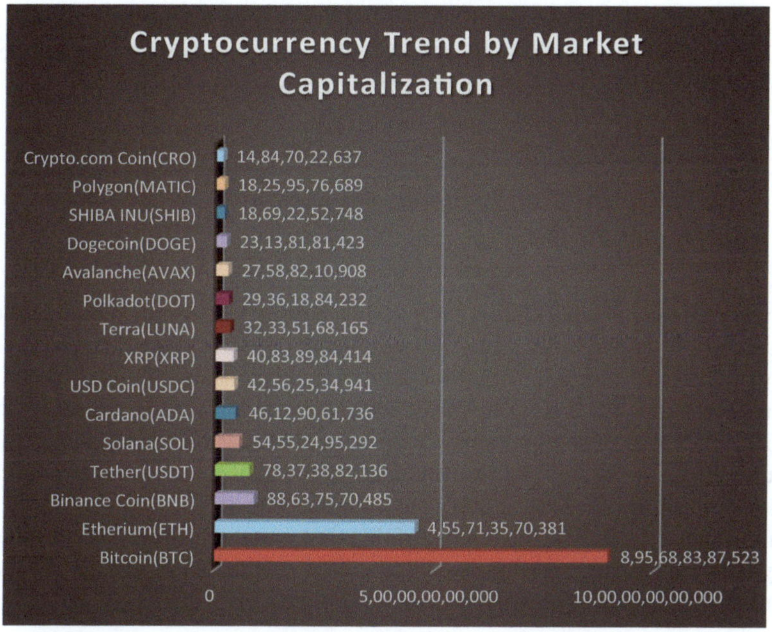

Fig. 15.4 Top 15 Cryptocurrencies by market capitalization (CoinMarketCap 2024)

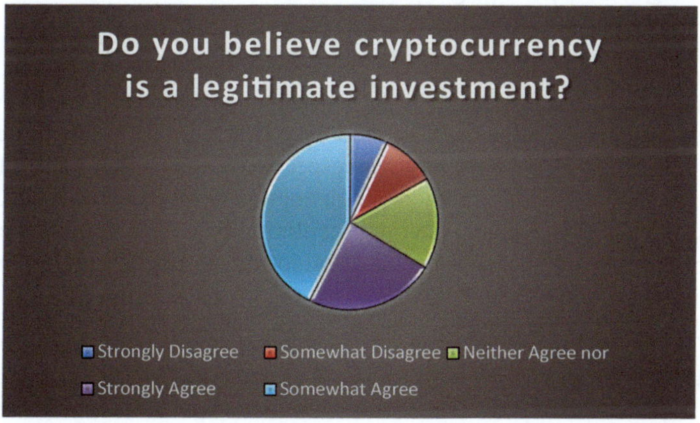

Fig. 15.5 According to a forbes Advisor Survey-UK (Forbes Advisor Australia 2024)

8.3 A Hybrid Future: Merging the Best of Both Worlds

A hybrid financial system seems a realistic future scenario. In this model, cryptocurrency would operate alongside traditional banking channels, offering faster and potentially cheaper options for specific transactions, such as smaller payments or transfers in less regulated regions. Traditional financial institutions would likely adapt and evolve, potentially integrating blockchain technology to streamline their own processes, creating a more efficient overall ecosystem.

8.4 Navigating the Road Ahead: Challenges and Opportunities

For cryptocurrency to truly disrupt the existing landscape, several challenges need to be addressed. Price volatility remains a significant concern, and evolving regulatory frameworks require careful navigation. Additionally, consumer education and trust-building measures are crucial to foster widespread adoption. Ultimately, whether cryptocurrency leads a revolutionary charge or coexists harmoniously with traditional structures hinges on its ability to deliver on its promises of clear advantages, enhanced security, and a transparent regulatory environment.

Only time will tell how this story unfolds, but one thing is certain: the future of money is being actively rewritten and both cryptocurrency and traditional financial institutions have roles to play in shaping it (Table 15.2).

9 The Algorithmic Tango: A Dance Between AI and Cryptocurrency

The world of technology is witnessing a fascinating convergence of two disruptive forces: artificial intelligence (AI) and cryptocurrency (Krishnan et al. 2021). Both are individually transforming their respective domains, but their intersection promises even greater transformation and disruption across diverse industries. This chapter delves into this dynamic interplay, exploring how AI is shaping the future of cryptocurrency and vice versa, forming an "algorithmic tango."

9.1 AI's Empowering Steps in the Crypto Space

AI offers several valuable tools that are redefining various aspects of the cryptocurrency ecosystem:

Market Insights and Predictive Analytics: AI algorithms can analyze vast datasets, identifying historical and real-time trends in the cryptocurrency market. This analysis can be used for informed investment decisions, potentially improving risk management and portfolio optimization. However, it's crucial to remember that AI cannot definitively predict the future, and solely relying on its predictions can be risky (ZeroCap, n.d.).

Table 15.2 Future implications of cryptocurrency in money transfer: Adapted from (Deloitte 2021, 2022; International Monetary Fund 2022; Ozili 2022; World Economic Forum 2023a)

Scenario	Potential outcome	Considerations
Integration and Collaboration	Traditional financial institutions incorporate blockchain-based solutions alongside existing systems	Requires willingness to adapt and invest in new technologies
Niche Adoption	Cryptocurrency becomes a preferred choice for specific use cases, such as smaller transactions, remittances, or specific corridors	Demands ongoing development and improvement of user experience and accessibility
Regulatory Clarity	Clear regulations establish a framework for responsible development and adoption of cryptocurrency in money transfer	Requires collaboration between governments, industry leaders, and regulatory bodies
Continuous Innovation	Ongoing advancements in technology address security concerns, improve scalability, and enhance user experience	Requires commitment to responsible development and ethical considerations
Financial Inclusion	Increased access to financial services, particularly in underserved communities, empowers individuals and stimulates economic growth	Requires addressing the digital divide and promoting financial literacy

Enhanced Security Measures: AI-powered systems can detect and prevent fraudulent activities in cryptocurrency transactions. They can analyze patterns of suspicious behavior, identify potential vulnerabilities in blockchain networks, and implement real-time security measures to combat cyber threats, leading to a more robust crypto environment (Luo et al. 2023).

Automated Trading Strategies: AI algorithms can be implemented to automate trading strategies, executing trades based on pre-defined parameters and real-time market conditions. This can help traders react faster to market fluctuations and potentially generate higher returns, but it is important to understand the inherent risks, as automated trading can also lead to significant losses if not carefully managed (Addy et al. 2024).

Decentralized Finance (DeFi) Engine: AI can play a crucial role in driving the development and adoption of DeFi applications. AI-powered tools can automate smart contract execution, manage liquidity pools, and provide users with personalized financial services within the DeFi space (Sadman et al. 2023).

9.2 Cryptocurrency Fueling AI's Advancement

While AI empowers various aspects of the cryptocurrency domain, the reverse is also true. Cryptocurrencies provide several unique advantages that can accelerate the development and adoption of AI:

Decentralized AI Infrastructure: Blockchain technology can be leveraged to create decentralized AI marketplaces, allowing individuals or organizations to share and access computing resources for AI tasks. This can democratize access to AI resources, especially for those who lack the infrastructure to run complex AI models (Casey 2024).

Secure and Transparent Data Storage: Blockchain technology can provide a secure and transparent platform for storing and managing the vast amounts of data needed for training AI models. This ensures data integrity and accountability, which are crucial considerations in developing and deploying AI systems (Luo et al. 2023).

Tokenized Incentive Mechanisms: Cryptocurrencies can be used to incentivize participation in AI projects and reward individuals who contribute valuable data or computational resources. This can help create a more collaborative and efficient ecosystem for AI development (Luo et al. 2023).

9.3 Navigating the Algorithmic Tango: Challenges and Considerations

Despite the exciting potential, several challenges and considerations need to be addressed at the intersection of AI and cryptocurrency:

Regulation and Governance: As both AI and cryptocurrency face evolving regulatory landscapes, ensuring clear and effective governance frameworks is crucial to foster innovation while mitigating potential risks (Arkansas Law Review 2023).

Bias and Transparency: AI algorithms can potentially inherit and amplify biases present in the data they are trained on. It is crucial to ensure transparency and fairness in the development and deployment of AI-powered crypto solutions (Brundage et al. 2020).

Accessibility and Education: Ensuring equitable access to AI-powered crypto tools and services is vital to prevent further marginalization and promote financial inclusion. Education and awareness initiatives are essential to empower users to make informed decisions in this dynamic landscape (Chen et al. 2022).

10 Conclusion

The traditional landscape of international money transfer, riddled with delays, intermediaries, and high fees, is ripe for innovation. Cryptocurrency emerges as a compelling force with the potential to disrupt the status quo, offering unique characteristics that hold significant promise.

Decentralization eliminates the need for intermediaries, streamlining processes and potentially reducing costs. Transparency enhances trust and security through immutable transaction records on the blockchain. Speed facilitates near-instantaneous settlements compared to the days or weeks typical of traditional systems. Potential cost-effectiveness arises from the reduced reliance on intermediaries and their associated fees.

These characteristics combined could pave the way for financial inclusion on a global scale. Cryptocurrency empowers the unbanked population, estimated at 2 billion individuals, by providing an alternative to traditional systems often inaccessible due to geographical limitations or documentation requirements. Additionally, seamless cross-border remittances become possible, enabling faster and more affordable transfers back to families in developing economies, stimulating local economies, and improving livelihoods (Deloitte 2021).

However, challenges remain that demand attention. Market volatility creates uncertainty for users, while security vulnerabilities require robust measures to mitigate risks. Evolving regulations pose a further hurdle, requiring collaboration between governments and industry leaders to establish frameworks that foster innovation while ensuring responsible development. Table 15.3 illustrates the ways cryptocurrency could revolutionize money transfer, offering benefits like decentralization and transparency, while also acknowledging challenges such as market volatility and user complexity.

The future of money transfer might not be a complete disruption but rather a dynamic interplay between established systems and emerging technologies. Cryptocurrency could become a viable alternative for specific use cases, particularly for smaller transactions, remittances, and corridors where existing methods are cumbersome and expensive. Traditional financial institutions, recognizing this potential, may need to proactively innovate by incorporating blockchain-based solutions to remain competitive.

Despite the uncertainties, cryptocurrency has captured the global imagination. Its potential to transform financial transactions—making them faster, cheaper, more inclusive, and truly global—warrants continued investigation, development, and responsible implementation. As awareness, acceptance, and regulations evolve, cryptocurrency could usher in a new era of financial inclusion and global connectivity.

The intersection of AI and cryptocurrency presents a future brimming with possibilities. By harnessing the strengths of both technologies, we can create a more secure, efficient, and accessible financial ecosystem. However, navigating this path requires responsible development, ethical considerations, and an ongoing commitment to addressing emerging challenges. As this dynamic field continues to evolve,

Table 15.3 Cryptocurrency: Potential for disruption in money transfer: Adapted from (Nakamoto 2008; Swan 2015; Deloitte 2021, 2022; International Monetary Fund 2022; Ozili 2022; World Economic Forum 2023a)

Feature	Advantage	Disadvantage
Decentralization	Eliminates intermediaries, streamlining processes and potentially reducing costs	Increases complexity for some users who may find traditional systems easier to navigate
Transparency	Enhances trust and security through immutable transaction records on the blockchain	Requires increased awareness of potential security risks associated with individual user responsibility
Speed	Facilitates near-instantaneous settlements compared to the days or weeks typical of traditional systems	May not be suitable for all types of transactions due to potential limitations in scalability and adoption
Potential Cost-Effectiveness	Reduced reliance on intermediaries can potentially lower fees associated with money transfer	Market volatility can create uncertainty and impact transaction costs
Financial Inclusion	Empowers the unbanked population by providing an alternative to traditional systems	Requires addressing issues of digital literacy and access to technology for widespread adoption
Cross-Border Remittances	Enables faster and more affordable transfers, potentially stimulating local economies in developing countries	Regulatory frameworks for cross-border cryptocurrency transactions continue to evolve, potentially impacting implementation

the "algorithmic tango" between AI and cryptocurrency will continue to reshape the world, demanding constant adaptation and innovation.

References

Addy WA, Ajayi-Nifise AO, Bello BG, Tula ST, Odeyemi O, Falaiye T (2024) Algorithmic trading and AI: a review of strategies and market impact. World J Adv Eng Technol Sci 11:258–267. https://doi.org/10.30574/wjaets.2024.11.1.0054

Arkansas Law Review (2023) Ark Law Rev 76(2), Article 1

Australian Government (2019) Regulation of cryptocurrency. Retrieved from Australian Government Treasury

BBC (2021) El Salvador makes Bitcoin legal tender. Retrieved from BBC News

Böhme R, Christin N, Edelman B, Moore T (2015) Bitcoin: economics, technology, and governance. J Econ Perspect 29(2):213–238

Brundage M, Amodei D, Kleinberg J, Whitley P (2020) The malicious use of artificial intelligence: forecasting, prevention, and mitigation. arXiv preprint http://arxiv.org/abs/2004.14823

Buterin V (2013) Ethereum: a next-generation smart contract and decentralized application platform. Ethereum. https://ethereum.org/en/whitepaper/

Casey MJ (2024) At Davos, crypto pushes case for decentralized AI. CoinDesk. https://www.coindesk.com/consensus-magazine/2024/01/23/at-davos-crypto-pushes-case-for-decentralized-ai/, 23 Jan 2024

Chainalysis (2020) The 2020 Geography of Cryptocurrency Report. Retrieved from Chainalysis

Chainalysis (2022) The 2022 cryptocurrency crime report. Retrieved from https://go.chainalysis.com/2023-crypto-crime-report.html

Chen Y, Lu Y, Bulysheva L, Kataev M Y (2022) Applications of blockchain in industry 4.0: A review. Information Systems Frontiers. https://doi.org/10.1007/s10796-022-10248-7

Chiu J, Koeppl TV (2017) The economics of cryptocurrencies—Bitcoin and beyond. Rev Financ Stud 30(3):670–711. https://doi.org/10.1093/rfs/hhx015

Christin N (2013) Traveling the silk road: a measurement analysis of a large anonymous online marketplace. In Proceedings of the 22nd international conference on World Wide Web, pp 213–224

Coincheck (2020) Incident report. Retrieved from https://coincheck.com/en/blog/4675

CoinMarketCap (2024) Cryptocurrency market capitalizations. Retrieved March 1, 2024, from https://coinmarketcap.com/

Congressional Research Service (2021a) Cryptocurrency: understanding the evolving landscape. Retrieved from https://fas.org/sgp/crs/misc/R46504.pdf

Congressional Research Service (2021b) Cryptocurrency: an evolving landscape. Retrieved from https://crsreports.congress.gov/product/details?prodcode=R45427, 1 Dec 2021

Deloitte (2021) Blockchain remittances. Retrieved from https://www2.deloitte.com/us/en/pages/about-deloitte/solutions/blockchain-and-digital-assets.html

Deloitte (2022) Blockchain & payments: hype or real?

Europol (2023) EU Serious and Organized Crime Threat Report (SOCTA). Retrieved from https://www.europol.europa.eu/publications-events/main-report

European Commission (2020) Regulation of Markets in Crypto-Assets. Retrieved from European Commission

Financial Action Task Force (FATF) (2019) Guidance for a risk-based approach to virtual assets and virtual asset service providers. Retrieved from https://www.fatf-gafi.org/publications/fatfrecommendations/documents/guidance-rba-virtual-assets.html

FATF (2020) Virtual assets and virtual asset service providers: key definitions and potential AML/CFT risks. Retrieved from https://www.fatf-gafi.org/media/fatf/documents/reports/Virtual-Assets-Definitions-Amendment-Recommendations.html

Forbes Advisor Australia (2024) Top cryptocurrency statistics and trends in 2024 [Forbes Advisor Australia]. Retrieved from https://www.forbes.com/advisor/au/investing/cryptocurrency/cryptocurrency-statistics/, 1 Mar 2024

Gensler G (2018) Testimony on virtual currencies: the oversight role of the U.S. securities and exchange commission and the U.S. commodity futures trading commission. U.S. Securities and Exchange Commission. https://www.sec.gov/news/testimony/testimony-virtual-currencies-oversight-role-us-securities-and-exchange-commission

Hassan S, Tsang P, Chang R (2018) The scams, the hackers, and the victims: A classified overview of cryptocurrency-related crime. IEEE Access 6:70638–70651

Hassan SR, Ijaz MF, Khan AN (2020) Medicalchain: a blockchain based platform for secure and accessible health records. Int J Adv Comput Sci Appl 11(4):374–380

Hileman G, Rauchs M (2020) Global cryptocurrency benchmarking study. Cambridge Centre for Alternative Finance. https://www.jbs.cam.ac.uk/faculty-research/centres/alternative-finance/publications/global-cryptocurrency-benchmarking-study/

HSBC (2018) HSBC completes first trade finance transaction using blockchain. Retrieved from https://www.hsbc.com/news-and-media/hsbc-stories/hsbc-completes-first-trade-finance-transaction-using-blockchain/
IBM (n.d.) IBM food trust. Retrieved from https://www.ibm.com/food-trust
International Monetary Fund (2022) Financial inclusion for all: promoting growth and development. Retrieved from https://meetings.imf.org/en/2023/Annual/Schedule/2023/10/14/imf-seminar-financial-inclusion-as-a-pathway-to-resilient-and-shared-growth
Kiayias A, Panagiotis O, Zikos N (2017) On the security of proof-of-stake blockchains. Retrieved from https://eprint.iacr.org/, 1 Sep 2017
KPMG (2023) Global fintech report 2023. Retrieved March 1, 2024, from https://kpmg.com/xx/en/home/industries/financial-services/pulse-of-fintech.html
Krishnan S, Balas VE, Julie EG, Yesudhas HR, Balaji S, Kumar R (Eds.) (2020) Handbook of research on blockchain technology. Academic Press
Krishnan S, Balas VE, Julie EG, Yesudhas HR, Kumar R (Eds.) (2021) Blockchain for smart cities. Elsevier
Luo B, Zhang Z, Wang Q, Ke A, Lu S, He B (2023) AI-powered fraud detection in decentralized finance: A project life cycle perspective. arXiv preprint. https://arxiv.org/abs/2308.15992
Nakamoto S (2008) Bitcoin: a peer-to-peer electronic cash system. Retrieved from https://bitcoin.org/bitcoin.pdf
Ozili P (2022) Cryptocurrency for financial inclusion and financial stability. Digit Policy, Regul GovAnce 24(5):564–570
Pew Research Center (2022) 16% of Americans say they have ever invested in, traded, or used cryptocurrency. Retrieved from https://www.pewresearch.org/fact-tank/2021/11/11/16-of-americans-say-they-have-ever-invested-in-traded-or-used-cryptocurrency/
PWC (2020) PwC Global crypto report. Retrieved from PwC
Ravindranath A (2021) How real estate tokenization platforms are transforming the property investment landscape. Retrieved from https://www.forbes.com/sites/forbesrealestatecouncil/2021/01/29/how-real-estate-tokenization-platforms-are-transforming-the-property-investment-landscape/?sh=55f7eeb1309b
Ron D, Shamir A (2013) Quantitative analysis of the full Bitcoin transaction graph. In International conference on financial cryptography and data security, pp 6–24
Sadman N, Ahsan MM, Rahman A, Siddique Z, Gupta KD (2023) Promise of AI in DeFi, a systematic review. MDPI 2(1):6. https://www.mdpi.com/2673-6470/2/1/6
Statistics & Data (2022) Top 10 cryptocurrency 2022 to invest. Retrieved March 1, 2024, from https://statisticsanddata.org/data/top-10-cryptocurrency-2022-to-invest/, 23 Dec 2022
Swan M (2015) Blockchain: Blueprint for a new economy. O'Reilly Media, Inc
Sygnum (2021) Sygnum partners with Kraken to offer institutional clients access to Kraken's leading cryptocurrency trading and custody services. Retrieved from https://www.sygnum.com/en/newsroom/2021-05-06-sygnum-partners-with-kraken-to-offer-institutional-clients-access-to-krakens-leading-cryptocurrency-trading-and-custody-services/
Vigna P, Casey MJ (2015) The age of cryptocurrency: how bitcoin and digital money are challenging the global economic order. St. Martin's Press
Wired (2014) Overstock.com becomes first major retailer to accept bitcoin worldwide. Retrieved from https://www.wired.com/2014/09/overstock-com-becomes-first-major-retailer-accept-bitcoin-worldwide/, September
World Bank (2023a) Global findex database. Retrieved from https://www.worldbank.org/en/publication/globalfindex
World Bank (2023b) Migration and remittance flows: Recent developments and outlook

World Economic Forum (2023a) Building a more inclusive and sustainable financial system with digital currency. Retrieved from https://www.weforum.org/agenda/2021/06/cryptocurrencies-financial-inclusion-help-shape-it/

World Economic Forum (2023b) Blockchain and the promise of financial inclusion: transforming lives through decentralized identity. Retrieved from https://www.weforum.org/reports/blockchain-and-the-promise-of-financial-inclusion-transforming-lives-through-decentralized-identity

Yermack D (2017) Is bitcoin a real currency? An economic appraisal. National Bureau of Economic Research. Retrieved from https://www.nber.org/papers/w29968, 1 Feb 2017

ZeroCap (n.d.) Crypto AI projects. ZeroCap insights. Retrieved from https://zerocap.com/insights/articles/crypto-ai-projects/

Achieving Cyber Security and Data Shield in Business Applications: Lensing 360 Degree Blockchain Protection in the Era of Digitization Landscapes

Bhupinder Singh and Christian Kaunert

Abstract As businesses increasingly rely on digital platforms and applications, the need for robust cybersecurity and comprehensive data protection measures has become paramount. Cybercriminals deploy sophisticated techniques to compromise data integrity, confidentiality, and availability. In this landscape, a holistic approach to cyber-security is imperative which transcending traditional measures to address the complexities of modern threats. Blockchain, the decentralized and tamper-resistant ledger technology has emerged as a game-changer in the realm of cyber-security. Its cryptographic mechanisms and distributed nature make it an ideal candidate for fortifying business applications against cyber threats. Beyond its association with cryptocurrencies, blockchain's inherent characteristics have profound implications for data protection. The immutable ledger provides a foundation for guaranteeing data integrity and authenticity, while smart contracts automate and secure business processes. These applications collectively contribute to a 360-degree protection paradigm, elevating the cyber-security posture of business applications. This chapter explores the paradigm of achieving 360-degree cyber-security and data shield in business applications, with a primary focus on the transformative potential of blockchain technology. From threat mitigation to data integrity assurance, the chapter sightsees the multifaceted applications of blockchain to fortify business applications against evolving cyber threats in the digitization landscapes.

B. Singh (✉)
Sharda University, Greater Noida, India
e-mail: bhupindersinghlaw19@gmail.com

C. Kaunert
Dublin City University, Dublin, Ireland

University of South Wales, Cardiff, Wales, UK

S. Krishnan et al. (eds.), *Unveiling Blockchain Enabled Business Applications*, Contributions to Management Science, https://doi.org/10.1007/978-981-95-2337-5_16

1 Introduction

Blockchain's impact on cyber-security is multifaceted. It introduces decentralized identity management, ensuring secure access controls. Hash functions ensure data security through irreversible encryption, while public and private key encryption guarantees secure data transmission. Zero-knowledge proofs add an extra layer of privacy protection, assuring users that sensitive information remains confidential even in a decentralized environment (Ray et al. 2024). Supply chain security is a critical aspect of business operations. Blockchain introduces transparency and traceability, mitigating risks associated with counterfeiting and ensuring the authenticity of goods. The decentralized nature of blockchain enables secure and verifiable transactions throughout the supply chain, bolstering resilience against cyber threats (Krishnan et al. 2020).

Information exchange inside the company is made easier when key data is presented visually, such as using bar charts, dashboards, or other statistical tools. An end-to-end digital thread, for example, may provide engineers and industrial design teams with useful information that helps them understand what's happening on the manufacturing line (Krishnan et al. 2021). The digital thread may be integrated with other programs, improving system compatibility as a whole. Manufacturers need to adopt a holistic strategy for managing data, acknowledging that gathering data entails more than just input. From the beginning, businesses should focus on collecting data from a range of sources, including humans on the factory floor, sensors, and other types of equipment (Krishnan et al. 2023). Throughout the production process, accuracy and consistency are guaranteed by the contextualization of machine, human, and procedural data with Computer-Aided Design (CAD) and Bill of Materials (BOM). A well-designed data strategy guarantees that the gathered data, following analysis, yields precise insights in line with predetermined and mutually agreed upon goals. Manufacturers are able to discover underlying causes, support preventative steps, and get warnings linked to important monitoring metrics by utilizing reports that integrate contextualized data (Raj et al. 2024).

1.1 Major Cyber-Attack Incidents and How Blockchain Can Mitigate Such Attacks

In the digital age, Cyberattacks have become one of the biggest threats that face organizations as well as government and individuals. A few of the most infamous incidents demonstrates that digital infrastructures are at risk and there is a fast need for solid security practices. The Wanna Cry ransomware attack is just one example of a particularly high-profile cyberattack, which took place in May 2017. The attack took down businesses, hospitals, and critical infrastructure in more than 100 countries via its disruption of over 200,000 computers. Governments and corporations across the globe ground to a halt when ransomware, taking advantage of an exploit in Microsoft

Windows, started encrypting files and asking for payment in Bitcoin. As the largest ransomware attack in history, it laid bare the shortcomings of legacy systems and how ill-equipped many were for such a wide scale assault (Yue and Shyu 2024).

Many other breaches that occurred during 2017, the largest took place when credit-reporting giant Equifax had its database accessed to reveal account information on over 147 million people. The attackers found a weakness in the site, likely employed a SQL injection attack or other form of exploit and redoubtable to get their hands on names, Social Security numbers birth dates addresses even driver's license I. Digits The incident shines a blinding spotlight on the risks of storing data centrally, as even strong security may not be enough to prevent universal exposure should there ever occur a single point of failure. Last year saw the SolarWinds hack of 2020, a wide-spread breach by experienced cyber attackers (Mithas et al. 2022). These hackers then planted malicious code into software developed by SolarWinds; this tampered-with product was later installed remotely onto thousands of clients in the form of American government agencies and Fortune 500 corporations. The supply chain attack effectively put the hackers at a position from where they could spy on these organizations for months, highlighting how vulnerable everything can become when trusted software providers are compromised (Kumar et al. 2019).

The cyber solution Blockchain technology, which is decentralized and immutable in nature could provide a concrete method to alleviate this type of cyber-attacks. Blockchain, on the other hand contrasts with typical centralised systems in that it works as a distributed ledger across thousands of nodes. Since there are lots of decentralized models, it means that one point could not become the vulnerability hackers could target to compromise a whole system (Felsberger et al. 2022). Blockchain, used for decentralized storage solutions (with its strengths in security and block updates), can augment the level of protection against ransomware attacks like WannaCry. Conversely, instead of saving data on vulnerable centralized servers; that same information can be dispersed, encrypted, and saved across a blockchain network. Even if a single node is hacked, the rest of data still remains secure. One of the other key issues in the Equifax breach was unauthorized access and tampering, which could be prevented by using a blockchain's immutable ledger. Once data is recorded in a blockchain system, it cannot be changes without the validation of all other network participants. This guarantee can allow hackers to tweak some sensitive information about us and reuse it (Rath et al. 2024).

That being so, blockchain is an add-on available for increasing supply chain security one of the more troubling issues that came out in relation to SolarWinds hack. Blockchain can be used to track and authenticate system updates, so if organizations you flag all other than Authorized code as invalid its hard for Hackers to introduce software with Malicious Code under the radar. Due to the transparent and traceable nature of blockchain, detection becomes more accountable in almost every part along with detectability as well. Ergo, while the cyberattacks are getting smarter and lethal with time that our current security systems can easily counter anyways, blockchain is here to provide a way out. The decentralization of data storage, the guarantee by immutability to maintain the version control standards and increasing transparency in a supply chain will make these cyber threats cease existence. As cyber threats keep on enhancing regularly, advance toward using blockchain in reconnoitering through security strategies to shield computerized assets and foundation (Asadollahi-Yazdi et al. 2020).

1.2 Contextualizing Digital Transformation Landscape in Business

In the age of digital transformation, manufacturers need to do more than just gather and have access to data. They need to know how to use it wisely in a particular context in order to make decisions that add value. The biggest challenges facing manufacturers today are effectively managing and deriving actionable insights from the vast amount of data that is accessible (Meyendorf et al. 2023). Manufacturers obtain data from a variety of sources, including people, platforms, and systems, each of which has a unique data format, compatibility, and structure. The process of integrating and harmonizing data from these many sources presents complex and time-consuming obstacles. Problems with data models, schema, data semantics, and naming standards frequently lead to data handling difficulties (Angelopoulos et al. 2019).

Even the highest quality data alone cannot bring about change, despite the fact that data is a crucial catalyst for change. In the era of digital transformation, manufacturers need to use data in a specific context to make decisions that add value; just aggregating and providing access to data is insufficient. The interpretation of manufacturing data becomes prone to mistakes or misconceptions in the absence of proper context. Inaccurate assessments and poor decision-making can have a detrimental effect on output, quality, and overall company outcomes when contextual knowledge is lacking. With employing modern digital technologies that are able to contextualize data, manufacturers may use that knowledge to make a variety of intelligent decisions. This might involve learning more about the effects of upgrading machinery or developing novel manufacturing techniques that maximize throughput (Lu et al. 2020).

1.3 Imperative for Robust Cyber-Security in the Age of Digitization: Blockchain Support Cyber-Security

Digital transformation is the reconsideration of how businesses operate in the digital age. In order to satisfy changing business and market demands, digital transformation entails using digital technology to either build new business processes or change current ones, as well as to shape corporate culture and improve consumer experiences. It extends beyond traditional roles like customer service, marketing, and sales and puts a lot of focus on how companies view and engage with their clients (Chander et al. 2022). With the use of digital technology, company processes and customer engagement may be reimagined by switching from antiquated tools like paper-based procedures to modern ones like smart apps. It is not necessary for new small enterprises to set up traditional procedures and then change them later. The organization may be made future-proof right from the start. Using digital tools to build a 21st-century firm is a more sustainable approach than sticking with outdated practices like handwritten ledgers and sticky notes. Having a digital mentality and preparing digitally puts a company in a flexible, adaptive, and growth-oriented position. Many businesses are reviewing their operations in the midst of the digital transformation process to make sure they are in line with the proper goals (Kasowaki and Ahmet 2024).

In a time of rapid technological advancement, the world's reliance on digital technology is growing, with data moving easily across a network of linked devices. There has never been a greater need for strong cyber-security safeguards to protect our data. Everything in our digital age is vulnerable, from personal information on an individual basis to national security. Exploring the field of cyber-security highlights its importance and makes it necessary to investigate the educational and employment opportunities in this area (Sima et al. 2020).

1.4 Objectives of the Chapter

This chapter has the following objectives to:

- fundamental knowledge of business applications in relation to cyber security.
- examine and clarify methods for putting complete blockchain security into practice.
- discussing problems and trends that the digital age is currently facing and offering solutions for smoothly integrating strong cyber-security measures into digitalized company operations.
- focusing on data security and safeguarding the sensitive data in business applications includes encryption techniques, access controls, and other relevant strategies. Figure 16.1 shows in the dimensions of various objectives of this chapter in four parameters.

1.5 Structure of the Chapter

This chapter expressly focusing on diverse arena of Achieving Cyber Security and Data Shield in Business Applications with 360 Degree Blockchain Protection in the Era of Digitization Landscapes. Section 2 elaborates the Current Threat Landscape: Analysis of Prevalent Cyber Threats Targeting Business Applications. Section 3 discusses the Threat Detection and Intelligence. Section 4 lays down the

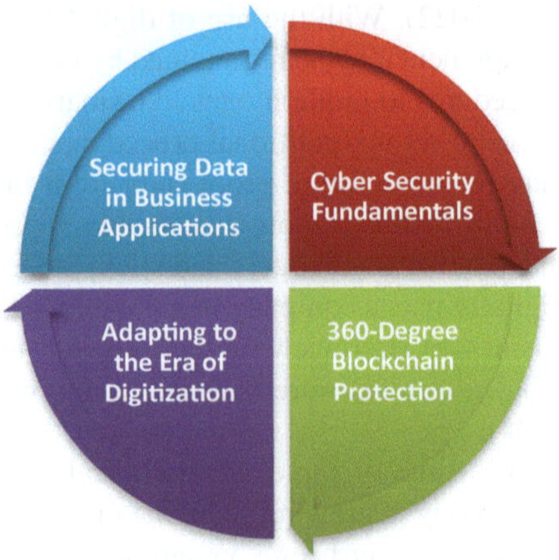

Fig. 16.1 Shows the objectives of the chapter. *Source* Authors' Original Contribution

Secure Coding Practices: Importance of Secure Coding in Preventing Vulnerabilities. Section 5 explains the Blockchain Technology Definition and Fundamental Principles. Section 6 scrutinizes the Cryptographic Mechanisms in Blockchain. Section 7 travels the Supply Chain Security with Blockchain. Section 8 elucidates the Network Security and Access Controls: Securing Network Infrastructure against Cyber Threats. Section 9 illuminates the Compliance and Regulatory Considerations. Finally, Sect. 10 conclude the Chapter with Future Scope.

2 Current Threat Landscape: Analysis of Prevalent Cyber Threats Targeting Business Applications

The field of cyber-security is ever-evolving, with new problems emerging as technology advances quickly. Ransomware attacks, social engineering assaults, data breaches, insider threats, AI-driven attacks, 5G, and IoT threats are some of the most common dangers of today. Attacks using ransomware have significantly increased in sophistication and frequency (Tseng et al. 2021). In some cases, the victim's data is encrypted by hackers who then demand a ransom to unlock the key. Attacks using ransomware have become more frequent, mostly affecting important infrastructure and large corporations. The use of "double extortion" tactics by cybercriminals, in which they threaten to reveal the stolen data unless the requested ransom is paid, is on the rise (Anastasi et al. 2021).

Social engineering attacks entail coercing people into disclosing private information, including credit card numbers or passwords. Phishing is a common type of social engineering assault in which attackers pose as reliable sources in order to trick victims into divulging personal information. Because social engineering assaults prey on human flaws rather than technological ones, countering them can be challenging.

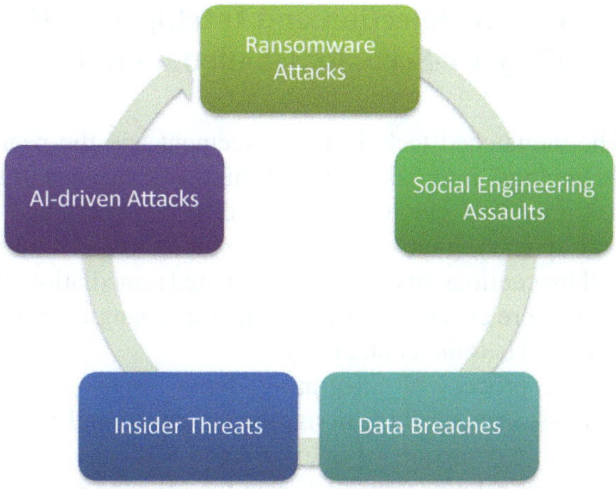

Fig. 16.2 Shows the prevalent cyber threats targeting business applications. *Source* Authors' Original Contribution

Unauthorized access to private information, frequently with the intention of stealing and selling it on the dark web, is known as a data breach (Dwivedi et al. 2021). These occurrences have the potential to cause significant financial losses and damage an organization's image. The insider threats don't always come from outside sources. With access to critical information, workers, contractors, or business partners may represent serious insider risks, especially if their resentment is strong. Because these people are usually trusted and have access credentials, it is difficult to mitigate these hazards. As artificial intelligence (AI) has grown, so too have AI-Driven Attacks, in which hackers use AI to increase their level of sophistication and avoid detection. Figure 16.2 specifies the 5 aspects where AI makes it possible for hacking procedures to be automated, giving thieves the ability to carry out assaults at a pace and scale never before possible (Singh Rajawat et al. 2021).

2.1 Emerging Trends in Cyber-Attacks and Their Potential Impacts

Cyber-attacks may have serious and detrimental effects on people and organizations alike. Businesses may experience operational disruptions, financial losses, intellectual property theft, and reputational damage as a result of these assaults. In extreme cases, a cyber-attack might cause such a large financial loss that the company is forced to close (Leng et al. 2020). Cyber-attacks can result in financial losses, identity theft, and the compromise of private information for an individual. The psychological fallout from a hack can make people distrustful of digital connections and feel vulnerable (Javaid et al. 2021).

3 Threat Detection and Intelligence: Dynamic Nature of Cyber-Security Challenges in Digital Era

The unmatched technical advancement and the pervasive incorporation of digital technology into all facets of human existence, the digital transformation age has presented previously unheard-of prospects and efficiency. But the quick and widespread digitalization has also brought up a variety of cyber-security issues that call for cautious analysis and calculated remediation. It dives deeply into the complex area where cyber-security and the ever-changing dynamics of digital transformation collide (Hassoun et al. 2023).

The process of integrating digital technologies into organizational business processes and resulting in significant operational changes is known as digital transformation. A company's user experience, business procedures, target markets, clientele, customer relationships and even the effects on varied cultures can all be impacted by this change. Emerging technologies including blockchain, cloud computing, artificial intelligence, big data and analytics, Internet of Things, and industrial Internet of things are critical facilitators for digital transformation. Businesses are speeding up their digital transformation initiatives because of the many benefits. But cybersecurity has grown to be a major concern, and companies need to safeguard their digital transformation tools and artifacts in order to maintain business continuity. It is imperative that businesses adopting digital transformation emphasize cyber-security measures and guarantee the security of their systems from potential threats (Aoun et al. 2021).

A significant vulnerability in cybersecurity is human mistake in setup, which is a big cause for concern (Borowski 2021a). Effectively overseeing appropriate system configuration can be extremely difficult, especially when large IT teams are engaged in the setup procedure. Reactive technologies might help teams recognize and resolve problems that occur during the replacement, modification, and upgrading of network systems as computer security grows more complex and innovative. Consider the complexity introduced by newer internet infrastructure like cloud computing layered atop older local frameworks. In enterprise systems, IT teams must ensure compatibility to secure these systems. Manual processes for assessing configuration security can lead to fatigue among teams as they alter endless updates alongside routine daily support tasks. Implementing intelligent, adaptive automation could provide teams with timely insights into newly discovered issues. This approach could offer guidance on available options or even automatically adjust settings as needed (Arden et al. 2021).

3.1 Role of Advanced Threat Detection Mechanisms

Cybercriminals may take advantage of holes in digital technology, thus it's critical for businesses to protect their technical solutions from online threats. Implementing access control, authentication, and encryption techniques can help accomplish cyber-security by shielding networks and data from malevolent or illegal access. Organizations can also think about purchasing cyber insurance plans to offer monetary security against damages brought on by a successful assault on their systems. Educating staff members about cyber-security assaults is also essential, since more knowledge promotes more dependable information security practices. Business companies need to understand cyber-security dangers and implement comprehensive plans to combat the sharp rise in cyber-attacks. Typically, the goals of these assaults are to interrupt regular corporate operations, extract financial advantages from users or evaluate, change, or delete important information. Cyber-security refers to methods of protecting networks and computers against criminal activity, including data loss and theft as well as illegal access (Rane 2023).

3.2 Utilizing Artificial Intelligence and Machine Learning for Proactive Threat Identification

AI-driven cyber-security is a disruptive force in the continuous war against cyber-attacks. Machine learning-driven, it enables enterprises to proactively detect and neutralize possible attacks, preventing damage before it starts. This method saves important time and money in addition to improving security. The modern digital age has brought about an unparalleled increase in the significance of cyber-security. Traditional security methods are less effective due to the growing complexity and scope of cyber threats. Although cyber-security and AI have been heralded as revolutionary and maybe near-term innovations, it's important to approach this idea with realistic expectations. The truth is that rather than a dramatic revolutionary shift, it may experience slow but steady advances in the near future. In order to examine the possible security implications of AI and machine learning, it is necessary to address the existing cyber-security concerns. Artificial Intelligence (AI) technologies encompass a wide range of procedures and elements that are widely recognized as standard (Awan et al. 2021).

4 Secure Coding Practices: Importance of Secure Coding in Preventing Vulnerabilities

The writing of executable programs in a language that a computer can understand is called coding or computer programming. When writing this source code, a software developer needs to consider a number of things, such as the application's architecture and design specifications; Efficiency and optimization of the code; Safety and security of codes, etc. With following certain best practices and guidelines, sometimes referred to as secure coding standards, safe coding is a technique that makes it easier to find and fix common software vulnerabilities. Using safe coding techniques is essential since it reduces often exploited software vulnerabilities and averts possible intrusions. Setting security as a top priority early on also reduces potential long-term expenses in the event that an attack exposes private user data. Even with the importance of secure code, there are still many software vulnerabilities. Utilizing the vulnerability list maintained by the National Institute of Standards and Technology (NIST), astounding multi-application vulnerabilities have been discovered in the past years alone (Kumar et al. 2020).

The process of building software that is protected against security threats is known as secure coding. It includes writing code in a high-level language while following strict guidelines in order to prevent vulnerabilities that might lead to data loss or harm to a particular system. Secure coding involves more than just creating, assembling, and deploying code into apps; it also involves setting up a secure development environment. This entails expanding upon a stable and secure IT infrastructure that includes services from reputable vendors, safe hardware, and software. The prevention of cyber-security incidents, which may result in the disclosure of private and sensitive data, begins from the very beginning of the software development lifecycle that is with the source code. Using safe coding techniques is important since it removes software weaknesses that are often attacked, which may prevent cyberattacks. Setting security as a top priority from the start also helps to lower potential long-term expenses that might arise if vulnerability exposes sensitive user data (Kurniawan et al. 2022).

5 Blockchain Technology—Definition and Fundamental Principles

Blockchain technology functions as a sophisticated database system that makes transparent information exchange within a corporate network possible. It works by storing data in blocks that are connected one after the other to create a chain. Since the network must agree in order to change or remove the chain, the data is still consistent in terms of chronology. As a result, an unchangeable or immutable ledger is established using blockchain technology to monitor transactions such as orders, payments, and accounts. In order to stop unwanted transactions and preserve consistency in the shared view of these transactions, the system has built-in safeguards (Jakka et al. 2022).

Financial transaction recordkeeping is hampered by traditional database technology. Usually, a trusted third party supervises and verifies transactions to prevent disagreements, which adds complexity and a single point of vulnerability. Blockchain solves these issues by establishing a decentralized, impenetrable transaction recording mechanism. For situations like as real estate sales, separate ledgers are kept for the buyer and the seller. Every transaction needs both parties approval, and the ledgers are automatically updated in real time. A single corrupt transaction in the past might jeopardize the entire ledger. Because of these special qualities, blockchain technology is being used extensively in many other fields, including the creation of virtual currencies like Bitcoin (Xing et al. 2021).

5.1 Decentralized and Tamper-Resistant Nature of Blockchain

Decentralization in the context of blockchain refers to the transfer of power and decision-making from a centralized entity a person, an organization, or a group of people to a dispersed network. Decentralized networks are primarily designed to reduce the amount of confidence that users must have in one another and to discourage the exercise of power or influence over the network that would jeopardize its integrity. In technological solutions, decentralization is a well-established idea. Three basic network designs are often taken into consideration: distributed, centralized, and decentralized. Even though decentralized networks are commonly used by blockchain technologies, it's crucial to remember that a blockchain application can't just be classified as decentralized or not; rather, decentralization is a spectrum that should be applied to every aspect of a blockchain application (Shi et al. 2020). An application can gain more justice and service by decentralizing the management and access to resources, however there are certain drawbacks, such as decreased transaction speed. In a perfect world, the increased stability and service levels these sacrifices provide would justify them as shown in Fig. 16.3 (Saeed et al. 2023).

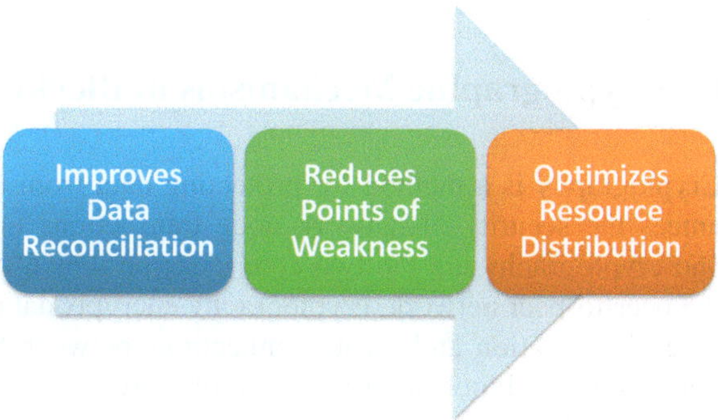

Fig. 16.3 Illuminates the benefits of decentralization. *Source* Authors' Original Contribution

5.2 *Exploring Blockchain's Potential to Reshape Cyber-Security Paradigms*

Because blockchain technology is decentralized, there is no single body that can manage the whole network which minimizes the risks that come with centralized points of failure. The key characteristics of blockchain are immutability which refers to the inability to change data once it is a part of the chain. This is accomplished by using cryptographic hashing which makes the data impervious to illegal changes or alteration (Kumar and Mallipeddi 2022).

5.3 *Blockchain Applications in Cyber-Security: Decentralized Identity Management for Secure Access Controls*

Blockchain identity management is the process of securely monitoring and controlling digital identities via the use of blockchain technology. It makes use of the immutability, transparency, and decentralization of the blockchain to establish trust, validate identities, and protect private data. Blockchain identity management gives people authority over their digital identities as opposed to the conventional dependence on centralized authorities or middlemen, guaranteeing increased security, privacy, and data ownership. Figure 16.4 highlights the multifarious benefits of the blockchain identity management concerning cyber-security in business applications (Bag and Pretorius 2022).

6 Cryptographic Mechanisms in Blockchain

Cryptography is a method for protecting data from unwanted access. It also guarantees the security of network node transactions inside the blockchain. Hashing and cryptography are the two core ideas on which the blockchain is built. Within the peer-to-peer network, cryptography encrypts data and hashing is used to secure block information and create connections between blocks in the blockchain. The main goals of blockchain cryptography are to safeguard users, transactions, and avoid double-spending. It is essential to the protection of different kinds of transactions on the blockchain network since it makes sure that the transaction data is only accessible, readable, and processed by those who are supposed to. The cryptography which comes from the Greek words "kryptos" which means "hidden" and "graphein" which means "to write" is the process of protecting data by converting it into a format that is unreadable by unauthorized parties. The use of mathematical methods, keys, and protocols allows for this transition. The four main goals of cryptography are non-repudiation, integrity, secrecy, and authenticity (Morgan et al. 2021).

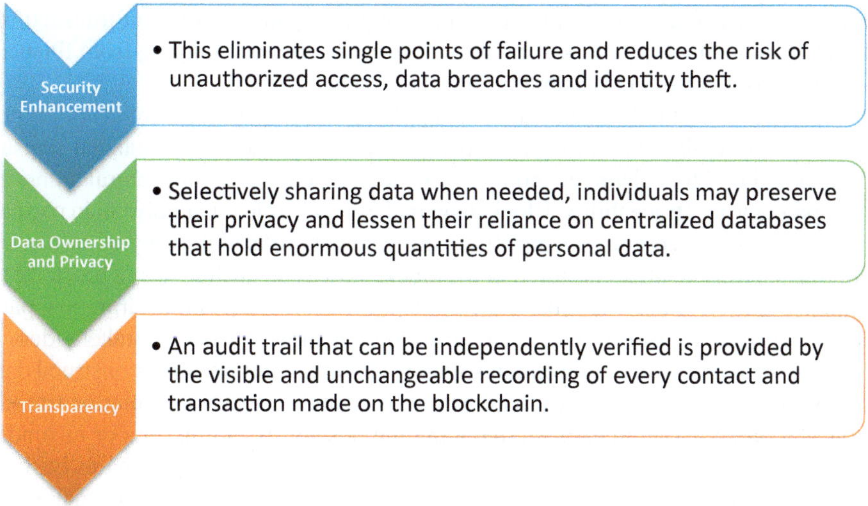

Fig. 16.4 Shows the benefits of the blockchain identify management. *Source* Authors' Original Contribution

Cryptography protects data from unwanted access and is essential to the blockchain because it allows transactions between two nodes in the network to be securely recorded. The two main ideas at the center of the blockchain are hashing and cryptography, as was previously established. In a peer-to-peer network, cryptography is used to encrypt communications, and hashing is used to protect block data and create connections between blocks in the blockchain. The primary goals of cryptography in the blockchain are to protect users, transactions, and transactions from being double-spent. It is essential for safeguarding different kinds of transactions on the blockchain network by making sure that the transaction data is only accessible, readable, and processed by the intended parties. Blockchain was developed using a number of cryptographic ideas. The development of cryptography places limits on how far blockchain can go in its progress (Singh et al. 2022).

7 Supply Chain Security with Blockchain

Blockchain solves problems like quality problems by making it easier to follow a product's route from raw material suppliers to final consumers. Blockchain enables producers of tuna in cans to track the product all the way back to the exact shipping vessel from which it came, recording every step of the product's journey on the transparent and unchangeable blockchain. With ensuring product traceability, recalls and lost income are reduced. Because blockchain technology guarantees transaction integrity, transparency, and traceability, it is especially well-suited for modern supply chain networks that include a large number of participants in frequent transactions. Blockchain technology has several uses in supply networks, ranging from financial transactions to product traceability (Nica and Stehel 2021).

Blockchain technology can control supply chain money flows by using cryptocurrencies like bitcoin, although not being extensively used yet. Financial transactions based on blockchain technology provide independence from central authority, transparency, and traceability. Blockchain makes it possible for supply chain partners to carry out smart contracts, simplifying intricate networks with many of parties and contracts. Every contract is stored in the blockchain as an unchangeable block that is impenetrable to tampering, guaranteeing its immutability and prohibiting any party from changing it (Borowski 2021a). Information moves between stakeholders in broad supply chains and blockchain solves problems with lost or misconstrued data. Transparency and traceability are improved by including every information transaction as a block in the blockchain, reducing the possibility of data loss or misunderstanding. There are advantages of blockchain technology for supply chains business as: Transparency: Blockchain improves supply chain transparency by making all transactions public to all parties. Finding Execution mistakes: Supply chain operations may identify and fix execution mistakes more quickly and efficiently, saving time and money to the immutable nature of blockchain transactions. Enhanced Security: The use of private keys to safeguard blockchain transactions makes them impervious to tampering and ensures a reliable, sequential audit record of all transactions. Increased Reactivity and Speed: Blockchain digitizes laborious operations, enabling instantaneous data transmission and guaranteeing supply chain networks (Ahmad et al. 2021).

8 Network Security and Access Controls: Securing Network Infrastructure Against Cyber Threats

Network security measures comprise a range of controls that are put in place to protect networks availability, confidentiality, and integrity. The fundamental information is easily accessible in the ever-evolving world of security measures, with an emphasis on preventing intruders from accessing networks. In this context, intrusion detection/prevention systems, firewalls, proxies, and gateways are essential tools (Wamba-Taguimdje et al. 2020).

Firewall: Based on pre-established rules, a firewall operates at levels 2–5 of the OSI model, allowing or blocking communication. A firewall's setup is crucial; it must be set up such that it only permits particular traffic to recognized services and by default blocks all traffic. The knowledge of the firewall administrator is essential for handling various settings (Popov et al. 2022).

Proxy: In the OSI model, a proxy firewall functions at layer 7 and handles traffic by removing its headers. It breaks the TLS connection at layer 4, allowing layer

7 examination of the data that was transferred. Compared to traditional firewalls, proxies provide more control over user traffic, but they also take more time and processing resources (Lu 2019).

Gateway: Traditionally, a gateway had firewall components and was located in the space between two networks. The line between a firewall and a modern gateway may become less clear because of the possibility of built-in web application firewalls (WAF) in gateways like those found in Microsoft Azure (Blobel 2020).

Systems for Intrusion Detection and Prevention: IDSs (intrusion detection systems) monitor network traffic in a passive manner while recording unusual activities. Based on where they are deployed, network-based IDS (NIDS) and host-based IDS (HIDS) are different. IDS capabilities are enhanced by intrusion prevention systems (IPSs) which actively block suspicious traffic (Borowski 2021b).

Virtual Private Network (VPN): Encryption is essential for safeguarding data while it is being sent. IPsec, SSH, and SSL/TLS are popular techniques for encrypting data while it's in transit. User and browser-based VPN connections are secured using SSL/TLS, now known as TLS. SSH replaces Telnet at layer 7, and network managers use it mostly for remote login. Encapsulating Security Payload (ESP) and Authentication Header (AH) are the two security protocols used by IPsec, a network layer protocol that offers encryption, integrity, and secrecy. Whereas ESP offers data origin authentication, integrity, and secrecy, AH guarantees data origin authentication and integrity. The implementation of security measures necessitates continuous learning, evolution, and threat-avoidance activities; incident response plans and teams are critical components of good cyber-security (Lim et al. 2021).

9 Compliance and Regulatory Considerations

Intellectual property (IP) protection is still very important and covers a wide range of assets including design papers, research and development data, manuals, and more. The two key issues are keeping information private and making sure only authorized people can access it. Crucial components used to properly manage access are data categorization and access control. Data Leak Prevention (DLP) technology may be used to successfully address concerns about unapproved data flow from a firm. DLP technology keeps an eye out for sensitive data in data flows like file transfers and emails. The DLP software takes preventive measures such as blocking or stopping the transfer and, depending on the business's wishes, may also encrypt the data, if it finds sensitive data such as a credit card number (Vogt 2021).

DRM technology is an additional tool for managing access to intellectual property. Common DRM-enabled programs include Netflix, Spotify, Kindle, iTunes, and Amazon Prime Video. Businesses like Cisco use DRM to control access to course materials when a consumer pays a class. DRM also enables customers to access content that they have paid for such as music, movies or books. Javelin and Lock-Lizard are two examples of DRM systems for content management; they provide access control settings depending on the preferences of the IP owner, deciding things like usage time, printing rights, and sharing permissions. Effective security issue identification and rectification is a prerequisite for implementing strong security measures. Logging is a basic place to start as almost every system creates logs. The choice of what to log is based on the risk tolerance of the company, the sensitivity of its assets, and its system vulnerabilities. Potential logs, customized to the unique requirements and goals of the company, might contain login attempts, network flows, packets, user actions, or even individual keystrokes. Proactively resolving security vulnerabilities is greatly aided by the use of Security Information and Event Management (SIEM) solutions in conjunction with log monitoring (Gadekar et al. 2022).

Organizations use rules and best practices found in the compliance and regulatory frameworks to meet regulatory obligations, improve processes, strengthen security, and accomplish a variety of commercial goals, such as going public or offering cloud solutions to government agencies. From the server room to the boardroom, these frameworks offer a common language. They are employed by Internal stakeholders, including auditors, should evaluate the controls in place inside their company; external auditors to evaluate and verify the organization's internal controls; external parties, such as prospective clients and investors, to evaluate the possible dangers of working with a company. Reaching compliance in a regulatory environment is a continuous process. Because environments change continuously, controls may no longer be as effective. It is imperative that frequent reporting and monitoring be done and each framework offers guidelines about the details of "regular monitoring" (Tao et al. 2021).

10 Conclusion and Future Scope

Blockchain technology is revolutionizing a variety of sectors with its unique characteristics and developing uses. Its ability to provide open, safe, and effective solutions is reorganizing current systems and encouraging the development of new ideas. Notwithstanding obstacles, the ongoing development of blockchain technology is broadening its potential uses and influence, bringing in a new era of digital transformation and cyber-security. The future developments in artificial intelligence (AI) for cyber-security are poised to revolutionize defense tactics as technology continues to progress. It is expected that advances in adversarial machine learning, proactive threat hunting, and predictive skills will influence cyber-security protection in the future. These advancements will advance to joint efforts by researchers, industry professionals, and legislators. In the continuous battle against cyber threats, the incorporation of AI and ML into cyber-security protection marks a major technological advancement. The proactive deployment of AI and ML technologies becomes critical for preserving an advantage over competitors in the rapidly changing digital ecosystem. To properly utilize the promise of these technologies and their applications, organizations must commit resources to research, cooperation, and ongoing innovation ensure a secure digital future.

References

Ahmad T, Zhang D, Huang C, Zhang H, Dai N, Song Y, Chen H (2021) Artificial intelligence in sustainable energy industry: Status Quo, challenges and opportunities. J Clean Prod 289:125834

Anastasi S, Madonna M, Monica L (2021) Implications of embedded artificial intelligence-machine learning on safety of machinery. Procedia Comp Sci 180:338–343

Angelopoulos A, Michailidis ET, Nomikos N, Trakadas P, Hatziefremidis A, Voliotis S, Zahariadis T (2019) Tackling faults in the industry 4.0 era—a survey of machine-learning solutions and key aspects. Sensors 20(1):109

Aoun A, Ilinca A, Ghandour M, Ibrahim H (2021). A review of Industry 4.0 characteristics and challenges, with potential improvements using blockchain technology. Comp Indus Eng 162:107746

Arden NS, Fisher AC, Tyner K, Lawrence XY, Lee SL, Kopcha M (2021) Industry 4.0 for pharmaceutical manufacturing: preparing for the smart factories of the future. Int J Pharm 602:120554

Asadollahi-Yazdi E, Couzon P, Nguyen NQ, Ouazene Y, Yalaoui F (2020) Industry 4.0: revolution or evolution? Am J Operat Res 10(06):241

Awan U, Sroufe R, Shahbaz M (2021) Industry 4.0 and the circular economy: a literature review and recommendations for future research. Bus Strat Environ 30(4):2038–2060

Bag S, Pretorius JHC (2022) Relationships between industry 4.0, sustainable manufacturing and circular economy: proposal of a research framework. Int J Organiz Analy 30(4):864–898

Blobel B (2020, September) Application of industry 4.0 concept to health care. In pHealth 2020: Proceedings of the 17th International Conference on Wearable Micro and Nano Technologies for Personalized Health, Vol 273. IOS Press, p 23

Borowski PF (2021a) Innovative processes in managing an enterprise from the energy and food sector in the era of industry 4.0. Processes 9(2):381

Borowski PF (2021b) Digitization, digital twins, blockchain, and industry 4.0 as elements of management process in enterprises in the energy sector. Energies 14(7):1885

Chander B, Pal S, De D, Buyya R (2022) Artificial intelligence-based internet of things for industry 5.0. In: Artificial intelligence-based internet of things systems, pp 3–45

Dinesh Arokia Raj A, Jha RR, Yadav M, Sam D, Jayanthi K (2024) Role of blockchain and watermarking toward cybersecurity. In: Multimedia watermarking: latest developments and trends. Springer Nature, Singapore, pp 103–123

Dwivedi YK, Hughes L, Ismagilova E, Aarts G, Coombs C, Crick T, et al (2021) Artificial Intelligence (AI): multidisciplinary perspectives on emerging challenges, opportunities, and agenda for research, practice and policy. Int J Inf Manage 57:101994

Felsberger A, Qaiser FH, Choudhary A, Reiner G (2022) The impact of Industry 4.0 on the reconciliation of dynamic capabilities: evidence from the European manufacturing industries. Prod Plann Cont 33(2–3):277–300

Gadekar R, Sarkar B, Gadekar A (2022) Key performance indicator based dynamic decision-making framework for sustainable Industry 4.0 implementation risks evaluation: reference to the Indian manufacturing industries. Annal Operat Res 318(1):189–249

Hassoun A, Aït-Kaddour A, Abu-Mahfouz AM, Rathod NB, Bader F, Barba FJ, et al (2023) The fourth industrial revolution in the food industry—Part I: Industry 4.0 technologies. Crit Rev Food Sci Nut 63(23):6547–6563

Jakka G, Yathiraju N, Ansari MF (2022) Artificial intelligence in terms of spotting malware and delivering cyber risk management. J Pos School Psychol 6(3):6156–6165

Javaid M, Haleem A, Singh RP, Khan S, Suman R (2021) Blockchain technology applications for Industry 4.0: a literature-based review. Blockcha Res Appl 2(4):100027

Kasowaki L, Ahmet S (2024) Shielding the virtual ramparts: understanding cybersecurity essentials (No. 11700). EasyChair

Krishnan S, Balas VE, Julie EG, Yesudhas HR, Balaji S, Kumar R (eds) (2020) Handbook of research on blockchain technology. Academic Press

Krishnan S, Balas VE, Julie EG, Yesudhas HR, Kumar R (eds) (2021) Blockchain for smart cities. Elsevier

Krishnan S, Balas VE, Kumar R (eds) (2023) Green blockchain technology for sustainable smart cities. Elsevier

Kumar S, Mallipeddi RR (2022) Impact of cybersecurity on operations and supply chain management: emerging trends and future research directions. Prod Oper Manag 31(12):4488–4500

Kumar K, Zindani D, Davim JP (2019) Industry 4.0: developments towards the fourth industrial revolution. Springer, Cham, Switzerland

Kumar SH, Talasila D, Gowrav MP, Gangadharappa HV (2020) Adaptations of Pharma 4.0 from Industry 4.0. Drug Inven Today 14(3)

Kurniawan TA, Maiurova A, Kustikova M, Bykovskaia E, Othman MH D, Goh HH (2022) Accelerating sustainability transition in St. Petersburg (Russia) through digitalization-based circular economy in waste recycling industry: a strategy to promote carbon neutrality in era of Industry 4.0. J Clean Prod 363:132452

Leng J, Ye S, Zhou M, Zhao JL, Liu Q, Guo W, et al (2020) Blockchain-secured smart manufacturing in industry 4.0: a survey. IEEE Trans Syst Man Cybern Syst 51(1):237–252

Lim CH, Lim S, How BS, Ng WPQ, Ngan SL, Leong WD, Lam HL (2021) A review of industry 4.0 revolution potential in a sustainable and renewable palm oil industry: HAZOP approach. Renew Sustain Ener Rev 135:110223

Lu Y (2019) Artificial intelligence: a survey on evolution, models, applications and future trends. J Manage Anal 6(1):1–29

Lu C, Lyu J, Zhang L, Gong A, Fan Y, Yan J, Li X (2020) Nuclear power plants with artificial intelligence in industry 4.0 era: top-level design and current applications—a systemic review. IEEE Acce 8:194315–194332

Meyendorf N, Ida N, Singh R, Vrana J (2023) NDE 4.0: progress, promise, and its role to industry 4.0. NDT & E International, 102957

Mithas S, Chen ZL, Saldanha TJ, De Oliveira Silveira A (2022) How will artificial intelligence and Industry 4.0 emerging technologies transform operations management? Prod Operat Manage 31(12):4475–4487

Morgan J, Halton M, Qiao Y, Breslin JG (2021) Industry 4.0 smart reconfigurable manufacturing machines. J Manuf Syst 59:481–506

Nica E, Stehel V (2021) Internet of things sensing networks, artificial intelligence-based decision-making algorithms, and real-time process monitoring in sustainable industry 4.0. J Self Govern Manage Econ 9(3):35–47

Popov VV, Kudryavtseva EV, Kumar Katiyar N, Shishkin A, Stepanov SI, Goel S (2022) Industry 4.0 and digitalisation in healthcare. Materials 15(6):2140

Rane N (2023) Transformers in Industry 4.0, Industry 5.0, and Society 5.0: roles and challenges

Rath KC, Khang A, Roy D (2024) The Role of Internet of Things (IoT) technology in Industry 4.0 economy. In: Advanced IoT technologies and applications in the Industry 4.0 digital economy. CRC Press, pp 1–28

Ray RK, Chowdhury FR, Hasan MR (2024) Blockchain applications in retail cybersecurity: enhancing supply chain integrity, secure transactions, and data protection. J Bus Manag Stud 6(1):206–214

Saeed S, Altamimi SA, Alkayyal NA, Alshehri E, Alabbad DA (2023) Digital transformation and cybersecurity challenges for businesses resilience: issues and recommendations. Sensors 23(15):6666

Shi Z, Xie Y, Xue W, Chen Y, Fu L, Xu X (2020) Smart factory in Industry 4.0. Syst Res Behav Sci 37(4):607–617

Sima V, Gheorghe IG, Subić J, Nancu D (2020) Influences of the industry 4.0 revolution on the human capital development and consumer behavior: a systematic review. Sustainability 12(10):4035

Singh SK, Sharma SK, Singla D, Gill SS (2022) Evolving requirements and application of SDN and IoT in the context of industry 4.0, blockchain and artificial intelligence. In: Software defined networks: architecture and applications, pp 427–496

Singh Rajawat A, Bedi P, Goyal SB, Shukla PK, Zaguia A, Jain A, Monirujjaman Khan M (2021) Reformist framework for improving human security for mobile robots in industry 4.0. Mob Inf Syst 2021:1–10

Tao F, Akhtar MS, Jiayuan Z (2021) The future of artificial intelligence in cybersecurity: a comprehensive survey. EAI Endor Trans Creat Technol 8(28):e3–e3

Tseng ML, Tran TPT, Ha HM, Bui TD, Lim MK (2021) Sustainable industrial and operation engineering trends and challenges toward Industry 4.0: a data driven analysis. J Indus Prod Eng 38(8):581–598

Vogt J (2021) Where is the human got to go? Artificial intelligence, machine learning, big data, digitalisation, and human–robot interaction in Industry 4.0 and 5.0: Review Comment on: Bauer M (2020) Preise kalkulieren mit KI-gestützter Onlineplattform BAM GmbH, Weiden, Bavaria, Germany. AI & SOCIETY 36(3):1083–1087

Wamba-Taguimdje SL, Fosso Wamba S, Kala Kamdjoug JR, Tchatchouang Wanko CE (2020) Influence of artificial intelligence (AI) on firm performance: the business value of AI-based transformation projects. Bus Process Manag J 26(7):1893–1924

Xing K, Cropley DH, Oppert ML, Singh C (2021) Readiness for digital innovation and industry 4.0 transformation: studies on manufacturing industries in the city of salisbury. In: Business innovation with new ICT in the Asia-Pacific: case studies, pp 155–176

Yue Y, Shyu JZ (2024) A paradigm shift in crisis management: the nexus of AGI-driven intelligence fusion networks and blockchain trustworthiness. J Contingen Crisis Manage 32(1):e12541

Blockchain Technology: Scalability Challenge, Solutions, and Future Directions

Bhargavi K. Chauhan and Shanti Verma

Abstract Blockchain technology has gained attention to not only tech realm but also makes attention from various sectors like education, health, insurance, finance, transportation, supply chain, agriculture, government, E-commerce etc.... The decentralized, trusted, secure, transparent, and immutable features of blockchain become particularly prominent when integrated with machine learning and the Internet of Things (IoT). However, it faces some challenges but out of them the scalability challenge is the most affecting challenge. As its popularity has prompted the research community to investigate and analyze its challenges. The objective of this study is to address the scalability challenge by examining its affecting parameters. The study further provides a comprehensive classification of recent solutions based on parameters such as methodology or techniques, architecture, types of network, and consensus mechanisms etc.... The conclusion includes recommendations and outlines potential avenues for future research that catering to the research community. This work aims to offer valuable classified information on the blockchain scalability challenge, solutions and different blockchain applications.

1 Introduction

Bitcoin is the first platform that uses blockchain technology, but its scope is not limited to cryptocurrencies. The main purpose of blockchain technology is to create a decentralized environment where no third party controls the data (Huang et al. 2022). Blockchain technology has many advantages such as immutability, transparency, traceability, and security etc.... (Zafar et al. 2022). Blockchain technology is based on a distributed ledger environment where individual nodes can participate equally in the transaction verification process and contain the same database without controlling

B. K. Chauhan (✉) · S. Verma
Department of Computer Science, LJ University, Ahmedabad, India
e-mail: bhargavi.chauhan49@gmail.com

S. Verma
e-mail: shanti.verma@ljku.edu.in

S. Krishnan et al. (eds.), *Unveiling Blockchain Enabled Business Applications*, Contributions to Management Science, https://doi.org/10.1007/978-981-95-2337-5_17

each other. The data in the database is cryptographically protected and once the data is stored, it becomes an immutable database (Besançon et al. 2022). The peer-to-peer (P2P) networking model is used for decentralized communication in which any node can broadcast the block. In the client/server network model, clients make service requests and servers fulfil those requests (Joshi et al. 2022). In the P2P network model, each network node can act as both a client and a server and be equally responsible for processing data (Chauhan and Patel 2021). Each node has to follow the consensus mechanism to verify the transaction. The consensus mechanism is a fault-tolerant mechanism used in blockchain technology that uses a self-regulating architecture (Qi 2022). There are many consensus mechanisms introduced after PoW (proof of work) used by Bitcoin; Proof of stake (PoS), Proof-of-space, Proof of Activity, Proof-of-authority, Proof of Burn, Proof of Elapsed time etc. (Joshi et al. 2022). Although blockchain technology has several advantages, it also has limitations like; Scalability, interoperability, sustainability, and governance (Vispute et al. 2021). Ethereum is the second-generation blockchain technology after Bitcoin which is first generation blockchain. It is a modified version of Bitcoin that introduces a proof-of-stake consensus mechanism and the concept of smart contracts. It is an open source platform that can be used to build any ecosystem into a decentralized application (Gupta et al. 2020).

Smart contracts in Ethereum are small blocks of code, and these reside on the blockchain and will have their addresses (Aggarwal and Kumar 2021). Executing a smart contract. The smart contract is executed when a transaction, which includes a smart contract address as the delivery address, is included to the blockchain (Solaiman et al. 2021). A smart contract may also call another smart contract. Smart contracts are further stateful; that is to say that smart contracts maintain a private storage (under them) that is retained throughout transactions (Papadis and Tassiulas 2020). Every individual can generate his or her own smart contract and append it to the blockchain, but a smart contract, once it joins the blockchain can not be changed (Khamis et al. 2021). Ethereum Virtual Machine (EVM), a user-created, sandboxed, runtime environment where all smart contracts are hosted. VM-based code executing within the EVM is not permitted access to the filesystem, and the network, or other processes (Liu et al. 2021). Typically, ethereum smart contracts are written in a higher-level language that is later compiled to the ethereum virtual machine bytecode and stored in the blockchain (McCorry et al. 2021). To make it easier to write smart contracts, Ethereum comes with Solidity described as a contract-oriented programming language and compiling to EVM bytecode (Vispute et al. 2021).

The basic components of blockchain system are; transactions, blocks (collection of transaction is known as block), network nodes (miners), consensus mechanism, and chain. Figure 17.1 shows the basic transaction life cycle of a blockchain, covering all the components of a blockchain system (Aggarwal and Kumar 2021). (Burdges et al. 2020). The components of blockchain are strong enough to decentralize any application and has the potential to transfer any sector. It supports mainly three types of network; public, private, and consortium (Upadhyay 2020). The implementation of various applications has been made easier by the introduction of many consensus mechanisms. All those consensus mechanisms help in optimizing the network as well

(Gupta et al. 2020). To overcome the limitations of blockchain network, many solutions have been introduced, leading to the creation of different blockchain designs (Atlam et al. 2020). The number of ledgers also plays a vital role in the performance of the blockchain network. İt supports single and multiple ledger design (Janssen et al. 2020). The Blockchain is confronted with a few challenges, the research community is constantly making their contribution to overcome that challenges (Allombert et al. 2019). Although blockchain technology has attracted the attention of many industries due to its great advantages, it also faces several challenges such as scalability, interoperability, high energy consumption, and complexity (Mohammed et al. 2020). Out of them scalability plays a vital role as it directly influenced the end user and this is the main reason it is gaining attention in the research community. The horizontal and vertical approaches have been developed to tackle the scalability challenge (Baird and Luykx 2020).

Block generating process

a. A new transaction is done in blockchain system which will be initially validated by all the nodes in the network.
b. The block is formed of the set of transactions. In short block, there is nothing but gathering the transactions.
c. The nodes(miners) use consensus mechanism to verify that block.
d. The verification that has been successfully completed produces the valid block and this valid block is added in the blockchain (Hashim et al. 2022; Singh et al. 2020; Allombert et al. 2019).

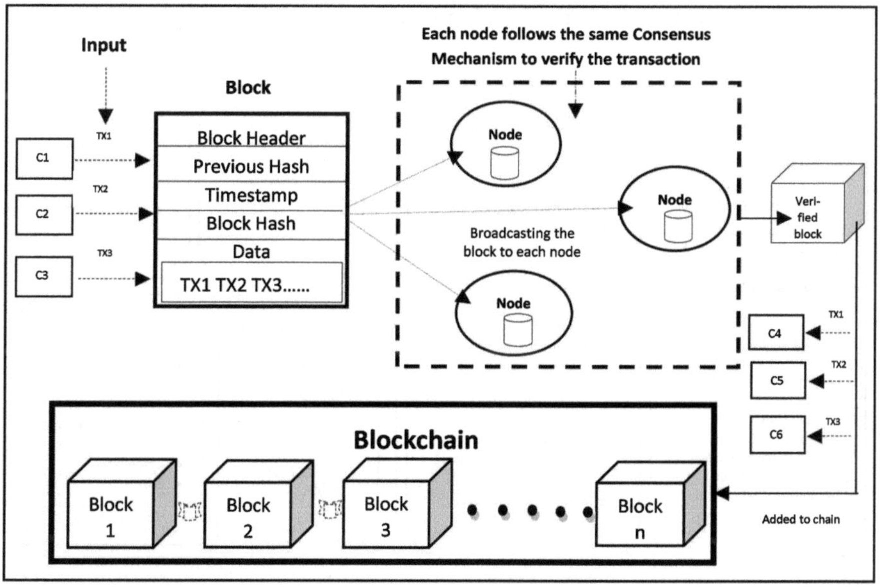

Fig. 17.1 Components of blockchain system

2 Blockchain Scalability

The main pillars of blockchain are decentralization, security, and scalability. All the connected nodes verify the transaction to make the network decentralized but this process slows down the network (Chauhan and Patel 2021). The network becomes slow down due to the high computational power needed to verify the transactions also. This makes it difficult to use blockchain networks in applications that require fast transaction processing speeds (Upadhyay 2020). The network becomes slow down if the more number of nodes participate in the transaction verification process (Gupta et al. 2020). The network nodes are not the only parameter that affects the scalability problem. There are many parameters that directly impact the scalability of chain (Brünjes and Gabbay 2020). Figure 17.2 shows some parameters which affect the scalability of chain (Upadhyay 2020).

The scalability issues on the blockchain can be divided into three categories namely: Throughput, Cost, and Capacity. Throughput: The problem on public blockchain 2.0 is waiting time on the transactions to be entered into the block because of the block size limit. Moreover, when the block generation time is not long then too many forks are created hence the block generation time cannot be artificially shortened. Due to these factors throughput is restrained (Liu et al. 2020). Cost: The user will need to pay to the minor, the transaction fee, once he creates the transaction and it becomes a burden, it is the cost problem. Capacity: The capacity can come too high to support a chain when all the transactions of the chain are stored in a chain. Scalability has become one of the biggest problems of blockchains because they should make all the transactions be verified by all of the nodes. Scalability is the big problem that restricts the development of the public blockchain (Allombert

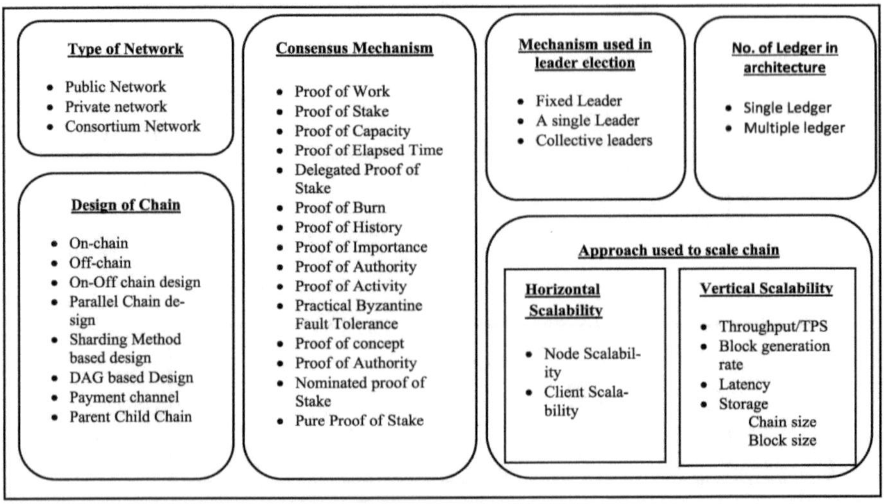

Fig. 17.2 Parameters affecting blockchain scalability

et al. 2019). The public network increases drastically in regard to the number of participants and data since no user has a limit as to whether he can join the network or not (Wang et al. 2022).

All the transactions and block verifications are going to be on the rise due to the increment in the number of users (Huang et al. 2022). This results in communication overhead that strains the network scalability. These systems are prone to slow down as the amount of transactions increases thus making them costly. This is why scaling is turning into an obstacle in a broader adoption of blockchain systems (Lavaur et al. 2023). The more nodes are sooner slowed which is because there is more time needed to transmit the information to the new node, thus there should be better designs which are made regarding scalability (Negka and Spathoulas 2021). The number of transactions conducted over a blockchain has gradually been increasing with time and with this rise in traffic the performance of the system has started to suffer as processing resources are now increasingly incapable of verifying more and more transactions at a time (Wang et al. 2023). Latency between nodes increases along with an increase in the number of new nodes in the network and thus decreases the speed of the validation process (Chen et al. 2022). The resultant effect is the low scalability of the blockchains in terms of their throughput transactions.

The Block interval time is changing with two discrete possibilities, either increase in the size of the block, decrease in size of the block (Zafar et al. 2022). Block size growth is linked with reduced velocity of propagation and with an increase in the risk of orphaned blocks, which thereby drives up the cost of maintaining the chain (Huang et al. 2022). Comparatively, a reduction in block size limits the amount of transactions that can be accommodated per block. Blockchain is based on the distributed database system where the data stored in such databases is present in peer-to-peer network (Besançon et al. 2022). It follows that every computation node has a copy of the ledger, and these copies together take up huge storage space. It creates difficulties in the management of such a geographically distributed distribution of data because the individual participants cannot exercise complete control over the data regimen as well as the storage of the data (Wang et al. 2023). A good data transfer process is part and parcel of scalability. Network performance is determined by latency, bandwidth, and throughput. Bandwidth is a term applied in describing the rate at which the highest amount of data can pass through a certain line of route. Latency is the amount of time it takes data to move between the client and the server and vice versa, and throughput refers to the amount of data that flows between the client and the server within a certain timeframe (Lavaur et al. 2022).

2.1 Type of Network

Blockchain primarily supports three types of networks: public networks, private networks, and consortium networks. The public networks have more nodes than private or consortium networks (Janssen et al. 2020). Once block broadcast to all the network nodes, each node validates the transaction of block. The verification process becomes slower in public network when we make comparison with private and consortium network. Bitcoin and Ethereum have public blockchain and having 7 tps and 15–20 tps respectively. On the other hand, multichain and tendermint having private chain with 2500 tps and 10,000 tps respectively (Kaur and Gandhi 2020).

2.2 Design of Chain

There are many scalable mechanisms, networks, architecture, solutions, methods, techniques, and models are introduced with implementation. However, it needs more research to scale the blockchain completely and already many researches are ongoing to do the same (Baird and Luykx 2020). The design of chain also affects the scalability parameters of the blockchain as each has a different network architecture, storage, consensus mechanism, methodology etc.... After discussing this factor, solutions are described below that covers this factor in detail (Janssen et al. 2020) (Table 17.1).

2.3 Consensus Mechanisms

After Proof of Work, numerous consensus mechanisms are being introduced day by day depending on the requirement of different platforms and applications. Every consensus mechanism has its own rules and logic to verify the transaction so every mechanism shows different throughput and latency (Lashkari and Musilek 2021). The combination of more than one consensus mechanism also shows different result. Bitcoin has PoW consensus mechanism and its results show 7 tps. Ethereum combines the PoW and PoS consensus mechanisms that achieve 15–20 tps. R3 corda uses practical byzantine fault tolerance mechanism and shows 15–1678 tps. Hyperledger comes with private and consortium network architecture that achieves 3500 tps, combining PBFT and PoET (Janssen et al. 2020). EOS combines PoS and PBFT and optimizes the throughput by 3996 tps. A chain has a public network that combines the PoW, PBFT, and PoS consensus mechanism and shows 1000 tps (Lashkari and Musilek 2021). Tendermint comes with PoC consensus mechanism and shows 10,000 tps. Table 17.2 presents various blockchain platforms, their consensus mechanism and throughput (Cortes-Goicoechea et al. 2021; Sivaraman et al. 2020).

2.4 Mechanism Used in Leader Election

The blockchain supports mainly three types of leader election model; a single leader, fixed leaders, and collective leaders (Khamis et al. 2021). In the single-leader model, a Proof-of-Authority consensus mechanism is used, where only one node has the right

Table 17.1 Blockchain scalability methods with their approach to scale

S. No	Scalability methods	Approach	Main purpose
1	State channel	Vertical	Enhance scalability and minimize transaction cost
2	Payment channels	Horizontal	Improve scalability and reduce fees, reduce the volume of transactions that have to be processed on the blockchain
3	Sidechains	Horizontal	Helps to reduce traffic on main chain as a separate blockchain running in parallel to the main blockchain
4	Cross chain	Horizontal	Multiple blockchains can operate independently and communicate with each other
5	Rollups	Vertical	Transaction processing and computation occur off chain and only relevant information transmitted to the main chain
6	Sharding	Horizontal	Sharing resources, data, or processing between multiple users or nodes within blockchain network
7	DAG	Horizontal	Enabling multiple transactions to be confirmed simultaneously across different branches or vertices of the DAG
8	Dynamic block size adjustment	Vertical	Block size adjusted based on network traffic, number of transaction and consensus rule
9	Block data	Horizontal and Vertical	Horizontal: enables parallel processing of transactions and increases throughputs Vertical: focuses on optimizing the efficiency of individual blocks, optimize the overall performance and storage capacity of the blockchain

(continued)

Table 17.1 (continued)

S. No	Scalability methods	Approach	Main purpose
10	Data Propagation	Horizontal	Enhance the resilience, reliability, and efficiency of the blockchain network by leveraging the collective computing power and bandwidth of its participants
11	Hybrid solution	Horizontal and Vertical	Combine horizontal scalability by distributing tasks or data among various nodes in the network, with vertical scalability achieved through optimizations within each components or layers of the blockchain architecture

to create a block, but the network becomes centralized. Fixed leader architecture uses a PBFT consensus mechanism with a fixed number of nodes to validate the transactions (Cortes-Goicoechea et al. 2021). The collective leader model is implemented on a public blockchain, where all network nodes are equally responsible for creating blocks and storing copies of the ledger (Solaiman et al. 2021). Single-reader and fixed-reader models have higher throughput compared to collective-reader models. This is because the collective leader model has a large number of nodes to validate transactions (Mohammed et al. 2020).

2.5 Approach Used to Scale the Chain

Horizontal approach in blockchain is implemented by adding network nodes or end users. The main goal of vertical approach is to optimize throughput, latency, block generation rate, and chain size of network (Allombert et al. 2019). Implementing a vertical scaling refers to adding more powerful hardware or resources to the existing network nodes. On the other hand, horizontal scaling focuses on optimizing overall performance of the network (Mohammed et al. 2020). Table 17.1 presents various blockchain scaling techniques and their ways of scaling the blockchain network.

2.6 Number of Ledger in the Architecture

Blocks or databases are maintained in a single ledger or multiple ledgers depending on the application requirements (Chauhan and Patel 2021). The single ledger is a

Table 17.2 Brief classification of some blockchain platforms

S. No	Platform/solution/protocol/network/Model	Scalability methodology used/technique used	Consensus mechanism used	Main purpose/features	Result: throughput/latency	References
1	**Raiden**	Off-chain Payment channel	Burning mechanism	Micropayments, interoperability, reduce transaction cost	15–30 s latency	Sguanci et al. (2021)
2	**StarkWare**	Off-chain Rollups	Zk-rollups	Zero knowledge proof technology	1000 tps	McCorry et al. (2021)
3	**Perun**	Off-chain State channel	Proof of Concept	Hub-and-Spoke model architecture, Virtual payment hub	10,000 tps	Dziembowski et al. (2019)
4	**Plasma**	Off-chain sidechain	Proof of Authority	Dapps, Hierarchical structure	10,000 tps	Vispute et al. (2021)
5	**Polygon**	Off-chain sidechain	Modified Proof of Stake	Developer friendly, interoperability, PoS bridge, DeFi, gaming, Dapps	65,000 tps	Besançon et al. (2022)
6	**Polkadot**	Off-chain Cross-chain	Nominated Proof of Stake	Parallel chain architecture, support cross chain message passing (XCMP), heterogeneous environment	1 million tps	Burdges et al. (2020)
7	**Loopring**	Off-chain Rollups	Zk-rollups	Independent operation, autonomous cryptocurrency exchange, low costing	2000 tps	Lavaur et al. (2023)
8	**Tezos**	Off-chain payment channel	Liquid proof of stake and Delegated Proof of stake	Self-amendment, upgradability of smart contract, Michelson smart contract language	1 min to confirm a Block	Allombert et al. (2019)
9	**Celer**	Off-chain State channel	Interoperability Protocol	Support sidechains, micropayments, Low cost fast transactions	30,000 tps	Sivaraman et al. (2020)
10	**BrokerChain**	On-chain sharding	Cross shard protocol	Resolve hot-shard issue, reduce cross-shard transaction, account-balance model	352 tps	Huang et al. (2022)
11	**S-Store**	On-chain sharding	Merkle B + tree mechanism	Reduce data migration, decrease Merkle tree reconstruction overheads		Qi (2022)

(continued)

Table 17.2 (continued)

S. No	Platform/solution/protocol/network/Model	Scalability methodology used/technique used	Consensus mechanism used	Main purpose/features	Result: throughput/latency	References
12	**Ethereum 2.0**	On-chain sharding	Proof of Stake	Updated version of Ethereum, More user friendly, Support interoperability	1 million tps	Cortes-Goicoechea et al. (2021)
13	**Hashgraph**	On-chain DAG	Gossip protocol and Hedera algorithm	Virtual voting, grouping transaction data structure	10,000 + tps	Baird and Luykx (2020)
14	**TIPS**	On-chain DAG	Transaction inclusion protocol	Reduce transaction collision, defend denial-of-service attacks	–	Chen et al. (2022)
15	**Cardano**	On-chain Dynamic Blocksize adjustment	Proof of stake, Ouroboros consensus algorithm	Hierarchical structure, treasury system, research-driven approach, interoperability	1 million tps	Brünjes and Gabbay (2020)
16	**RINA**	On-chain Data propagation		Provide secure and configurable environment	–	Pandey et al. (2021)
17	**Harmony**	Layer 0 Hybrid model	Effective proof of stake	Combination of sharding and consensus mechanism	–	Joshi et al. (2022)

unique ledger containing all chain blocks where all the data are secured and maintained in the single ledger. In a multi-ledger architecture, data is divided into different ledgers. Each ledger has different privacy policies, rules, and data access permission. The entire database is distributed across multiple ledgers (Chauhan and Patel 2021). The single ledger architecture is implemented on the platforms like bitcoin, ethereum, litecoin, monero, and many more. The multi-ledger architecture shows greater scalability compared to the single ledger architecture. The multi-ledger architecture is implemented on the platforms like hyperledger fabric, corda, ripple, stellar etc.... (Pandey et al. 2021; Liu et al. 2021; Atlam et al. 2020).

3 Blockchain Scalability Solutions

3.1 Off-Chain (Layer-2) Solutions

Off-chain or Layer 2 is an approach that handles the transaction process outside of the main chain. Figure 17.3 shows several off-chain solutions categorized into state channels, payment channels, sidechains, cross-chains, and rollups. Each category has multiple solutions with different consensus mechanisms, methods, network types, etc.... (Khamis et al. 2021; Liu et al. 2021).

3.1.1 State Channels

The main purpose of state channel is to optimize interactions between participants. The Raiden Network focuses on Ethereum-based low-cost and fast micropayment transactions by building payment channel between participants (Zafar et al. 2022). The Celer Network is an advanced version of Raiden that supports sidechain elements

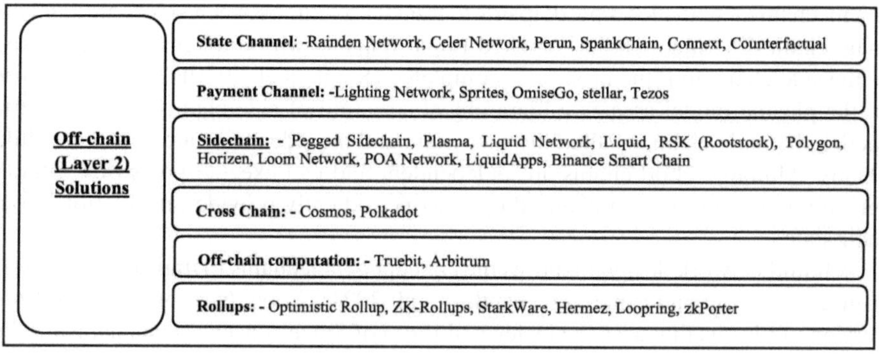

Fig. 17.3 Off-chain (layer-2) scalability solutions

and provides developers with a wide range of tools to build a variety of applications including; decentralized finance (DeFi), games, and many more. Perun ensures fast transaction processes by creating a virtual payment channel on the network (Dziembowski et al. 2019).

SpankChain is a special chain design aimed at payments in the adult industry. Developers can build their own payment channels using counterfactual model which includes rich functionality for developing state channel-based applications (Negka and Spathoulas 2021; Dziembowski et al. 2019).

3.1.2 Payment Channels

Payment channels are specialized state channels focused on optimizing the payments of participants (Solaiman et al. 2021). The lightning network provides low-cost, fast micro transactions based on Bitcoin and try to reduce the scalability of Bitcoin. A sprite is similar to lighting networks, but its main aim is to improve the scalability of Ethereum (Sivaraman et al. 2020). OmiseGo supports the concept of interoperability and provides the decentralized financial infrastructure that exchanging financial services as well. The stellar comes with its own consensus algorithm that facilitates the cross-border payment. Tezos uses liquid proof of stake consensus mechanism with smart contract execution and developers can also use this solution to build channels (Sivaraman et al. 2020; Papadis and Tassiulas 2020).

3.1.3 Sidechains

The sidechains are separate or independent chain that allows the participants to perform multiple transactions outside of the main chain. The key features of sidechains are interoperability, higher throughput, reduces traffic on main chain, security, and customization of smart contract (Aggarwal and Kumar 2021). The pegged sidechain is connected to the main chain via a two-way peg connection. This is a mechanism that allows participants to transfer tokens between sidechains and the main chain. The main chain is locked to participants whose tokens are transferred to the side chain until the process is completed (Singh et al. 2020). The architecture of the plasma model creates a hierarchical structure of side chains. In the plasma model, the main chain is called the root chain and the sidechains are called the child chains. Although child chains are independent and can execute processes without intervention from the root chain, they must periodically provide information about their state.

A liquid network design that uses the concept of sidechains to improve the performance of Bitcoin and Ethereum (Pandey et al. 2021). Loom Network is an updated version of Plasma that uses the Delegated Proof of Stake consensus mechanism. The main aim of this network is to provide the infrastructure and tools to the developers. Polygon was previously known as Matic and its main goal is to design a scalable version of Ethereum. It uses the same consensus mechanism as Ethereum and uses a combination of proof of work and proof of stake. Horizen is an updated version of ZenCash and a key feature of this chain is to improve privacy and anonymity for participants (Lashkari and Musilek 2021). Table 17.2 presents a classification of different solutions that attempt to solve the blockchain scalability problem and give their contributions. Different categories of solutions are presented with the consensus mechanisms used, their features and results detailing their throughput.

Each solution uses different architectures and mechanisms, resulting in different results in terms of throughput and latency (Kaur and Gandhi 2020).

3.1.4 Rollups

The main purpose of rollup is to reduce congestion on the main chain by executing transaction processes outside of the main chain. Once the transaction process is complete, the compressed data is passed to the main chain which validates the transaction before adding it to the chain. There are mainly two categories under rollups; Optimistic Rollups and ZK-Rollups (Zero-Knowledge Rollups) (Cortes-Goicoechea et al. 2021).

The optimistic rollup mechanism assumes that transactions are valid by default and notifies the main chain if a problem occurs during validation (Liu et al. 2020). While ZK rollup propagates the transaction's compressed cryptographic proof to the main chain (Liu et al. 2020). StarkWare gives the secured and efficient proof-computation using "zero-knowledge scalable transparent arguments of knowledge". Both Hermez and Loopring are Ethereum-based zero-knowledge models that use the ZK-Rollup rollup type but Loopring is only used for decentralized exchange purposes (Lavaur et al. 2023, 2022).

3.2 On-Chain (Layer-1) Solutions

The on-chain scalability solutions directly deal with the main chain and aim to address the scalability issue. Figure 17.4 shows some common categories of on-chain scalability solutions with different protocols, networks, or mechanisms (Sivaraman et al. 2020).

3.2.1 Sharding

Sharding technology has received a lot of attention as it provides a natural approach to scaling blockchain (Zafar et al. 2022). Many solutions have been introduced and implemented using this method and try to optimize the scalability of blockchain. The main purpose of this technique is to distribute the workload among parallel chains (Huang et al. 2022). Mainly Sharding technique has 6 types; transaction sharding,

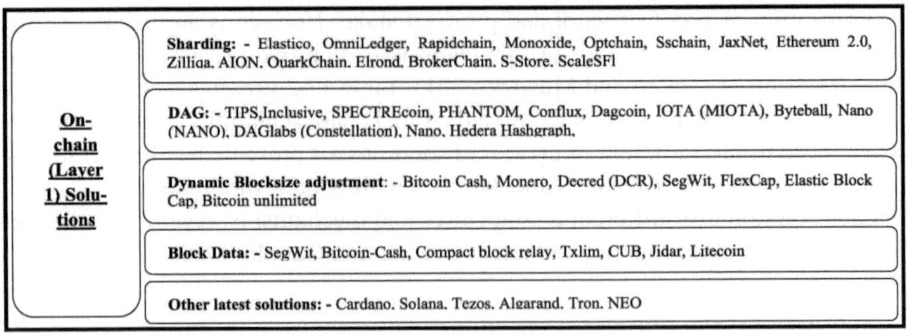

Fig. 17.4 On-chain (layer-1) solutions

network sharding, database sharding, state sharding and computational sharding, and hierarchical sharding.

Each type has different techniques and different architecture which shows different results (Sguanci et al. 2021). A network sharding model divides the entire network into multiple sub-networks and each sub-network performing a different task (Qi 2022). In the transactional sharding model, the number of transactions is divided into multiple shards and each shard contains a different transaction for validation. To reduce overload on individual ledgers, the entire ledger is split into multiple ledgers with database sharding. The main focus of computational sharding is to partition the tasks of a smart contract and assign different tasks to different shards (Cortes-Goicoechea et al. 2021). Hierarchical sharding using parent–child model technique in which each shard has a different number of sub-shards, and each sub-shard has its own sub-shards and so on (Hashim et al. 2022; Liu et al. 2021).

3.2.2 Dag

A DAG is also known as Tangle in which each transaction confirms the previous transactions and forming a directed acyclic graph (Kamble et al. 2021). This mechanism allows the parallel processing of transactions and provides great potential to scale the chain. However, this technology is still in its infancy stage and has not yet been proven fully scalable (Joshi et al. 2022).

3.2.3 Dynamic Blocksize Adjustment

Another approach of on-chain scalability is to make the blocksize dynamic based on demand to scale the chain. The network adjusts the blocksize with different volumes during the process of transactions (Wang et al. 2023).

3.3 *Layer 0 Solutions*

The main purpose of the Layer 0 solutions is to change/modify the architecture or fundamental changes in the protocol layer of the blockchain. Figure 17.5 shows some solutions of layer 0 which is divided into mainly two categories; data propagation-based solutions and hybrid solutions (Solaiman et al. 2021; Kamble et al. 2021).

3.3.1 Data Propagation

Once the transactions added in the block, the block is broadcast to all network nodes for verification purpose and valid block added in the blockchain. This process of data sharing is known as data propagation (Liu et al. 2021).

Erlay is a data propagation protocol try to minimize the bandwidth use requirement of Bitcoin network modes. RINA (Recursive Inter Network Architecture) is the latest protocol of data propagation based solutions which use principal of IPC (inter process communication) of networking and recurses the services of network (Wang et al. 2022).

3.3.2 Hybrid Solutions

Hybrid approach is combining the elements of on-chain and off-chain scalability with supporting the interoperability between different Blockchains. Polkadot uses hybrid approach that use the combination of proof of stake and nominated proof of stake useful to transfer the message and data in a trust free environment (McCorry et al. 2021). The main purpose of Near protocol is to provide the user friendly environment to the participants. Dragonchain, a platform combines the public and private elements focuses on flexibility and interoperability which use proof of time and proof of stake consensus mechanism (Solaiman et al. 2021).

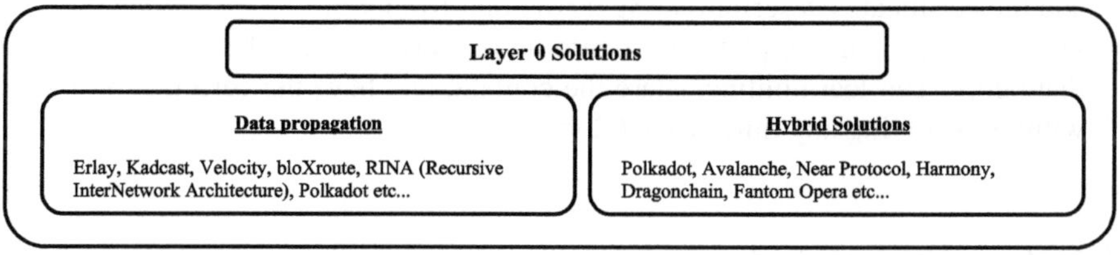

Fig. 17.5 Layer 0 solutions

4 Blockchain Applications

4.1 Healthcare Sector

- Patient data management: Patient data is very sensitive and protecting it should be at the forefront. Blockchain can be a trustworthy approach to the protection of this information.
- Medical credentials tracking: The decentralized and the cryptographic nature of blockchain allows the safe storage and sharing of patient medical records, providing additional security to such sensitive information being stolen.
- Management of drugs in the chain system: There are issues of lack of transparency, inability to track the drugs, loss of trust, and circulation of expired products in the drug industry. Blockchain technology has also become a solution that answers most of these questions.
- Healthcare claims processing and payment: The blockchain technology is able to revolutionize the claims payment process not only due to the non tamper proof, decentralized nature of such technology but also due to the consensus protocols and other features of the technology (Zafar et al. 2022; Negka and Spathoulas 2021).

4.2 Education Sector

- Students records keeping management system: Blockchain technology will enable the unchangeable records of the students like academic achievements, degree certificates, transcripts, and extracurricular activities. These documents are safely kept and easy to access thus enabling sharing to other universities and other similar institutions.
- Transparent certification process: Blockchain can make the certification process transparent by maintaining a note of every step taken in the certification process and the certificates description (Aggarwal and Kumar 2021).
- Decentralized learning platform: Learning platform can also be decentralized so that people can learn online, share the content, and have the certification issued without involving any third party (Gupta et al. 2020).

4.3 Government Sector

- Digital identity management: They could be in the form of the government using the blockchain platform that provides the processing of the secured and tamper-proof digital identities to citizens.

- Decentralized voting system: Voting system using the blockchain technology has the possibility to enhance transparency and integrity of the elections by providing secure and immutable records of an election.
- Land and property registry: Through the application of the blockchain technology, the registries of any piece of land and property can be brought out into the open and they become tamper free that minimizes the chance of fraud and possessive disagreements.
- Tax and revenue collection: The blockchain technology provides clear and auditable tracking of financial transactions and this would make the process of tax and revenue collection easier. This has the ability to augment government collection of revenue, and reduce tax evasion and enhance the collection of taxes (Zafar et al. 2022; Atlam et al. 2020).

5 Future Directions

The blockchain technology opens numerous possibilities for the different sectors with multiple use cases. In healthcare sector; Patient data is highly sensitive and ensuring its security is paramount. Blockchain technology offers a reliable solution for safeguarding this data (Solaiman et al. 2021). The decentralized and cryptographic characteristics of blockchain enable the secure storage and transmission of patient medical records, enhancing protection against unauthorized access to this sensitive data.

The pharmaceutical industry faces challenges including lack of transparency, tracking issues, trust deficits, and the circulation of expired products (Gupta et al. 2020). Blockchain technology has emerged as a solution which addressing many of these issues. Blockchain technology possesses the potential to revolutionize the payment and claims adjudication process not solely because of its decentralized and tamper-proof attributes but also owing to its consensus protocols and other unique characteristics.

In education sector; Blockchain technology facilitates the immutable records management of the students including academic achievements, degree certificates, transcripts, and extracurricular engagements. These records are securely stored and conveniently accessible, facilitating sharing with other universities and related entities. Blockchain can convert the certification process transparent by recording each stage of the certification journey and the certificates detail. The learning platform can be decentralized, facilitating online learning, content sharing, and certification issuance without the need of any third parties (Burdges et al. 2020).

In Government sector; Government can use blockchain platform that facilitates the secured and tamper-proof digital identities for citizens. The voting system based on the blockchain technology has the potential to improve the election transparency and integrity by offering safe and unchangeable records of vote. By utilizing the blockchain technology, land and property registries can be made transparent and tamper-free that reducing the possibility of fraud and ownership disputes. Blockchain technology offers transparent and auditable financial transaction record, which can simplify tax and revenue management procedures. This can increase government revenue collection, decrease tax evasion, and improve tax compliance (Upadhyay 2020).

Day by day the research community gives their contribution to optimize the challenges of blockchain technology. The number of research increases that combines the concept of machine learning and internet of things with blockchain technology (Krishnan et al. 2020). ML is a subset of artificial intelligence that provides numerous algorithms and models that help to make analysis, prediction, and accurate decisions based on historical data. ML features include automation, pattern recognition, healthcare diagnostics, sentiment analysis, reduce of human error, innovative applications, and many more. IOT provides the connectivity, interoperability, remote monitoring, real-time processing, and cloud integration etc.... The advantages of blockchain are decentralization, security, immutability, data integrity, and transparency (Papadis and Tassiulas 2020).

The combination of BT, IOT, and ML enhanced the security, efficiency, and transparency. IoT is helpful to collect real-time data from interconnected devices. BT secures this data transparently and with integrity. ML processes this accurate data for valuable analysis that will be helpful for accurate decision-making (Baird and Luykx 2020). The combination provides the smart contract automation, immutable data records, decentralized trust, security enhancement, predictive analytics, immutable prediction records, supply chain transparency etc.

6 Conclusion

This paper describes the fundamentals of blockchain, scalability challenges due to various influencing factors, classification of scalability solutions, and future directions. Many scalability solutions have been introduced, but not all of them are sufficient to fully scale blockchain. The research community needs to continue their contribution to solving the challenges of blockchain technology. The machine learning and IoT are also helpful to solve the challenges of blockchain but it requires further findings and contributions. Extensive research contributions are still needed to explore the full potential of blockchain and its related applications.

References

Aggarwal S, Kumar N (2021) Core components of blockchain. In: Advances in computers (Vol. 121). Elsevier, pp 193–209

Allombert V, Bourgoin M, Tesson J (2019) Introduction to the tezos blockchain. In: 2019 International Conference on High Performance Computing & Simulation (HPCS), July. IEEE, pp 1–10

Atlam HF, Azad MA, Alzahrani AG, Wills G (2020) A review of blockchain in internet of things and AI. Big Data Cogn Comput 4(4):28

Baird L, Luykx A (2020) The hashgraph protocol: efficient asynchronous BFT for high-throughput distributed ledgers. In: 2020 International Conference on Omni-layer Intelligent Systems (COINS), August. IEEE, pp 1–7

Besançon L, Da Silva CF, Ghodous P, Gelas JP (2022) A blockchain ontology for DApps development. IEEE Acc 10:49905–49933

Brünjes L, Gabbay MJ (2020) UTxO-vs account-based smart contract blockchain programming paradigms. In: Leveraging Applications of Formal Methods, Verification and Validation: Applications: 9th International Symposium on Leveraging Applications of Formal Methods, ISoLA 2020, Rhodes, Greece, October 20–30, 2020, Proceedings, Part III 9. Springer International Publishing, pp 73–88

Burdges J, Cevallos A, Czaban P, Habermeier R, Hosseini S, Lama F, Wood G (2020) Overview of polkadot and its design considerations. arXiv preprint arXiv:2005.13456

Chauhan BK, Patel DB (2022) A systematic review of blockchain technology to find current scalability issues and solutions. In: Proceedings of Second Doctoral Symposium on Computational Intelligence: DoSCI 2021. Springer Singapore, 2022

Chen C, Chen X, Fang Z (2022) TIPS: transaction inclusion protocol with signaling in DAG-based b. IEEE J Sel Areas Commun 40(12):3685–3701

Cortes-Goicoechea M, Franceschini L, Bautista-Gomez L (2021) Resource analysis of Ethereum 2.0 clients. In: 2021 3rd Conference on Blockchain Research & Applications for Innovative Networks and Services (BRAINS), September. IEEE, pp 1–8

Dziembowski S, Eckey L, Faust S, Malinowski D (2019) Perun: virtual payment hubs over cryptocurrencies. In: 2019 IEEE Symposium on Security and Privacy (SP), May. IEEE, pp 106–123

Gupta S, Sinha S, Bhushan B (2020) Emergence of blockchain technology: fundamentals, working and its various implementations. In: Proceedings of the international conference on innovative computing & communications (ICICC), April

Hashim F, Shuaib K, Zaki N (2022) Sharding for scalable blockchain networks. SN Comput Sci 4(1):2

Huang H, Peng X, Zhan J, Zhang S, Lin Y, Zheng Z, Guo S (2022) Brokerchain: a cross-shard blockchain protocol for account/balance-based state sharding. In: IEEE INFOCOM 2022-IEEE Conference on Computer Communications, May, pp 1968–1977. IEEE

Janssen M, Weerakkody V, Ismagilova E, Sivarajah U, Irani Z (2020) A framework for analysing blockchain technology adoption: Integrating institutional, market and technical factors. Int J Inf Manage 50:302–309

Joshi S, Pise AA, Shrivastava M, Revathy C, Kumar H, Alsetoohy O, Akwafo R (2022) Adoption of blockchain technology for privacy and security in the context of industry 4.0. In: Wireless communications and mobile computing

Kamble SS, Gunasekaran A, Kumar V, Belhadi A, Foropon C (2021) A machine learning based approach for predicting blockchain adoption in supply Chain. Technol Forecast Soc Chang 163:120465

Kaur G, Gandhi C (2020) Scalability in blockchain: challenges and solutions. In: Handbook of research on blockchain technology. Academic Press, pp 373–406

Khamis J, Rottenstreich O (2021) Demand-aware channel topologies for off-chain payments. In: 2021 International Conference on COMmunication Systems & NETworkS (COMSNETS), January. IEEE, pp 272–280

Krishnan S, Balas VE, Julie EG, Yesudhas HR, Balaji S, Kumar R (eds) (2020) Handbook of research on blockchain technology. Academic Press

Krishnan S, Balas VE, Julie EG, Yesudhas HR, Kumar R (eds) (2021) Blockchain for smart cities. Elsevier

Krishnan S, Balas VE, Kumar R (eds) (2023) Green blockchain technology for sustainable smart cities. Elsevier

Krishnan S, Ganesan LP (2025) Integrated dairy production and cattle healthcare management using blockchain NFTs and smart contracts. Systems 13:65
Kullig N, Lämmel P, Tcholtchev N (2020) Prototype implementation and evaluation of a blockchain component on IoT devices. Proc Comput Sci 175:379–386
Lashkari B, Musilek P (2021) A comprehensive review of blockchain consensus mechanisms. IEEE Acc 9:43620–43652
Lavaur T, Lacan J, Chanel CP (2022) Enabling blockchain services for IoE with Zk-rollups. Sensors 22(17):6493
Lavaur T, Detchart J, Lacan J, Chanel CP (2023) Modular zk-rollup on-demand. J Netw Comput Appl:103678
Liu Y, Yu FR, Li X, Ji H, Leung VC (2020) Blockchain and machine learning for communications and networking systems. IEEE Commun Surv Tutor 22(2):1392–1431
Liu C, Bodorik P, Jutla D (2021) A tool for moving blockchain computations off-chain. In: Proceedings of the 3rd ACM International Symposium on Blockchain and Secure Critical Infrastructure, May, pp 103–109
McCorry P, Buckland C, Yee B, Song D (2021) Sok: validating bridges as a scaling solution for blockchains. Cryptology ePrint Archive
Mohammed A, Potdar V, Yang L (2020) Key factors affecting blockchain adoption in organizations. In: Big Data and Security: First International Conference, ICBDS 2019, Nanjing, China, December 20–22, 2019, Revised Selected Papers 1. Springer, Singapore, pp 455–467
Negka LD, Spathoulas GP (2021) Blockchain state channels: a state of the art. IEEE Acc 9:160277–160298
Pandey AA, Fernandez TF, Bansal R, Tyagi AK (2021) Maintaining scalability in blockchain. In: International Conference on Intelligent Systems Design and Applications, December. Springer International Publishing, Cham, pp 34–45
Papadis N, Tassiulas L (2020) Blockchain-based payment channel networks: challenges and recent advances. IEEE Acc 8:227596–227609
Qi X (2022) S-store: a scalable data store towards permissioned blockchain sharding. In: IEEE INFOCOM 2022-IEEE Conference on Computer Communications, May. IEEE, pp 1978–1987
Sguanci C, Spatafora R, Vergani AM (2021) Layer 2 blockchain scaling: a survey. arXiv preprint arXiv:2107.10881
Singh A, Click K, Parizi RM, Zhang Q, Dehghantanha A, Choo KKR (2020) Sidechain technologies in blockchain networks: an examination and state-of-the-art review. J Netw Comput Appl 149:102471
Sivaraman V, Venkatakrishnan SB, Ruan K, Negi P, Yang L, Mittal R, Alizadeh M (2020) High throughput cryptocurrency routing in payment channel networks. In: 17th USENIX Symposium on Networked Systems Design and Implementation (NSDI 20), pp 777–796
Solaiman E, Wike T, Sfyrakis I (2021) Implementation and evaluation of smart contracts using a hybrid on-and off-blockchain architecture. Concurr Comput Pract Exper 33(1):e5811
Upadhyay N (2020) Demystifying blockchain: a critical analysis of challenges, applications and opportunities. Int J Inf Manage 54:102120
Vispute A, Patel S, Patil Y, Wagh S, Shirole M (2021) Scaling blockchain by autonomous sidechains. In: Proceeding of Fifth International Conference on Microelectronics, Computing and Communication Systems: MCCS 2020. Springer, Singapore, pp 459–473
Wang X, Jiang X, Liu Y, Wang J, Sun Y (2022) Data propagation for low latency blockchain systems. IEEE J Sel Areas Commun 40(12):3631–3644
Wang Q, Yu J, Chen S, Xiang Y (2023) Sok: Dag-based blockchain systems. ACM Comput Surv 55(12):1–38
Zafar S, Bhatti KM, Shabbir M, Hashmat F, Akbar AH (2022) Integration of blockchain and Internet of Things: challenges and solutions. Ann Telecommun:1–20

Harnessing Blockchain: Challenges and Innovations in the Digital Era

G. Arun Sampaul Thomas, S. Muthukaruppasamy, K. Saravanan, T. S. Karthik, and Seifedine Kadry

Abstract People and businesses alike are always on the lookout for new and improved methods to safeguard information and streamline processes, considering the rapid development of digital advances. Many people believe that blockchain technology will be the key to this promising future. This technology underpins Bitcoin and other cryptocurrencies, yet its uses extend far beyond mere digital currency. The financial, supply chain, healthcare, and government sectors have already experienced its disruptive influence. To fully utilize blockchain technology, it is essential to address both its challenges and its advancements. Blockchain's decentralized structure creates a more democratic data system by eliminating the need for a single data authority. Because the blockchain is secure and everyone can see the data, it's impossible to change or remove any information, which guarantees transparency. Because of this, blockchain technology has promise for many uses. This article, "Blockchain Beyond Cryptocurrency: Decentralized Industries and Transforming Systems," discusses how the technology eliminates middlemen, like banks, to speed up and improve the security of financial transactions. Improved supply chain visibility reduces the prevalence of counterfeit goods and increases product authenticity. A safe and efficient exchange of patient data leads to improved health outcomes. While blockchain technology exhibits potential, it also faces significant challenges, as evidenced in The Future of Blockchain and Evaluating the Potential and Problems articles. Problems with scalability, energy consumption, legal restrictions, and technological complexity are preventing broad use. High-volume transactions, for

G. Arun Sampaul Thomas (✉)
Department of AI, Anurag University, Hyderabad, Telangana, India
e-mail: arunsam.infotech@gmail.com

S. Muthukaruppasamy
Department of EEE, Velammal Institute of Technology, Tiruvallur, Tamil Nadu, India

K. Saravanan
DCSE, College of Engineering, Guindy, Anna University, Chennai, Tamil Nadu, India

T. S. Karthik
Department of ECE, SRMIST, Chennai, Tamil Nadu, India

S. Kadry
Department of Applied Data Science, Noroff University College, Kristiansand, Norway

S. Krishnan et al. (eds.), *Unveiling Blockchain Enabled Business Applications*, Contributions to Management Science, https://doi.org/10.1007/978-981-95-2337-5_18

example, are not well-suited to the Bitcoin network because of its limited transaction processing power. We must overcome these challenges, first in the financial services industry and then elsewhere, to fully reap the benefits of blockchain. Businesses must be flexible and responsive to customer demands, according to an expert review, which is important for reducing blockchain implementation risks. We are still years away from seeing blockchain in action, despite the far-reaching implications technology will have for the private and public sectors. Getting blockchain technology into the hands of as many people as possible is a top priority right now, so there must be constant innovation in this area.

1 Introduction

Before learning about blockchain, you must understand databases and how they work. Oracle defines a database as a computer-stored collection of structured data. A database organizes information or data. Blockchain requires simply knowing that databases store data in a tabular format and are electronic. In recent years, blockchain has emerged as a breakthrough technology that could transform several sectors. Blockchain adoption faces many challenges before the financial services industry can benefit. Businesses should monitor and adapt to market expectations. Blockchain technology will have a major impact on the public and private sectors, but that day is still far off. The list of challenges for blockchain adoption indicates the need for technical advancements. Blockchain technology currently aims to be available to everyone. This article will serve as a comprehensive guide to learning the fundamentals of blockchain technology, regardless of whether the blockchain concerns cryptocurrencies such as Bitcoin or if you are simply curious about its essential concepts (Blockchain Revolution and Its Basic 2023).

The blockchain is a highly precise database that stores information in a way that is not at all like any other database. A blockchain stores data differently from a conventional database. We compile the information into entities known as blocks. We predetermine the storage capacity of each block, and upon completion of its use, we chain it with the previous block. The term "blockchain" refers to the virtual blockchain generated by this action (7 Major Challenges for the Blockchain Adoption in 2023 2023).

A blockchain, a decentralized and disseminated digital ledger, disperses transactions across several network-based computers, also known as nodes. This expertise enables transparent, secure, and unchangeable peer-to-peer transactions, eliminating the need for intermediaries like banks or central authorities. Every blockchain-based transaction is saved in a "block," associated with the blocks that previously came, establishing a chronological hierarchy of data. Blockchain technology, one of the most innovative digital innovations, can disrupt many businesses. A blockchain primarily records transactions. In this chapter, we investigate the core concepts and origins of blockchain technology, its significance in attaining decentralization, transparency, and security, and its initial use case with cryptocurrencies (Tapscott et al. 2016).

As a result of the proliferation of digital developments, individuals and businesses are consistently searching for innovative approaches to safeguard their data and simplify their operations. People have hailed a technology known as blockchain as the key to unlocking this extremely bright future. While blockchain currently serves as the foundation for Bitcoin and other cryptocurrencies, its potential applications extend beyond digital currency. Several different industries, including supply-chain management, healthcare, banking, and government sector-based transactions, have already observed the disruptive implications of this phenomenon (Evaluating the Potential and Challenges in Blockchain 2023).

Decentralization in blockchain technology, in contrast to more conventionally centralized systems, functions on a decentralized network, meaning no specific being has rule over it. This approach achieves the elimination of single points of failure and promotes trust among participants. All network participants copy and distribute the blockchain ledger, ensuring universal access to the same information. We refer to this as the distributed ledger architecture. This transparency also fosters trust and responsibility.

Mechanisms for Reaching Consensus Utilizing a consensus mechanism allows for the validation of new blocks and adding them to the blockchain. Delegated Proof of Stake (DPoS), Proof of Work (PoW), and Proof of Stake (PoS) are all examples of well-known methods. Without the need for a centralized authority, these techniques guarantee that all network members agree with one another. In blockchain technology, cryptography relies on cryptographic techniques to ensure transaction safety and data consistency. Public-key cryptography gives participants the ability to protect their digital assets and have identities that are completely unique to them.

Public blockchains, like Bitcoin and Ethereum, are open to anyone, thus allowing anyone to contribute, execute, and authorize blocks. They offer transparency and censorship resistance, but they face scalability challenges. Private blockchains are restricted to a specific group or organization. They provide higher scalability, privacy, and control over the network but sacrifice decentralization to some extent. A group of organizations, rather than a single entity, governs consortium blockchains. They balance decentralization and control, making them suitable for collaborative industry applications.

However, there will always be a period of excitement associated with any new technology. It is a time-consuming process to eliminate all the obstacles and then put that energy to work to power the contemporary world. In this case, the blockchain is no different. Even though there are a significant number of opportunities, it will still take some time to overcome all the obstacles successfully. To assist those who are new to blockchain technology in gaining a better understanding of the technology, this guide will discuss the fundamentals of blockchain and its challenges, along with impending solutions to those problems (Top 10 Blockchain Adoption Challenges 2021)

2 Major Blockchain Adoption Challenges

Blockchain is defined by the following characteristics of challenges,

a. **Challenging to grasp and implement**

The benefits of blockchain technology are difficult for the average person to grasp because of how complicated it is. One must familiarize themselves with the concepts of distributed ledgers and encryption before putting this ground-breaking technology to use. The adoption of blockchain technology is hindered because financial institutions can provide secure payment gateways and other services at more affordable prices than blockchain technology.

b. **Missing Knowledge and Abilities**

To manage blockchain technology, we need competent people to join our team, in addition to software and hardware. To the best of our knowledge, blockchain technology is far from finished. Nowadays, only a few people can operate such complex machinery. Conversely, this type of trained staff is in high demand. Accordingly, we must offer greater wages if we wish to recruit competent people. Consequently, we will have to spend a little more money to hire suitable people.

c. **Privacy and Trust Issues**

Privacy and blockchain aren't usually compatible. Full anonymity is an issue because the public ledger powers the system. And yet, is privacy essential for the success of any company. Boundaries need to be clearly stated for many businesses that handle personal information. People put their personal information in their hands. Consequently, their privacy would be compromised if recorded in a public ledger. Thus, changing the registers on the public ledger is necessary to limit data access. Limiting access to it to paying clients would solve this problem.

d. **The Implementation's Cost Is Higher**

A common motivation for using blockchain technology is the desire to eliminate middlemen and save money when transferring funds. Despite its obvious advantages, blockchain technology is still in its early stages, making it challenging to integrate with current systems. In addition, the increased cost prevents public and private entities from embracing it.

e. **Cooperation Between Systems**

Interoperability is a major issue that requires fixing because it is a barrier to organizations and industries using this technology. Most blockchains function autonomously and do not engage with peer networks since they cannot transmit or receive data from other blockchain-based systems. This problem has prompted numerous initiatives. For instance, Ark employs the Smart Bridges architecture, which connects disparate networks. Worldwide interoperability, transmission, and transfer are all supposedly provided by this effort.

f. **Inadequate Oversight**

The absence of legislation is one of the main obstacles to implementing blockchain technology in a company. Numerous organizations are using blockchain technology as a tool for transactions. This is going to be crucial for a lot of products, as we will see. However, currently, there are no hard-and-fast regulations regarding it. Consequently, no one follows any rules when it comes to blockchain technology. But this is when things get tricky. While blockchain's transparency is a plus, the technology's security features are still lacking. Whether or not it poses any danger to us is something we cannot say with certainty. To tackle these issues, governments and other businesses must adopt rules for blockchain technology (Ivanova et al. 2023)

g. **Greater Use of Energy**

Before adding a transaction to the network, the technology uses a proof-of-work technique to confirm and build confidence. Processing, validating, and, most importantly, securing the entire network through the solution of difficult mathematical riddles requires a lot of computing power from this system. Ethereum's co-founder has suggested an alternative to proof-of-work called proof-of-stake as a solution to this issue. With this method, players won't have to use as much energy to solve difficult tasks.

2.1 Top 10 Blockchain Major Adoption Challenges

Blockchain technology continues to be the most popular buzzword in the world of technology. According to the protagonists, it will bring about a revolution in the financial industry, which includes real estate, healthcare, and the legal system. Figure 18.1 illustrates the primary source of all these hypes, which is the capability to function in a remote environment while being housed in a tamper-proof facility (Top 10 Blockchain Adoption Challenges 2021; Filippi and Santos 2022)

1. **Inefficient Technological Design**

Implementing blockchain poses a significant hurdle. While blockchain technology offers numerous advantages, it has deficiencies in certain technological aspects. A code defect or loophole is a crucial aspect in this context. Bitcoin pioneered this aspect, but the entire system exhibits a noticeable lack of efficiency in its architecture. Indeed, Ethereum attempted to address the deficiencies of Bitcoin, although it remains insufficient. Consider decentralized application development as an illustration. Ethereum enables developers to deploy decentralized applications (dApps) using its platform. Until now, numerous decentralized applications (dApps) have been developed using these platforms. However, most of them appear to suffer from inaccurate programming and vulnerabilities. Users can exploit these vulnerabilities and infiltrate the system expediently. Therefore, the discussion of security is not effectively functioning in this context. The situation will undoubtedly become more convenient by resolving this obstacle to blockchain adoption.

Fig. 18.1 Top 10 major blockchain challenges. *Source* https://101blockchains.com/blockchain-adoption-challenges/

2. **The Criminal Connection**

The anonymity aspect of blockchain technology has attracted professionals and individuals involved in illegal activities. What is the reason? The network operates in a decentralized manner, ensuring that our identity remains unknown to anyone. As a result, bitcoin has become the predominant choice for currency within illicit markets and the hidden corners of the internet known as the dark web. It is detrimental to cultivating a positive reputation. The negative reputation of this name is causing significant hesitation among individuals considering an in-depth examination of the entire system. Furthermore, it is inherent for individuals to desire to distance themselves from any involvement in illicit activities. Presently, criminals employ these coins to procure restricted, illicit equipment and modes of payment. In addition, they demand cryptocurrency as payment in exchange for the ransom. The sole means to address this issue is to completely terminate the criminal association and ensure more effective deployment of blockchain technology.

3. **Low Scalability**

Scalability is another obstacle that arises when utilizing blockchain technology. Blockchains are efficient for a limited number of users. However, what occurs when a large-scale integration occurs? Ethereum and Bitcoin currently boast the largest user base on their respective networks, and they are facing significant challenges in managing this influx. As the number of users on the network increases, the processing time for transitions becomes longer. Consequently, the transactions incur higher costs than usual, further limiting the network's number of users. The entire transaction can take several days to process. Ultimately, this problem of using blockchain technology is diminishing its profitability. While some blockchain systems have demonstrated quicker performance, they tend to experience reduced speed as the number of users accessing the system increases. Therefore, it is imperative to promptly address this dilemma since it is causing the entire technological sector to become lackluster.

4. **High Energy Consumption**

Energy consumption poses a significant obstacle to the widespread use of blockchain technology. The majority of blockchain technologies adhere to the infrastructure of Bitcoin and employ Proof of Work as their consensus method. Nevertheless, Proof of Work is not as commendable as it appears. Maintaining the system's functionality will require sufficient processing capacity. Currently, miners are utilizing 0.2% of the overall electricity. If the rate of increase continues, miners will consume more power than the globe can sustain. Consequently, it has now emerged as one of the main obstacles to this network. Several organizations are actively attempting to circumvent the use of blockchain due to this difficulty. Hence, it is imperative to maintain control over the situation, as it represents a significant obstacle to the adoption of blockchain technology. Blockchain can employ alternative consensus mechanisms to authenticate transactions. Energy-efficient consensus algorithms. Adopting this approach is the sole means by which we can truly restore the positive impact of blockchain technology. (Arun Sampaul Thomas and Harold Robinson 2020a)

5. **Lack of Privacy**

Blockchain and privacy are incompatible. The public ledger system drives the system; hence, complete privacy is not the primary priority. However, is it possible for any organization to operate effectively without privacy. It is essential for companies dealing with privacy to establish clear boundaries. Their clientele entrusts them with confidential data. Thus, if all the data is stored in a publicly accessible ledger, its privacy would be compromised. Hence, it is imperative to modify the registers to restrict access to the data. Restricting access to only customers will be a viable solution in this case. For bitcoin and other cryptocurrencies, this is a crucial necessity. However, this gives rise to apprehensions among governments and enterprises. Both governments and enterprises need to safeguard and control access to their data for diverse purposes. Consequently, blockchain technology is currently limited in handling sensitive data until a viable solution is found. Either a private or federated blockchain would be suitable for this scenario. We would obtain restricted access, ensuring the confidentiality of all our sensitive data, as intended.

6. **No Regulation**

Implementing blockchain in an organization poses a significant difficulty. Several organizations are adopting blockchain technology as a method of conducting transactions. We shall encounter numerous products that are reliant on this. However, there is a lack of explicit regulations pertaining to this matter. Therefore, there are no prescribed regulations that individuals adhere to in relation to the blockchain. However, this is the point at which the problem arises. While blockchain ensures transparency, one of its advantages, it does not provide any security. We cannot ascertain whether it will be safe for us or not. To overcome these obstacles, governments and heavily regulated industries must establish regulatory frameworks for blockchain technology.

7. **Security Problems**

Security is an essential subject of great importance in this context. It is widely acknowledged that every blockchain technology emphasizes its security features. However, like every other technological advancement, blockchain also possesses certain vulnerabilities in terms of security. The occurrence of a 51% assault on the network is a significant vulnerability in terms of network security. During this cyber assault, hackers might get control of the network and manipulate it to their advantage. They could modify the transaction process and limit others' ability to generate a block. To address this issue, enhancing the security measures at the protocol layer is necessary. We have already observed certain vulnerabilities in its security at this point. Nevertheless, just a few instances provide effective protocols capable of managing this situation. The long-term safety of these products is uncertain.

8. **Lack of Adequate Skill Sets**

Aside from software and hardware, it is imperative to locate competent staff to oversee the management of blockchain technology. Blockchain technology is a recently developed and continuously advancing field. Currently, only a small number of individuals possess the necessary expertise to aid with such advanced technologies. Conversely, there is a significant need for this highly skilled workforce. If we desire to recruit highly skilled individuals, we must be prepared to provide substantial compensation. Hiring the appropriate individuals will incur expenses. Like other technological advancements, the blockchain will persistently undergo further development. Indeed, there are certain difficulties, but they do not pose as hindrances. It is necessary to implement new norms and standards. Soon enough, we are contemplating the adoption of blockchain technology for our organization. Therefore, it is important to be cautious of the obstacles that arise when implementing blockchain technology.

9. **Blockchains Can Be Slow**

The blockchain is intricate. Hence, the processing time for any transaction is prolonged. Additionally, the encryption of the system further hampers its speed. Despite their assertion of being more expedient than conventional payment systems, there are instances where they are unable to fulfil this promise. The process of finalizing a transaction may require several hours to complete. Therefore, purchasing a cup of coffee will result in inconvenience. It is best suited for substantial transactions if timing is not critical. There is a certain degree of danger involved. Furthermore, was it not intended to eliminate the inherent vulnerability of blockchains.

In theory, this idea can be applied to blockchain networks that utilize functions other than storing value, such as recording transactions or interactions within the Internet of Things (IoT) ecosystem. These routes, and even computer files, might experience sluggishness and inefficiency. It is not always the case. The network experiences deceleration solely when it is burdened with an excessive number of users. As it increases in size, its speed decreases. The problems associated with the implementation of blockchain technology may soon become a significant obstacle. Therefore, it is imperative that we anticipate a prompt resolution. Nevertheless, certain emerging technologies provide superior velocity in comparison to conventional ones. The establishment has a vested stake in the failure of the blockchain. Frankly speaking, it has the potential to vanish more rapidly than we expect. Therefore, it is imperative to promptly identify a resolution for these obstacles hindering the widespread acceptance and use of blockchain technology (Arun Sampaul Thomas et al. 2023; Arun Sampaul Thomas and Harold Robinson 2020b).

Banks generate substantial profits by utilizing intermediaries. Consequently, the expenses are significantly lower than the typical amount. However, if blockchain can surpass this aspect, it will be significantly more cost-effective than the current alternative. Organizations must take these issues into account while using blockchain technology. Banks exert significant lobbying influence over governments and legislators. If the traditional banking sector deems it advantageous, they have the potential to diminish the influence of blockchain technology. Additionally, they have the capacity to greatly diminish its practicality and restrict its accessibility.

10. **Public Perception**

One major obstacle to adopting blockchain technology is the absence of a comprehensive knowledge base. Most of the general population remains uninformed about this technology's existence and potential use. For blockchain to succeed, it must garner widespread support. Despite the ground-breaking nature of the technology, it remains insufficient to captivate a larger user base. Furthermore, the absence of effective promotion for this market segment contributes to its lack of popularity. If we are not engaged in this matter, we would be completely unaware of its existence. Presently, blockchain technology is nearly synonymous with Bitcoin. The prevailing perception among individuals is that bitcoin is the exclusive blockchain network. Others are unaware of it, except for the cryptocurrency.

Simultaneously, the value of Bitcoin persists in ascending to unparalleled magnitudes. Cryptocurrency continues to be linked to illicit activities such as money laundering, black market trading, and other unlawful operations. Before achieving widespread acceptance, the public must comprehend the distinctions among bitcoins, alternative cryptocurrencies, and blockchain technology. This will aid in mitigating the adverse connotations associated with cryptocurrency and allow the technology to showcase its brilliance. The outcome will be a heightened inclination to utilize the technology. Although these applications demonstrate the vast capabilities of blockchain technology, obstacles such as scalability, energy consumption, regulatory frameworks, and interoperability continue to exist. The progression of blockchain technology entails overcoming these obstacles to facilitate smooth incorporation and extensive acceptance in various sectors.

2.2 Addressing Challenges and Advancements in Blockchain Technology

Several barriers stand in the way of the broad implementation of blockchain technology; nonetheless, these issues have prompted ongoing innovation and developments aimed at overcoming restrictions and improving the platform's capabilities (Blockchain Beyond Cryptocurrency 2023).

For blockchain networks, scalability continues to be a big, challenging obstacle. Existing systems are experiencing limited processing speed and capacity because of the growing volume of transactions. Implementing solutions like sharding, layer-two protocols, and blockchain interoperability initiatives is an attempt to overcome these scalability difficulties (Tapscott and Tapscott 2022; Filippi and Santos 2022).

Critics have questioned the environmental sustainability of certain consensus techniques, particularly Proof of Work (PoW) in cryptocurrencies like Bitcoin, due to their high energy consumption. The goal of emerging compromise models such as Proof of Authority (PoA) and Proof of Stake (PoS) is to lower the amount of energy consumed while simultaneously maintaining the network's security. Regarding the widespread success of blockchain technology, the lack of standardized regulatory frameworks presents some challenges. The process of incorporating blockchain technology while maintaining compliance with pre-existing regulations is so challenging that governments and regulatory organizations are now working to navigate it. To encourage general acceptance, it is essential to have legal frameworks that are both clear and consistent.

Compatibility between various blockchain networks and traditional systems remains a significant challenge. We are continuously working to develop protocols and standards that will enable seamless data interchange and communication between various blockchain platforms and traditional databases. Despite the presented hurdles, the trajectory of blockchain technology continues to evolve. Technological advancements are removing the current restrictions and opening the door for a blockchain ecosystem that is more scalable, sustainable, and interconnected (Akter et al. 2023; Arun Sampaul Thomas and Muthukaruppasamy 2023)

3 Future of Blockchain and Evaluating the Potential and Challenges

Blockchain must overcome a variety of adoption-related obstacles before it can eventually become a part of society. Scalability is one of these, along with the time required to validate transactions, the cost, and security. Recently, hacking incidents in Bitcoin-based businesses and start-ups have garnered significant attention. However, we are unaware that most of these situations are the result of malicious users. An example of this would be a software wallet, which can be considered the Bitcoin equivalent of a virtual bank account. Installing this wallet on our personal computer is required to access and save all our credits. On the other hand, the fact that there is no encryption is a flaw in this system. Everyone can access our personal computers and steal our credits.

The threat of a 51% attack is still another possibility. If more than 51% of users have the power, people can make any changes they want to the network. However, this is an event that is extremely unlikely to occur and will not be of great use to the attackers. Furthermore, we still lack a complete understanding of the regulations governing settlement processes involving bitcoin protocols. These hurdles to the deployment of blockchain technology also represent a threat (Tapscott and Tapscott 2021).

Regulatory authorities in certain nations are taking more steps to address this issue, which is good. Financial organizations can use blockchain technology to build and manage internal networks without restrictions. This will affect third-party blockchain applications like payments and transfers. Both the US and UK central agencies want a reasonable approach. The latest news from major financial institutions supports this pragmatic approach. Blockchain is the only concern, although many counter-platforms exist. Hyperledger, for instance, offers many enterprise-focused projects. Ethereum is also developing permissioned platforms, which might solve most blockchain difficulties for enterprises. Blockchain technology has enormous potential but challenges scalability, energy consumption, regulation, and interoperability. However, ongoing research and development is addressing these issues. Sharding, sidechains, and layer-2 scaling solutions are helping blockchain transcend its current limitations, indicating its bright future. Permissioned blockchain could benefit many enterprises. Many applications and projects are running smoothly. Thus, every blockchain issue will have a solution and an opportunity (Chaum 2021).

Many companies are keen to embrace this technology and are investing heavily in it. Therefore, society and the economy will soon undergo significant changes.

Blockchain can disrupt our present financing structure.

- Transform interbank payments.
- Ensure peer-to-peer payments.
- Offer opportunities to replace current exchange mechanisms.
- Preserve financial data, client records, and agreements.

Blockchains are distributed digital ledgers that securely record transactions over a network of computers. Blockchains are also known as blockchains. Its most compelling selling points are decentralization, openness, and safety principles. Through the process of decentralization, it is possible to create a data system that is more democratic in the absence of a centralized data authority. Within a blockchain context, the term "transparency" refers to the fact that all participants can freely view all the data. Finally, the data recorded on a blockchain is extremely difficult to change or remove, providing an additional layer of defensiveness (Tapscott and Tapscott 2021).

Due to these characteristics, blockchain technology is interesting for various applications. Blockchain technology has the potential to simplify and safeguard financial transactions, thereby amputating the prerequisite for intercessors such as banks, share markets, and so on. There is a possibility that blockchain technology could improve visibility in supply chains, which will, in turn, contribute to the authenticity of products and reduce the prevalence of counterfeit items. If patient information is shared securely, it improves the delivery of healthcare and the results.

Even though it has a lot of potential, blockchain technology still facing still faces many challenges. Scalability, energy consumption, regulatory constraints, and technological complexity are major obstacles that make widespread adoption difficult to achieve. These days, blockchain networks are experiencing significant difficulties in terms of scalability. Since it can only process a limited number of transactions per second, Bitcoin, for instance, is not an efficient option for high-volume transactions. Another concern is the amount of energy that is consumed (Banerjee et al. 2021; Krishnan et al. 2020).

Blockchain technology is a transformational challenge in the field of digital technologies, and it can alter the course of various sectors across the board. Despite this, it is facing several significant challenges, much like every other technology still in its early stages. These problems must be confronted head-on for them to transform from a potentially fruitful notion into a revolutionary instrument that is widely embraced. In this article, we take a comprehensive look at the three most critical challenges, which are security, scalability, and adoption, to consider the elucidations being developed to overcome these challenges (Embracing the Blockchain Revolution in Conquering Scalability 2023).

The Scalability Mystery of the blockchain is struggling to process massive transactions quickly. Most blockchain platforms have limited transaction processing capacity, which hinders high-volume transaction processing firms. Fortunately, creative methods can solve this scaling problem. Two promising approaches:

Sidechains are linked to the main blockchain. Sidechains can handle more transaction types and process transactions faster than the core blockchain. Sharding: splitting the blockchain into smaller pieces for concurrent processing. This division speeds up transaction processing, allowing more transactions per second. Strong Security with no central authority makes blockchain's distributed ledger impervious to tampering. However, recent high-profile blockchain hacks have raised security worries. A multifaceted strategy can improve security with strong encryption: Strong encryption protects data and makes system compromise considerably harder. Secure

Consensus Methods help reach blockchain consensus. Blockchain security requires careful selection of consensus algorithms due to their security levels (Iansiti and Rögös 2019; Krishnan et al. 2021).

Adoption is difficult because blockchain is a new technology and concept for many. The following initiatives can boost blockchain adoption. Education is to spread blockchain's benefits, which can sway firms to adopt it. Standards Development and Standards assure blockchain platform compatibility, enabling wider adoption. Apart from scalability, security, and acceptance, blockchain faces other growing challenges:

Regulation-based Blockchain's unclear regulatory environment hinders company adoption. The energy consumption of some blockchain networks, especially those that mine for security, is a major environmental issue. Skilled blockchain professionals are scarce, preventing organizations from adopting and implementing blockchain technology.

Despite challenges, the blockchain sector is thriving and is expected to develop exponentially. Grand View Research predicts an 87.7% CAGR for the global blockchain market from 2023 to 2030 (Blockchain Revolution and Its Basic 2023). Thus, the blockchain market might reach $23 billion by 2023. Blockchain's impact on various industries is expected to grow as these problems are overcome, and the technology improves. There is a great deal of potential for blockchain technology to revolutionize various businesses. However, many obstacles must be overcome before widespread acceptance occurs. As we work toward the development of robust answers to these difficulties, we are getting nearer to realizing the complete promise of this game-changing technology. Today's leaders are tasked with navigating these intricacies, ensuring that they can profit from the potential that blockchain technology brings while simultaneously managing the associated risks. Consequently, it is of the utmost importance for executives who wish to advance their organizations into the era of blockchain technology to comprehend and overcome these problems (Blockchain Revolution and Its Basic 2023; Arun Sampaul Thomas and Muthukaruppasamy 2023).

There is always a prospect for blockchain technology to be used in a diversity of new fields. Concepts such as NFTs (Non-Fungible Tokens), Decentralized Finance (DeFi), tokenization of assets, and Decentralized Autonomous Organizations (DAOs) are examples of cutting-edge developments that have the potential to change industries and interactions. A persistent quest for innovation and growth is exemplified by the journey that blockchain technology has taken beyond the original limits it faced. The final half of this blog will summarize the larger ramifications and potential applications of blockchain technology beyond the realm of Bitcoin. This will be accomplished as we navigate the landscape of difficulties and advancements (Banerjee et al. 2021; Krishnan et al. 2023).

4 Blockchain Beyond Cryptocurrency: Decentralized Industries and Transforming Systems

Few technological innovations have aroused as much interest and potential as blockchain technology. Blockchain has evolved from its humble beginnings as Bitcoin's underlying technology to become a global powerhouse that can revolutionize entire industries and systems.

4.1 Use Cases and Industries

The disruptive power of blockchain technology could impact many different industries. Some of these areas include healthcare, supply chain management, and finance. Included among the most notable use cases are:

a. **Cryptocurrencies and Digital Assets:** Bitcoin and other cryptocurrencies enable secure, decentralized digital transactions and store value.
b. **Supply Chain Management:** Blockchain improves traceability, transparency, and efficiency in supply chains by tracking products from their origin to the end consumer.
c. **Healthcare:** Blockchain can securely store and share medical records, facilitate interoperability, and ensure the integrity of clinical trials.
d. **Voting Systems:** Blockchain-based voting systems enhance transparency, eliminate fraud, and increase public trust in democratic processes.

In its most basic form, blockchain is a disseminated ledger system that allows for the safe recording and confirmation of trades via a distributed computer network. Cryptography connects an immutable chain of blocks, each containing a single transaction, to all previous blocks. With this core design, there is no need for a governing body to guarantee security, immutability, and transparency. Blockchain technology, while rooted in Bitcoin and cryptocurrencies, has applications that extend beyond financial transactions. New possibilities for innovation in many fields have arisen because of advancements in rudimentary expertise.

The decentralization process leads to the absence of a central authority, meaning no solitary unit controls the network, enhancing transparency and reducing the risk of manipulation or fraud. Immutability utilizes untransferable recorded data on the blockchain to provide secure and tamper-proof transaction evidence. Transparency means that all participants in the blockchain network have access to the same information, fostering trust and accountability.

Blockchain's adaptability has resulted in many uses outside the financial sector. Its secure architecture and decentralized nature have spurred innovation across a variety of industries, including healthcare, voting systems, supply chain management, and more. Blockchain technology has sparked a paradigm shift in numerous sectors, expanding its original use beyond the realm of cryptocurrency. Their inventive solutions to long-standing problems, enabled by their decentralized, secure, and transparent nature, have transformed businesses' operations and interactions (Filippi and Santos 2022).

Blockchain technology has had a huge impact on SCM. Businesses can use its immutable ledger to trace products from their inception to their destination. Blockchain technology enables "smart contracts," which automate the execution of contracts according to specified criteria. This innovation simplifies operations across sectors by removing middlemen and guaranteeing trust in contractual agreements. One strong method for managing IP rights is blockchain's immutability. It provides a transparent and secure platform for recording patents, copyrights, and trademarks to facilitate ownership verification and content provenance. Blockchain technology provides a safe environment for the healthcare industry to store and exchange patient records. This ensures the confidentiality of patient information and permits all healthcare practitioners quick access to up-to-date medical records. Blockchain technology has the potential to completely transform voting systems. It allays fears of election fraud by making records easily accessible and impossible to alter, encouraging faith in and engagement with democratic systems (Iansiti and Rögös 2019).

The revolutionary potential is resonating across sectors, governmental structures, and the fundamental nature of digital interactions as it expands beyond the domain of cryptocurrencies. In the years to come, blockchain technology has the potential to radically alter our world. Web 3.0, which aims to create a decentralized internet where users have more agency over their data and digital interactions, relies on blockchain technology. In addition to promoting improved privacy, this trend toward decentralization paves the way for novel decentralized ecosystems and applications (Mytton 2023; O'Connor et al. 2022).

Central banks are considering the idea of blockchain-based Central Bank Digital Currencies (CBDCs) worldwide. Central authorities govern CBDCs, which offer programmable and secure digital currencies. They also seek to simplify financial transactions, lower costs, and increase financial inclusion. Businesses in a wide range of industries are beginning to see the potential of blockchain technology. Industry consortia and collaborative initiatives are developing custom blockchain solutions, accelerating adoption and innovation in certain industries. As time goes on, the list of possible uses for blockchain grows (Blockchain Beyond Cryptocurrency 2023)

The expansion of blockchain technology beyond the realm of cryptocurrencies is a social movement toward more openness, confidence, and safety rather than just a technical phenomenon. To fully realize the potential of this transformation, inventors, regulators, and stakeholders must work together.

Decentralization eliminates the necessity for a central authority, enabling direct transactions between participants. This reduces the risk of single points of failure and enhances resistance to censorship and unauthorized control. Security is a cornerstone of blockchain technology. Cryptographic techniques such as hashing and digital signatures secure transactions and link them to previous transactions, forming an unalterable chain. This enhances data integrity and guards against unauthorized modifications. Transparency ensures that all participants have access to the same information, reducing the potential for fraud and increasing accountability. It also enables efficient auditing and tracking of transactions (Tapscott et al. 2016).

a. **Cryptocurrencies as a Use Case**

Blockchain-enabled cryptocurrency creation. Cryptography secures transactions and controls new units in cryptocurrencies. Bitcoin, the first cryptocurrency, showed how blockchain could enable peer-to-peer digital payments.

b. **Decentralization: Bitcoin (Cryptocurrency)**

Bitcoin uses mining to verify transactions on a decentralized network of nodes. The first miner to solve challenging mathematical problems adds a brand-new block of transactions to the blockchain. Decentralized consensus eliminates the requirement for central transaction validation.

c. **Transparency: Supply Chain Management**

In food and luxury goods, blockchain tracks products from origin to customer. Blockchain transactions record each supply chain step, creating a transparent and tamper-proof product history. Consumers can scan a QR code to check product legitimacy and ethical sources.

d. **Immutability: Land Title Registry**

Traditionally, land title records are susceptible to fraud and disputes. With blockchain, land titles can be recorded as transactions on a blockchain. Once recorded, these records cannot be altered or tampered with, ensuring a reliable and immutable record of land ownership.

e. **Cryptographic Techniques**

Blockchain technology makes use of digital signatures to ensure that transactions are legitimate. A user's digital signature is appended to a transaction when they start it. Verifying the signature's authenticity using the sender's public key allows other participants to ensure that the transaction originated with the rightful owner. Ethereum (Smart Contracts and Tokens) introduced the idea of smart contracts, which are coded agreements that automatically carry out their terms. When certain criteria are satisfied, the provisions of these contracts are automatically executed and enforced. Ethereum also made it possible to create tokens, which stand for assets or functions inside a blockchain network.

After years of academic debate, blockchain technology has now found a home in the real world, transforming whole sectors and ways of doing business. Discover how IBM, Walmart, and Ripple use blockchain technology to their advantage in areas like decentralization, transparency, and cryptocurrency.

f. **IBM's Food Trust: Revolutionizing Supply Chain Transparency**

IBM's Food Trust platform is a prime example of how blockchain can enhance transparency in complex supply chains. In collaboration with major retailers and food producers, IBM uses blockchain to trace the journey of food products from farm to table. This empowers consumers to verify the origin, quality, and safety of their food. Each step recorded on the blockchain, from production to distribution, greatly reduces the risk of fraud and contamination (Unveiling the Power of Blockchain 2023; Arun Sampaul Thomas and Harold Robinson 2020a).

g. **Walmart's Blockchain-Based Traceability: Trust in Every Aisle**

Walmart, a global retail giant, has also integrated blockchain technology into its food supply chain. In cases of food recalls or quality issues, traditional methods of identifying affected products can be time-consuming and imprecise. Walmart's blockchain-based traceability system enables the company to pinpoint the source of a product's issue within minutes, preventing widespread recalls and reducing consumer health risks.

h. **Ripple's XRP: Reinventing Cross-Border Payments**

Aiming to revolutionize cross-border payments, Ripple's XRP builds upon Bitcoin, the pioneer of cryptocurrencies. International money transfers using more conventional methods are frequently error-prone, costly, and time-consuming. Financial organizations now have a faster and cheaper way to transfer cross-border payments thanks to Ripple's blockchain-based platform. Because blockchain transactions are public, both the sender and the receiver can see how far along the transaction is. These examples demonstrate how blockchain technology already has a tangible effect on many industries. Supply chain efficiency, customer trust, and financial services are all being enhanced by corporations through decentralization, transparency, and cryptocurrency. More inventive ways blockchain is changing industries and business practices will be covered as we progress through this series. Few ideas and principles of blockchain technology are put into practice in these cases. We will explore blockchain's consensus methods, various blockchain types, and their growing industrial uses in future blogs (Unveiling the Power of Blockchain 2023).

5 Expert Review with Importance in Mitigating Blockchain Implementation Challenges

Blockchain is a game-changing technology that can improve productivity, security, and transparency in any industry. The need for a seamless rollout is growing as more and more businesses see blockchain's potential. Nevertheless, there may be barriers to adoption due to the difficulty of comprehending blockchain development. If you want to know how to make the most of blockchain technology, you should investigate expert blockchain consulting. Blockchain, an irreversible and trustless platform for sharing data and exchanging currency, has shaken up traditional business methods. Its cost reductions, operational efficiency, and data security have piqued the interest of enterprises across industries. To get these advantages, thorough preparation and growth are necessary (Evaluating the Potential and Challenges in Blockchain 2023).

5.1 *Specific Challenges that Businesses May Encounter During the Blockchain Implementation Process*

Blockchain is a game-changing technology that can improve productivity, security, and transparency in any industry. The need for a seamless rollout is growing as more and more businesses see blockchain's potential. Nevertheless, there may be barriers to adoption due to the difficulty of comprehending blockchain development. Full use of blockchain technology may be yours with the assistance of an expert blockchain consultant. Blockchain, an irreversible and trustless platform for sharing data and exchanging currency, has shaken up traditional business methods. Its cost reductions, operational efficiency, and data security have piqued the interest of enterprises across industries. To get these advantages, thorough preparation and growth are necessary (Top 10 Blockchain Adoption Challenges 2021).

To use blockchain, a distributed ledger system that relies on consensus, one needs expertise in computer science, networking, and cryptography. Because of poor resource allocation, businesses run the risk of wasting money on inefficient implementations and security holes. Expert blockchain consultants are highly prized. They can break down complex concepts into simpler terms, provide an overview of blockchain technology, and help companies see how it may improve their operations. Through their seminars and training, consultants may bring executives and engineers together in support of blockchain integration.

5.2 *Identifying and Addressing Business-Specific Blockchain Deployment Challenges*

Improving the blockchain's scalability is an important challenge. Processing more transactions without sacrificing speed or security becomes more important as the network grows. Experts in blockchain technology may assess a company's needs and implement tailored scaling strategies. It is critical to secure sensitive data and valuable assets on blockchain networks. Security and auditing measures can achieve this. Blockchain consultants conduct security audits and penetration tests to identify vulnerabilities and offer cyber-security recommendations.

The fact that there are several different blockchain systems and protocols causes problems with interoperability. To guarantee data flow and interoperability, consultants construct defined protocols and link existing systems with blockchain technology. Smart contracts, which are agreements that can execute themselves, are crucial to the majority of blockchain applications. But because of their complexity and potential for bugs, they necessitate comprehensive testing and auditing. Experts in blockchain analyse code and improve the safety and efficiency of smart contracts.

Adherence to Bitcoin regulations is not always clear. Consulting firms aid businesses in meeting financial, privacy, and data protection regulations. Blockchain Consulting's Role in Facilitating Tailored Products: Given the unique nature of each company, there is no universal solution for implementing blockchain technology. We tailor blockchain technology consulting services to meet the specific needs of each customer. Consultants, as experts in blockchain, contribute best practices at the table. Organizations benefit from their knowledge of development frameworks, consensus processes, and blockchain protocols, which they use to discover optimal solutions (Top 10 Blockchain Adoption Challenges 2021; Arun Sampaul Thomas et al. 2023; Arun Sampaul Thomas and Harold Robinson 2020b).

To lessen the likelihood of setbacks and maximize the likelihood of a project's success, blockchain consulting services can detect potential issues in their early stages. Although deploying a blockchain could seem expensive at first, it could end up saving money with the help of knowledgeable advisors. Through careful planning and execution, businesses may enhance the return on investment (ROI) from their blockchain investments.

Blockchain technology has the potential to impact businesses worldwide. Adopting blockchain technology is not a picnic; it requires thorough planning and expert oversight. Through blockchain consulting services, organizations can overcome challenges, take advantage of opportunities, and fully utilize the decentralized and secure capabilities of blockchain technology. A new era of innovation and productivity could be on the horizon for blockchain users who surround themselves with capable advisors. Step into the next level of efficiency and innovation with the help of expert advisory services and the best blockchain technologies. To begin the blockchain technology revolution for your company, contact us right away.

6 Conclusion

Despite these challenges, blockchain technology is slowly finding its way into many other industries, such as education, retail, logistics, the media, entertainment, and many more. However, there are still several hurdles to jump to see a true and actual change in how financial services use new technology. It appears that businesses must pay close attention to this technology and be prepared to adjust their services to achieve the desired outcomes. The technology known as blockchain has the potential to revolutionize various facets of our lives, including financial transactions, supply chain management, and other areas. Understanding the fundamentals of blockchain technology allows us to investigate the range of possible applications for this technology and keep ourselves updated on the ever-changing landscape. This disruptive technology is transforming industries and empowering individuals globally, making it crucial to embrace the decentralized revolution and closely monitor its progress. While blockchain technology is sure to make a splash in the commercial and public spheres, the future is still a way off, despite what many think. The trajectory of blockchain represents technological progress and a societal shift. It ushers in an era characterized by the convergence of decentralization, transparency, and trust, which will redefine our digital world. Given the previously mentioned list of challenges related to blockchain technology, new technological developments are necessary.

Given the current state of blockchain technology, our primary focus should be on making this cutting-edge technology accessible to a wider audience. Blockchain-based technology has significantly transformed our comprehension of transaction and data management processes. Beyond the realm of cryptocurrencies, several applications have emerged because of their fundamental ideas, which include decentralization, transparency, and security. As we continue our examination of blockchain technology, we will delve into numerous facets of this technology. We will also discuss this technology's impact on various industries, including banking, government, and more.

References

Akter M et al (2023) Blockchain technology for sustainable development goals: a review of trends and opportunities. Sustainability

Arun Sampaul Thomas G, Harold Robinson Y (2020a) IoT, big data, blockchain and machine learning besides its transmutation with modern technological applications. In Internet of Things and big data applications—Part of the Intelligent Systems Reference Library book series, vol 180. Springer, Cham, pp 47–63, ISSN: 978-3-030-39118-8

Arun Sampaul Thomas G, Harold Robinson Y (2020b) Real-time health system (RTHS) centered Internet of Things (IoT) in healthcare industry: benefits, use cases and advancements in 2020. Springer's Multimedia Technologies on the Internet of Things Environment (Scopus Indexed). ISSN: 978-981-15-7965-3

Arun Sampaul Thomas G, Muthukaruppasamy S (2023) Green IoT use case approaches for blockchain technology taking industry 5.0 to the next level. In green blockchain technology for sustainable smart cities, Elsevier, pp 119–141. ISSN: 978-0-323-95407-5

Arun Sampaul Thomas G, Muthukaruppasamy S, Saravanan K, Muleta N (2023). Revolutionizing smart grids with big data analytics: a case study on integrating renewable energy and predicting faults. In: Kumar Sharma D, Sharma R, Jeon G, Kumar R (eds) Data analytics for smart grids applications—a key to smart city development. Intelligent Systems Reference Library, vol 247. Springer, Cham. https://doi.org/10.1007/978-3-031-46092-0_11

Banerjee A et al (2021) Blockchain technology for decentralized innovation: opportunities and challenges. Technol Forecast Soc Change

Blockchain Beyond Cryptocurrency (2023) https://medium.com/@kunalchhablani14/blockchain-beyond-cryptocurrency-revolutionizing-industries-and-transforming-systems-865790ae696b

Blockchain Revolution and Its Basic (2023) https://www.linkedin.com/pulse/blockchain-basics-introduction-revolutionary-bhoskar-pmp-csm/

Chaum D (2021) Navigating the challenges of trust in blockchain-based systems. Financ Crypt Data Security

De Filippi P, Santos A (2022) Navigating the future of decentralized innovation: challenges and opportunities in the blockchain era. Technolo Forecast Soc Change

Embracing the Blockchain Revolution in Conquering Scalability (2023) https://timesofindia.indiatimes.com/blogs/a-window-to-the-tech-world/embracing-the-blockchain-revolution-in-2023-conquering-scalability-security-and-adoption-challenges/

Evaluating the Potential and Challenges in Blockchain (2023) https://medium.com/coinmonks/is-blockchain-the-future-evaluating-the-potential-and-challenges-e653a7c096ce

Iansiti M, Rögös K (2019) The trust machine: the story of blockchain and the future of the internet.

Ivanova A et al (2023) Blockchain for social good: a survey of potential applications and challenges. Comput Human Behavior.

Krishnan S, Balas VE, Julie EG, Yesudhas HR, Balaji S, Kumar R (eds) (2020) Handbook of research on blockchain technology. Academic Press, London

Krishnan S, Balas VE, Julie EG, Yesudhas HR, Kumar R (eds) (2021) Blockchain for smart cities. Elsevier, Amsterdam

Krishnan S, Balas VE, Kumar R (eds) (2023) Green blockchain technology for sustainable smart cities. Elsevier, Amsterdam

Mytton S (2023) Building trust in the age of blockchain: a framework for governance and accountability. J Bus Ethics

O'Connor M et al (2022) Building trust in blockchain: a multi-stakeholder approach. MIT Sloan Manage Rev

7 Major Challenges for the Blockchain Adoption in 2023 (2023) https://wazirx.com/blog/7-major-challenges-for-blockchain-adoption-in-2023/

Tapscott D, Tapscott A (2021) The blockchain revolution: how it will change the way we live, work, and trust

Tapscott D, Tapscott A (2022) Blockchain revolution: how digital ledgers are transforming business in finance, healthcare, and more

Tapscott A et al (2016) Blockchain revolution: how blockchain technology will transform business, government, and our lives.

Top 10 Blockchain Adoption Challenges (2021) https://101blockchains.com/blockchain-adoption-challenges/

Unveiling the Power of Blockchain (2023) https://www.linkedin.com/pulse/unveiling-power-blockchain-introduction-transparency-vipin/

MIX
Papier aus verantwortungsvollen Quellen
Paper from responsible sources
FSC® C105338

If you have any concerns about our products,
you can contact us on
ProductSafety@springernature.com

In case Publisher is established outside the EU,
the EU authorized representative is:
Springer Nature Customer Service Center GmbH
Europaplatz 3, 69115 Heidelberg, Germany

Printed by Libri Plureos GmbH
in Hamburg, Germany